Introduction to the mechanics of plastic forming of metals

**Monographs and textbooks on mechanics of solids and fluids**

editor in chief: G. Æ. Oravas

**Mechanics of plastic solids**

editor: J. Schroeder

# Introduction to the mechanics of plastic forming of metals

Wojciech Szczepiński

*Polish Academy of Sciences*
*Institute of Fundamental Technological Research*
*Warszawa*

SIJTHOFF & NOORDHOFF INTERNATIONAL PUBLISHERS

Alphen aan den Rijn, The Netherlands

PWN—POLISH SCIENTIFIC PUBLISHERS

Warszawa

First English edition based on *Wstęp do analizy procesów obróbki plastycznej* published in 1967 by Państwowe Wydawnictwo Naukowe, Warszawa

Translation: Wojciech Szczepiński

ISBN-13: 978-94-009-9549-9      e-ISBN-13: 978-94-009-9547-5
DOI: 10.1007/978-94-009-9547-5

Sijthoff & Noordhoff International Publishers B.V.,
Alphen aan den Rijn, The Netherlands

# Contents

*Contents*

*Contents*

# *Preface*

The plastic forming of metals is a branch of technology, in which great progress has been achieved mainly by collecting practical data resulting from long-time experience. The contribution of theoretical analysis to this progress is not significant. The common argument was, that such an analysis is too difficult and often even impossible. However, the constantly increasing significance of coldwork processing in industrial practice and the progress in the automatic control of these processes, demanding a broad knowledge of the influence of various parameters, and also the need to improve the mechanical behaviour of coldworked elements have proved beyond doubt that better understanding of the theory of plastic forming processes is necessary. This general trend has become evident in several recent publications in recent years, even though these represent only the engineering, approximate approach to theoretical analysis.

Recent significant progress in the mathematical theory of plasticity allows to build an analysis of plastic forming processes on a more sound basis. In particular there have been developed graphical methods by means of which slip-line and hodograph meshes are constructed for plane strain and axially symmetric problems. These graphical methods yield not only the required forces but also the mode of plastic deformation, even for very complex problems. Numerical procedures also give good results if the digital computer technique is employed.

The aim of the book is to present as simply as possible those methods of the mathematical theory of plasticity which may be applied to the analysis of plastic forming processes. Recent achievements in this analysis are presented, including some original theoretical results and their experimental verification.

Since a large range of problems has been covered in the book, there remained no space to include tables or diagrams which would be directly applicable in engineering practice. Our attempt was to give a systematic presentation of the modern methods used in the analysis of the plastic forming processes, and we hope that the book will be useful for graduate students of mechanical engineering and for mechanical engineers concerned with theoretical knowledge.

In order to give the book a uniform character we shall not deal here with the approximate methods of analysis, although such approximate methods are at the present used broadly in the engineering practice due to their simplicity. Their description may be found in a number of well-known monographs quoted at the end of the book as Supplementary references.

# Mechanical properties of metals

## 1.1 The plastic behaviour of metals

Most of the commercial metals subject to plastic forming processes have a polycrystalline structure. This means that they are composed of a large number of relatively small crystalline grains randomly oriented in space. Each grain displays anisotropy of mechanical properties depending on crystallographic directions. Polycrystalline metals in the annealed condition display, however, approximately isotropic properties, resulting from random orientation of a large number of fine grains. They generally offer more resistance to deformation than single crystals. As an example, in Fig. 1 are presented stress-strain curves for zinc, for which this effect is

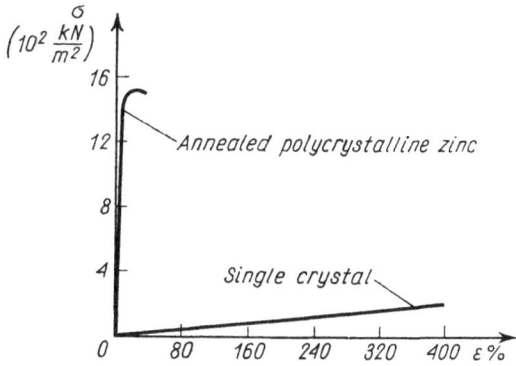

Fig. 1. Stress-strain curves for zinc (after C. F. Elam, see [4])

particularly evident (see [4]). Many factors are held responsible for the different formability of monocrystals and polycrystalline metals, the main role being attributed to the complex effects connected with grain boundaries in an polycrystalline aggregate.

## 1 Mechanical properties of metals

Applying uniaxial tension to standard specimens is the most frequently used method for testing the mechanical behaviour of metals. Also compression of relatively short solid cylinders and torsion of thin-walled tubular specimens is used in routine practice. From the tension test one obtains the diagram of loading force $P$ versus elongation $\Delta l$ from which the stress-strain curve can be constructed. Two measures of strain are usually employed. If the length of a tensile specimen is increased from $l_0$ to $l$, the amount of deformation may be measured either as the conventional strain $e = (l - l_0)/l_0$ or as the logarithmic (natural) strain, obtained by adding increments of strain referred to the instantaneous length of the specimen $\varepsilon = \int_{l_0}^{l} dl/l = \ln(l/l_0)$. There exists the simple relationship $\varepsilon = \ln(1 + e)$ between these two measures. As a measure of stress one usually takes the loading force $P$ divided by the area of the initial cross-section of the specimen, i.e. $\sigma_0 = P/A_0$. Such a measure of stress does not correspond to the actual stress since the area of the cross-section decreases gradually during the test. The decreasing of the cross-section area is particularly strong at the final stage of the test, when a local necking occurs at a certain segment of the specimen. Necking is accompanied by a characteristic drop of the conventional stress $\sigma_0 = P/A_0$. However, if we take the quotient $\sigma_r = P/A_{min}$ of the loading force $P$ by the area $A_{min}$ of the instantaneous narrowest cross-section area as a measure of stress, then the stress-strain curve increases continuously up to the occurrence of fracture (Fig. 2).

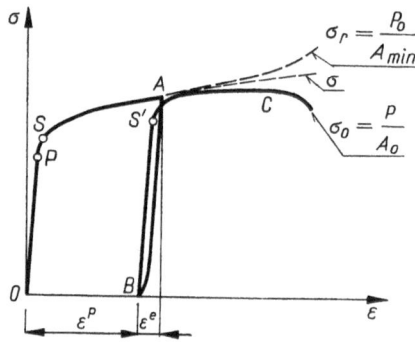

Fig. 2. Stress-strain curve for metals

Frequently, the slope of the $\sigma_r$-curve increases after the neck is formed, which is connected with the triaxial state of stress created in this region. A detailed analysis of this phenomenon is given in specialist monographs

2

(see e.g. [92]) on material testing, and it will not be presented here because of its small significance for the metal forming theory. If the triaxiality of the stress state in the neck is accounted for, one obtains the corrected curve $\sigma$ which represents the properties of deformed material.

Let us now consider the main properties of the tension curve in the $\sigma, \varepsilon$ coordinate system. The initial portion $OP$ of the diagram is linear. The end point $P$ of that portion represents the stress which is called the *proportional limit*. Deformations within this range are elastic, and they vanish when the load is removed. The elastic type of deformation slightly overruns the sector $OP$, thus the stress at which plastic (permanent) deformations begin to appear is greater than the proportional limit. This stress corresponds to the point $S$ on the tension curve, whose ordinate determines the elastic limit. The exact measurement of the elastic limit is rather difficult and depends mainly on the accuracy of the measuring device. Hence, for the sake of clarity the notion of conventional elastic limit has been introduced. The *conventional elastic limit* represents the stress level accompanied by a certain very small permanent deformation, most commonly equal to 0.01 or 0.02 percent. Thus the conventional elastic limit is denoted as $\sigma_{0.01}$ or $\sigma_{0.02}$, respectively. To give rise to further plastic deformation the load must be increased. This very important effect is referred to as the *strain-hardening* of the material. The slope of that portion of the stress-strain diagram $c_1 = \mathrm{d}\sigma/\mathrm{d}\varepsilon$ is called the *hardening modulus*. The physical meaning of that term is clear if the unloading and consecutive reloading of the specimen, previously deformed until a certain point $A$, is considered. During unloading from $A$ to $B$ a certain small part of deformation vanishes; it represents the elastic part of strain $\varepsilon^e$. The remainder constitutes the plastic part of strain $\varepsilon^p$. If the specimen is loaded again, deformation up to a certain point $S'$ is elastic. Since the stress level corresponding to $S'$ is higher than that corresponding to $S$, the plastically deformed material increases its elastic limit. Under the still increasing loading force the material begins again to deform plastically. It is important to note that after a short transitory sector of strong curvature the diagram becomes an extension of the initial part of the stress-strain curve obtained before unloading. If the loading had been continued uninterruptedly after the point $A$ is reached, the two lines practically would coincide.

Some metals as for example mild steel, certain aluminium alloys, polycrystalline molybdenum and cadmium display certain characteristic

3

features of the stress-strain diagram (Fig. 3). The stress after reaching the
point *G* sudenly drops. The stress $\sigma_u$ is called *upper yield point*. Then the
material undergoes plastic deformation at an almost constant value of
stress $\sigma_l$, called *lower yield point*. The remaining part of the diagram is
similar to that on Fig. 2. The magnitude of the upper yield point depends
mainly on the conditions under which the tension test is run, whereas the
lower yield point has a fixed value.

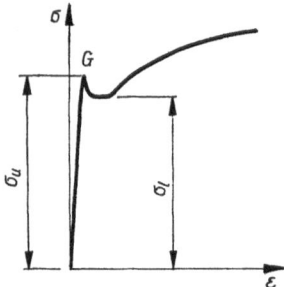

Fig. 3. Stress-strain curve for mild steel, certain aluminium alloys, polycrystalline
molybdenum and cadmium

Certain difficulties arise concerning determination of the yield point
for materials whose tension diagram is of the type schematically shown in
Fig. 2. In engineering practice the conventional yield point $\sigma_{0.2}$ is used,
equal to stress accompanied by 0.2 percent of permanent deformation.

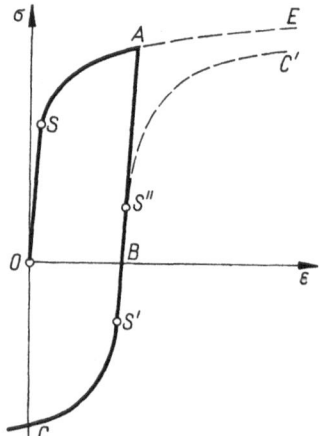

Fig. 4. Schematic diagram of the Bauschinger effect. Curve *BC* is shifted to the
position *BC'*

Generally, metals under compression have the same stress-strain curve as under tension, provided the signs of stress and strain are changed. However, materials behave in a different manner if previous tensile plastic deformation is followed by compression. Let tensile loading until the point $A$ (Fig. 4) and unloading until $B$ be followed by compression causing a shortening of the specimen. If now the segment $BC$ is shifted to take up the position $BC'$, then it turns out that it will run below the tension curve $OAE$. Even more remarkable is the difference between ordinates of the point $S''$ and of the corresponding point on the tension-following-previous-tension curve (Fig. 2). This phenomenon is known as the *Bauschinger effect*, and is observed also in the reversible torsion test and other more complex loadings of plastically prestrained metals. The Bauschinger effect is of great significance in the formulation of the strain-hardening hypotheses, which will be discussed in more detail in Chapter 3.

The stress-strain curve strongly depends on the test temperature. By rising the temperature sufficiently one obtains considerable lowering of the yield point and of the curve itself. This phenomenon is widely exploited in the practice of plastic forming of metals. In Fig. 5 are presented tension

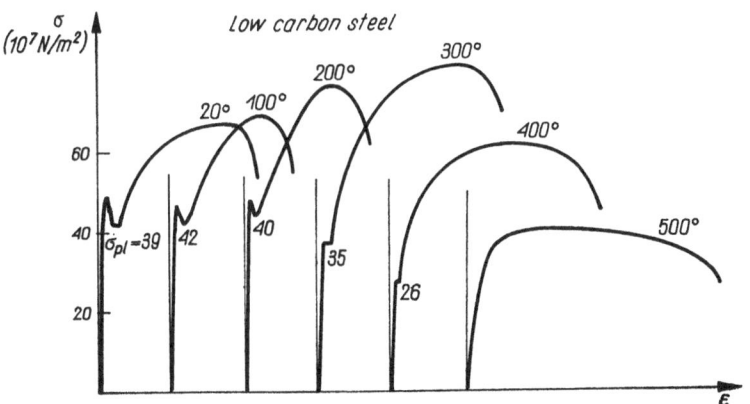

Fig. 5. Tension diagrams for mild steel at different temperatures

diagrams for mild steel at different temperatures. At temperatures of 100° and 200°C a slight rise of the yield point is observed, but beginning from 300°C the yield point decreases remarkably. As shown in Fig. 6, copper for any temperature increment displays lowering of stresses required to deform it plastically.

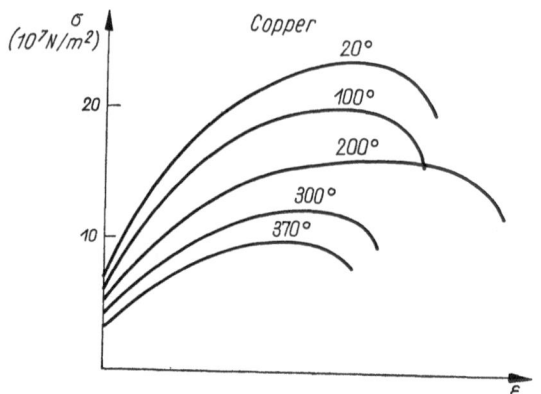

Fig. 6. Tension diagrams for copper at different temperatures

The tension test is not particularly advantageous for determining the plastic properties of metals, which need to be known in the theoretical analysis of various plastic forming processes. The main disadvantage of this test consists in the loss of stability connected with necking occurring at the relatively small plastic deformations. The triaxiality of the stress state in the neck adds to the intricacy of interpretating the test results. Although the end part of the stress-strain diagram can be corrected by means of respective calculations [21, 12] and so the effect of triaxiality of the stress state cán be eliminated, the fracture of the specimen at relatively small deformations remains the main weakness of the tension test. A substantial amount of total elongation is concentrated in the necking region, and the so-called *uniform deformation* reached, before necking begins to form, does not usually exceed the magnitude of several percent. In plastic forming processes deformations are in most cases, particularly when compressive stresses predominate, remarkably large. For that reason, of great practical significance are methods of material testing which permit to obtain the stress-strain diagram for as large as possible strains.

Being confined in space, we are not able to describe here all the various methods of testing, which may be found in specialist works. Two methods, however, will be shortly mentioned below for their ingenious simplicity. In [35] the specimens having the form of a flat ring shown in the bottom part of Fig. 7, are twisted between two punches. Moreover, the specimen is compressed between the two punches by means of an axial force acting along the axis of the testing device. An internal mandrel and

outer ring prevent the outflow of the material in the radial direction. On
the contact surfaces of the two punches a number of cuts is made in order
to prevent mutual sliding on the specimen-punch interface. Before twisting
starts, the specimen is compressed by a large axial force. Then both

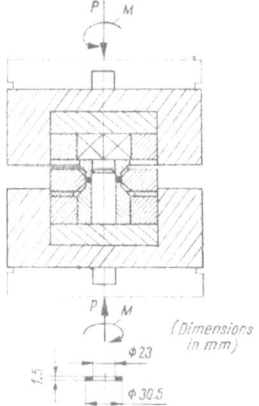

Fig. 7. Testing device for twisting specimens in the form of flat ring (after
S. Erbel [35])

punches begin to rotate slowly in opposite directions twisting the specimen compressed between them. During the test the diagram twisting
moment versus angle of twist is recorded automatically. If the magnitude of the compressive force is properly chosen, very large shear strains
up to several thousands percent can be obtained. One can suspect,
since the deformations involved are so large, that there is taking place a
continuously repeating process of local decohesion followed by immediate
bonding due to large compressive stress. An example of the twisting moment
versus deformation diagram obtained for copper is presented in Fig. 8. It
is clearly seen that above a certain value of deformation the material

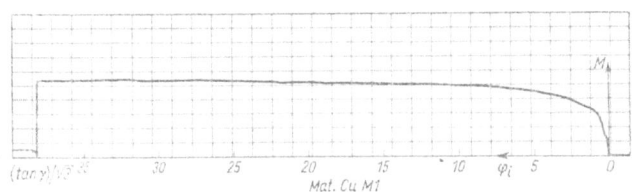

Fig. 8. Diagram of twisting moment versus shear angle (after S. Erbel [35])

7

does not display any strain-hardening effect and it behaves as a perfectly plastic material.

The method used in [83] consists in the twisting of a thin disc of outer diameter 70 mm and inner diameter 16 mm. The disc is clamped at the outer and inner edges. When the two clamping devices rotate slowly in opposite directions the rectilinear radii drawn on the specimen's surface before test begin to bend. For a given value of the twisting moment the stresses are distributed linearly along the radius. Thus on the basis of the curves obtained on the specimen's surface one can determine the stress-strain relation. Figure 9 shows photographs of the lines on the specimen before and after deformation.

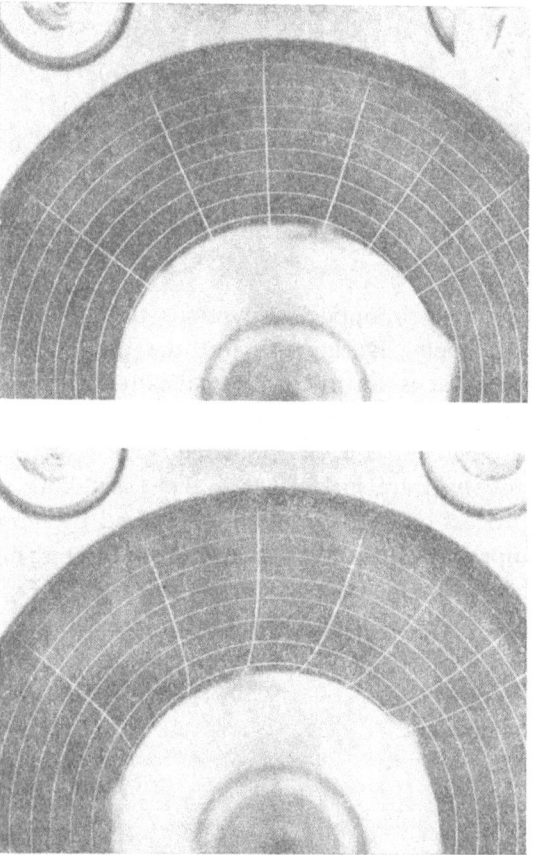

Fig. 9. Deformation of a twisted disc (after Z. Marciniak [83])

## 1.2 Effect of strain rate on resistance of metals to deformation

The effect of strain rate on the stress-strain curve is one of the important factors in the analysis of plastic forming processes. In most practical plastic forming processes this effect is at room temperatures rather insignificant. However, in high energy forming, and first of all in the cases of explosive forming, the effect of strain rate is of great practical significance.

Recently the effect of strain rate on the resistance to deformation has been investigated in numerous works. The book by J. F. Bell [5] gives an excellent treatment of this subject. Figure 10 shows results of tests per-

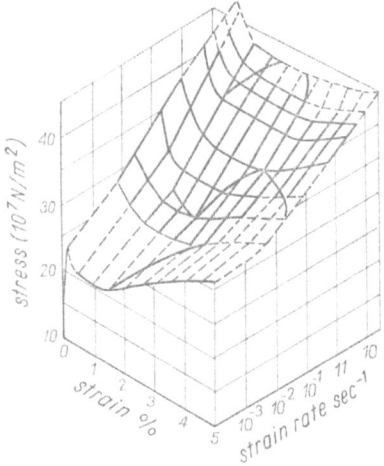

Fig. 10. Variation of yield stresses with strain and strain rate for mild steel (after K. J. Marsh and J. D. Campbell [86])

formed in [86] on mild steel specimens compressed with different velocities. Sections of this block-diagram by planes corresponding to specified constant strain rates correspond to stress-strain diagrams obtained from tests conducted with a fixed rate of strain.

However, as more thorough investigations [15, 16, 77] show, the value of stress is not uniquely determined by the actual value of strain and strain rate, but depends also on the history of the strain rate variation during the process of deformation. Deviations from the surface obtained by means of a sequence of tests performed with constant rates of stress are not greater than 10 percent, even if the strain rate suddenly changes

9

its value from a very slow to a very high one. Thus the assumption that the surface of the kind presented in Fig. 10 determines the stress in all cases is for practical purposes quite admissible.

For the high energy forming methods, of particularly practical significance are investigations of the effect of the very high strain rates on the stress-strain curve. Such investigations for carbon steel were performed in [17]. Results are presented in Fig. 11 which shows the diagram of the

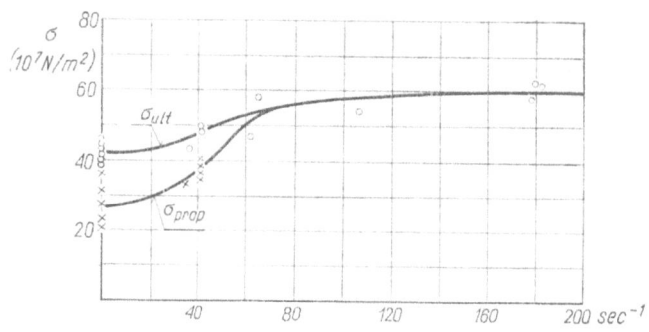

Fig. 11. Variation of the yield locus with strain rate for mild steel (after D. S. Clark and P. E. Duwez [17])

proportional limit and fracture stress against the strain rate. For strain rates exceeding 80 sec$^{-1}$ the proportional limit is twice larger than the proportional limit measured at low strain rates. Note that the two curves merge at high strain rates, which indicates that the stress-strain curve runs horizontally, or in other words, that there is no strain-hardening.

For metals deformed at higher temperature the effect of strain rate is remarkably stronger than at room temperature. For a nickel-steel the increase of the resistance to deformation at the temperature 500°C is quite remarkable even at strain rates of the magnitude used in standard processes of plastic forming. Therefore in such cases this effect cannot be neglected in the theoretical analysis.

### 1.3   Idealized stress-strain curves

In numerous problems of the theory of plastic forming taking the real stress-strain relation into account leads to solutions of considerable mathematical complexity. For that reason it is common practice to devise simplified stress-strain diagrams, idealizing properties of the material but, on the other hand, allowing to obtain effective solutions. The most common simplification consists in neglecting the elastic deformations. This does not deviate too far from the physical reality, because in most plastic forming processes plastic deformations are of two orders greater than elastic deformations. Further simplifications concern the stress-strain curve in the plastic range. The most far-going simplification consists in assuming a rigid-perfectly material (Fig. 12a). For this idealized model of the material,

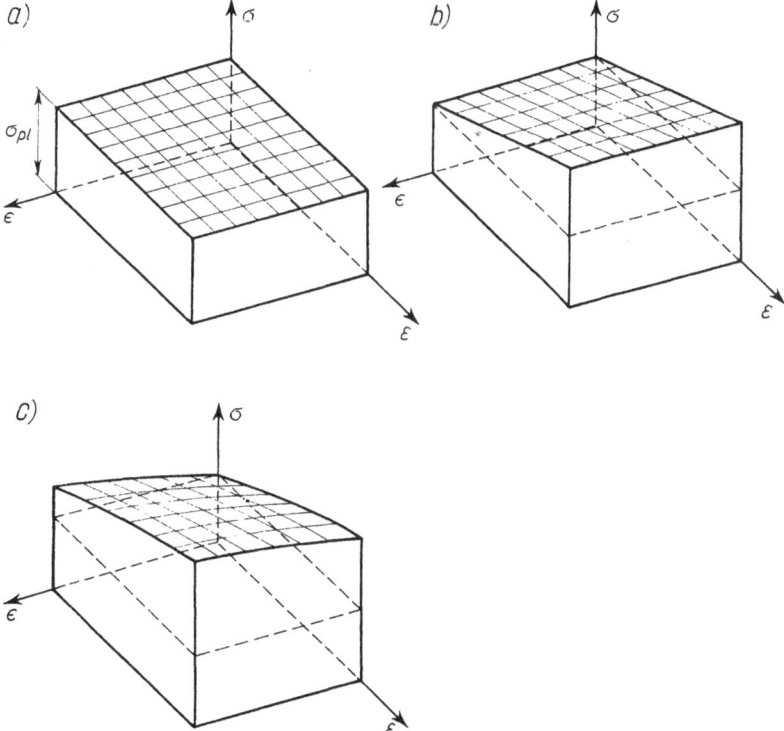

Fig. 12. Idealized models of plastic material; a) rigid-perfectly plastic model, b) rigid-plastic material with constant hardening modulus, c) rigid-plastic material with strain-hardening and strain-rate sensitivity; $\varepsilon$ denotes the strain and $\in$ is the strain rate

a tensile specimen is completely rigid until the axial stress equals $\sigma_{pl}$, whereupon the material flows at a constant stress. The block-diagram shown in Fig. 12a indicates that there is also no effect of the strain rate

11

on the resistance to deformation for such a model of the material. If the load is removed, the material does not contract elastically. Thus the entire deformation remains as the permanent one. The model of a rigid-plastic material proved very useful in practical applications; it allowed to solve a large number of problems connected, among others, with the plastic forming of metals. Neglecting strain-hardening in this model in numerous practical applications, particularly when the metal undergoes plastic deformation at high temperature, also does not remarkably deviate from the reality.

If the strain-hardening effect cannot be neglected, other types of model of the material are used. The simplest of them is shown in Fig. 12b. It corresponds to a linear stress-strain relation with a constant hardening modulus. In some particular cases solutions can be obtained involving a more general model of the rigid-plastic material, as for example the one presented in Fig. 12c.

# Stresses, strains and flow velocities

## 2.1 The state of stress

Let us assume the Cartesian coordinate system $x, y, z$. The state of stress at an arbitrary point of the body is determined by nine stress components measured as force per unit area. Positive directions of stress components are shown in Fig. 13. Let $\sigma$ denote the normal stress components and $\tau$

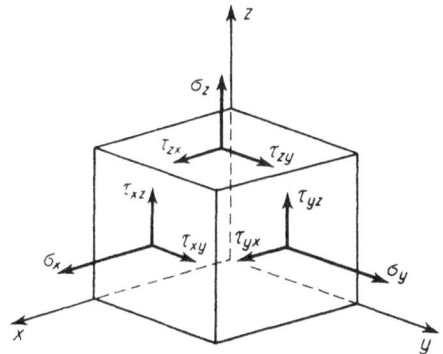

Fig. 13. Positive directions of stress components

the shear stress components. The conditions of equilibrium, found by taking moments about axes through the centre of the elementary cube, give the equalities $\tau_{xy} = \tau_{yx}$, $\tau_{zx} = \tau_{xz}$, $\tau_{zy} = \tau_{yz}$. Thus the number of independent stress components reduces to six. They form a symmetric tensor

$$\begin{bmatrix} \sigma_x & \tau_{xy} & \tau_{xz} \\ \tau_{yx} & \sigma_y & \tau_{yz} \\ \tau_{zx} & \tau_{zy} & \sigma_z \end{bmatrix}$$

which, using suffix notation, may be briefly written as $\sigma_{ij}$ $(i, j = 1, 2, 3)$. [1]
In suffix notation the coordinate axes will be denoted as follows: $x = x_1$,
$y = x_2$, $z = x_3$. The components of the stress tensor will be denoted
according to the scheme

$$\begin{bmatrix} \sigma_{11} & \sigma_{12} & \sigma_{13} \\ \sigma_{21} & \sigma_{22} & \sigma_{23} \\ \sigma_{31} & \sigma_{32} & \sigma_{33} \end{bmatrix}.$$

By comparing the two tables, one can find the corresponding notation
for the same stress components.

At every point of the body one can distinguish three orthogonal
planes, on which only normal stresses are acting, whereas shear stress
components vanish. Such planes are called the *principal planes*, and the
normal stresses acting there are defined as the *principal stresses*. In order
to find the values of the principal stresses let us consider the conditions
of equilibrium of an elementary tetrahedron (Fig. 14) whose three faces

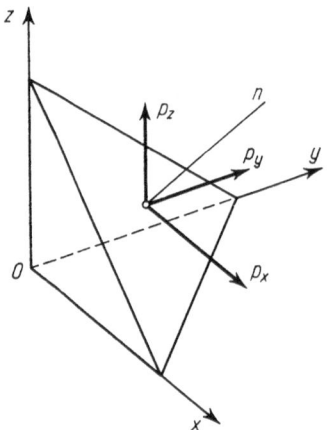

Fig. 14. Elementary tetrahedron

contain the coordinate axes, and the fourth is formed by a plane with the
normal $n$ arbitrarily inclined to the coordinate system. The direction
cosines of the normal will be denoted by $a_x, a_y, a_z$, or briefly in suffix
notation by $a_i$ $(i = 1, 2, 3)$. The resultant stress $p$ acting on this face may

[1] See Appendix 1 for an explanation of suffix notation and the summation convention.

be resolved into three components written as $p_x, p_y, p_z$, or in different way as $p_j$ $(j = 1, 2, 3)$. From the condition of equilibrium of the tetrahedron we obtain three equalities which can be written in the following form:[1]

$$p_j = \sigma_{ij} a_i. \tag{2.1}$$

The normal stress component is equal to the sum of projections of the components $p_j$ on the $n$-direction

$$p_n = p_j a_j.$$

The tangential stress, which is the second component of the resultant stress, is obtained from the relation

$$p_t^2 = p^2 - p_n^2 = p_x^2 + p_y^2 + p_z^2 - p_n^2.$$

On the plane, on which the sought for principal stress $\sigma$ is acting, the resultant stress must be directed along the normal $n$, because only in such a case the shear stresses vanish. This leads to the condition

$$p_j = a_j \sigma.$$

By substituting these relations in equations (2.1) we arrive at the homogeneous set of linear equations with the three sought for direction cosines of the normal $n$ to the principal plane

$$\sigma_{ij} a_i - \sigma a_j = 0.$$

It is a well-known property of such systems of equations that they have non-zero solutions only in the case where the characteristic determinant is equal to zero

$$\begin{vmatrix} \sigma_x - \sigma & \tau_{xy} & \tau_{xz} \\ \tau_{yx} & \sigma_y - \sigma & \tau_{yz} \\ \tau_{zx} & \tau_{zy} & \sigma_z - \sigma \end{vmatrix} = 0.$$

By resolving the left-hand side we obtain an equation of the third order with respect to $\sigma$,

$$\sigma^3 - I_1 \sigma^2 + I_2 \sigma + I_3 = 0, \tag{2.2}$$

---

[1] The summation convention is used here (see Appendix 1). Equation (2.1) is equivalent to three equations, the first of which takes the form

$$p_x = \sigma_x a_x + \tau_{yx} a_y + \tau_{zx} a_z.$$

where

$$I_1 = \sigma_x + \sigma_y + \sigma_z,$$

$$I_2 = \sigma_x \sigma_y + \sigma_y \sigma_z + \sigma_z \sigma_x - \tau_{xy}^2 - \tau_{yz}^2 - \tau_{zx}^2,$$

$$I_3 = \begin{vmatrix} \sigma_x & \tau_{yx} & \tau_{zx} \\ \tau_{xy} & \sigma_y & \tau_{zy} \\ \tau_{xz} & \tau_{yz} & \sigma_z \end{vmatrix}.$$

Equation (2.2) has always three real roots which determine the values of the principal stresses $\sigma_1, \sigma_2, \sigma_3$. Of course, these roots cannot depend on the choice of the coordinate system. This means that also the coefficients $I_1, I_2, I_3$ must be independent from the rotation of the $x, y$ and $z$ axes. Thus they are the invariants of the stress tensor. By assuming that the coordinate axes $x, y, z$ coincide with the principal directions at a specific point, one can express the stress invariants in terms of principal stresses

$$I_1 = \sigma_1 + \sigma_2 + \sigma_3,$$

$$I_2 = \sigma_1 \sigma_2 + \sigma_2 \sigma_3 + \sigma_3 \sigma_1,$$

$$I_3 = \sigma_1 \sigma_2 \sigma_3.$$

The magnitude

$$\sigma_m = \tfrac{1}{3} I_1 = \tfrac{1}{3}(\sigma_1 + \sigma_2 + \sigma_3) = \tfrac{1}{3}(\sigma_x + \sigma_y + \sigma_z)$$

represents the mean stress. Experimental tests show that the value of the mean stress practically has no influence on the plastic yielding of metals. For that reason it is advantageous to assume that the state of stress at a given point is composed of two parts, namely of the isotropic triaxial compression or tension by stresses equal to the mean stress $\sigma_m$ and of that part which remains after the first part is subtracted. It means that the stress tensor may be resolved into two tensors: a spherical one representing the mean stress, and a stress deviatoric

$$\begin{bmatrix} \sigma_x - \sigma_m & \tau_{xy} & \tau_{xz} \\ \tau_{yx} & \sigma_y - \sigma_m & \tau_{yz} \\ \tau_{zx} & \tau_{zy} & \sigma_z - \sigma_m \end{bmatrix}, \quad \text{or in another way,}$$

$$S_{ij} = \begin{bmatrix} S_x & S_{xy} & S_{xz} \\ S_{yx} & S_y & S_{yz} \\ S_{zx} & S_{zy} & S_z \end{bmatrix}. \tag{2.3}$$

In suffix notation the stress deviatoric $s_{ij}$ is related to the stress tensor by the expression

$$s_{ij} = \sigma_{ij} - \delta_{ij}\sigma_m, \qquad (2.4)$$

or in another way,

$$s_{ij} = \sigma_{ij} - \tfrac{1}{3}\delta_{ij}\sigma_{kk}, \qquad (2.4a)$$

since according to the summation convention, we have

$$\sigma_m = \tfrac{1}{3}\sigma_{kk}.$$

In (2.4) $\delta_{ij}$ is Kronecker's symbol, defined by

$$\delta_{ij} = \begin{cases} 1 & \text{for} \quad i = j, \\ 0 & \text{for} \quad i \neq j. \end{cases}$$

The role of stress deviatoric in the theory of plastic deformations is fundamental. The principal directions of the stress deviatoric coincide with those of the stress tensor. Its invariants written in terms of principal stresses have the form

$$
\begin{aligned}
I_1' &= 0, \\
I_2' &= \tfrac{1}{6}[(\sigma_1-\sigma_2)^2 + (\sigma_2-\sigma_3)^2 + (\sigma_3-\sigma_1)^2], \qquad (2.5)\\
I_3' &= (\sigma_1-\sigma_m)(\sigma_2-\sigma_m)(\sigma_3-\sigma_m).
\end{aligned}
$$

The second invariant $I_2'$ of the stress deviatoric plays an important role in the theory of plasticity. It represents the square of a magnitude called the *equivalent stress* or the *intensity of shear stresses* $\sigma_i = \sqrt{I_2'}$. On account of its great significance the formula for $\sigma_i$ in the arbitrarily oriented coordinate system $x, y, z$ is given below:

$$\sigma_i^2 = \tfrac{1}{6}[(\sigma_x-\sigma_y)^2 + (\sigma_y-\sigma_z)^2 + (\sigma_z-\sigma_x)^2] + \tau_{xy}^2 + \tau_{yz}^2 + \tau_{zx}^2. \qquad (2.6)$$

In cases of elementary modes of loading which are used in material testing, this expression has a simple physical meaning. For pure shear we have $\tau_{xy} = \tau$, the remaining stress components being equal to zero. Hence from (2.6) we obtain $\sigma_i = \tau$. For simple uniaxial tension, i.e. for $\sigma_x = \sigma$, and other stress components having zero values, we have $\sigma_i = \sigma/\sqrt{3}$.

Knowing the principal directions and values of principal stresses at a given point, one can easily find the magnitude of the shear stress and the normal stress for a plane arbitrarily inclined to the principal directions. The orientation of the plane in space is determined by the normal $n$, whose direction cosines with respect to the principal axes $1, 2, 3$ are $a_1, a_2, a_3$, respectively. Figure 15 shows an elementary tetrahedron, three faces of which coincide with the principal planes, with the stresses $\sigma_1$, $\sigma_2$, $\sigma_3$ acting on them. The fourth face represents the chosen inclined arbitrarily plane under consideration. Let us resolve the sought for resultant stress acting

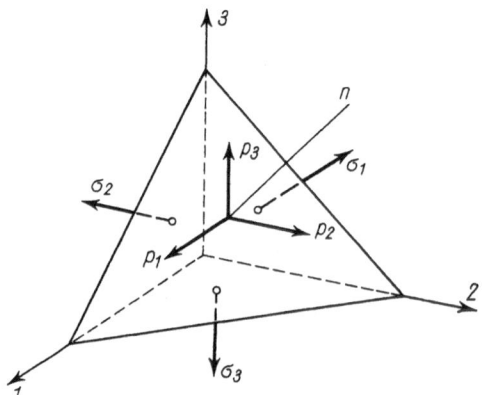

Fig. 15. Elementary tetrahedron

on this plane into three components $p_1, p_2, p_3$. The conditions of equilibrium of the tetrahedron require that these components should have the values $p_1 = \sigma_1 a_1$, $p_2 = \sigma_2 a_2$, $p_3 = \sigma_3 a_3$. The normal component of the resultant stress $p$ can be obtained by projecting its components $p_1, p_2, p_3$ on the normal direction $n$, and by adding the three projections

$$p_n = \sigma_1 a_1^2 + \sigma_2 a_2^2 + \sigma_3 a_3^2. \tag{2.7}$$

The tangential component is given by

$$p_t^2 = p^2 - p_n^2. \tag{2.7a}$$

Let us now investigate the values of stresses acting on certain planes in some special positions with respect to the principal directions. Such planes are shown in Fig. 16. The principal planes are denoted by I. The planes II have normals identically inclined to all three principal axes. Then the direction cosines of the normals are $a_1 = a_2 = a_3 = 1/\sqrt{3}$. These planes are called *octahedral planes* since they coincide with the faces of a regular octahedron. The stresses acting on the octahedral planes are called

18

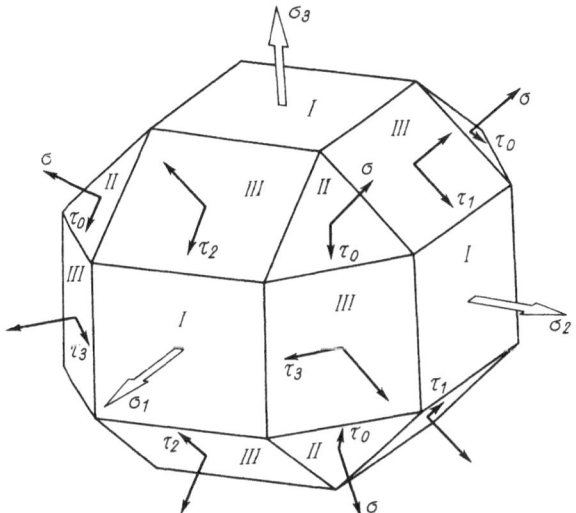

Fig. 16. Planes specifically inclined to principal directions: I—principal planes,
II—octahedral planes, III—planes of maximum shear stresses

*octahedral stresses.* The value of the normal octahedral stress is obtained
directly from (2.7),

$$p_n^0 = \tfrac{1}{3}(\sigma_1 + \sigma_2 + \sigma_3) = \sigma_m,$$

which equals one third of the first invariant of the stress tensor, or in other
words, the mean stress. The octahedral shear stress $\tau_0$ may be obtained
from the formula $\tau_0 = \sqrt{p^2 - \sigma_m^2}$, which after some rearrangements gives

$$\tau_0 = \tfrac{1}{3}\sqrt{[(\sigma_1 - \sigma_2)^2 + (\sigma_2 - \sigma_3)^2 + (\sigma_3 - \sigma_1)^2]}.$$

It can be seen that this stress component is related in a simple way to
the second invariant of the stress deviatoric and to the intensity of shear
stresses,

$$\tau_0 = \sqrt{\tfrac{2}{3} I_2'} = \sqrt{\tfrac{2}{3}} \sigma_i.$$

Thus the intensity of shear stresses is proportional to the octahedral shear
stress.

The planes III are of great significance. Each of them is parallel to
one of the principal directions and intersects two other principal axes at

an angle of 45°. The shear stresses acting on these planes are determined by the relations

$$\tau_1 = \frac{\sigma_2 - \sigma_3}{2},$$

$$\tau_2 = \frac{\sigma_1 - \sigma_3}{2},$$

$$\tau_3 = \frac{\sigma_1 - \sigma_2}{2}.$$

They are the largest of all shear stresses acting on any plane parallel to the respective principal axis. The one among the stresses $\tau_1$, $\tau_2$, $\tau_3$ which is greater than two others, constitutes the maximum shear stress at the point under consideration. If for instance $\sigma_1 > \sigma_2 > \sigma_3$, the shear stress $\tau_2$ will be the maximum one (Fig. 16). The planes III are called the *planes of maximum shear stresses*.

The state of stress on a cross-section arbitrarily inclined to the principal axes can be found by means of the *Mohr's graphical stress representation*. This method is particularly advantageous if stresses on the planes parallel to one of the principal directions and arbitrarily inclined to the two others are to be found. An example of such planes represents the planes III in Fig. 16. Let us examine the state of stress on the plane parallel to the principal axis *3*. The normal to this plane makes the angle $\alpha$ with the *1*-axis. From (2.7) we obtain after simple rearrangements (Fig. 17)

$$\sigma_\alpha = \frac{\sigma_1 + \sigma_2}{2} + \frac{\sigma_1 - \sigma_2}{2} \cos 2\alpha,$$

$$\tau_\alpha = -\frac{\sigma_1 - \sigma_2}{2} \sin 2\alpha.$$

(2.8)

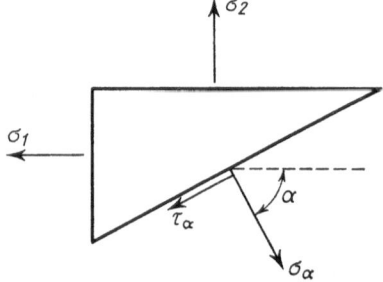

Fig. 17. Stresses on an arbitrary plane parallel to principal direction *3*

Similar relations can be obtained for the planes parallel to the principal axes *2* and *3*. Relations (2.8) have a simple geometrical interpretation in the stress plane (Fig. 18a). The points having coordinates $\sigma_\alpha$ and $\tau_\alpha$ cor-

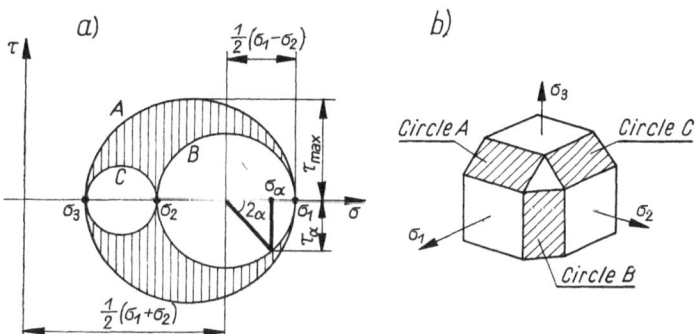

Fig. 18. Geometrical representation of a stress state by means of Mohr circles

responding to different values of the angle $\alpha$ form a circle *B* of diameter $\sigma_1 - \sigma_2$. Thus, making use of this circle, one can directly obtain the state of stress existing on any arbitrary plane parallel to the principal axis *3* and having the normal making an angle $\alpha$ with the principal axis *1*. The state of stress is determined by the coordinates of the point at which the radius inclined at the angle $2\alpha$ to the horizontal axis $\sigma$ intersects the Mohr circle. Similarly, the circle *A* represents the states of stress on the planes parallel to the principal direction *2*, and the circle *C* the states of stress on the planes parallel to the direction *1* (Fig. 18b). Points lying inside the shaded region in Fig. 18a correspond to the states of stress on the planes which are not parallel to any of the principal directions. According to the previous statement, the largest shear stresses on each of the three Mohr circles correspond to the angle $2\alpha = 90°$, thus to the angle $\alpha = 45°$ in the physical space. The maximum shear stress is equal to the radius of the greatest of the Mohr circles.

## 2.2  The state of strain

In order to describe the change of the shape of the body undergoing plastic deformations, let us compare the positions of its material points before and after deformation. According to the remark in Section 1.3, we will assume in all further considerations that the elastic part of deformation does not exist. In other words, deformations will be identified with *plastic deformations* only. Let us assume that a specific point $P$, determined by its initial coordinates $x, y, z$ suffers, due to the process of plastic deformation, a small displacement which is determined by a vector **u** with the components $u_x, u_y, u_z$ (Fig. 19). Let us analyse how the lengths of the

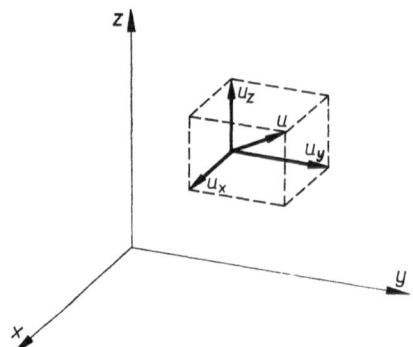

Fig. 19. Displacement vector

edges of an elementary parallelepiped have changed, assuming that initially the edges were parallel to the axes $x, y, z$ and their lengths were $dx, dy$ and $dz$, respectively. The change suffered by the edge $dx$ may be written as $(\partial u_x/\partial x)dx$. Hence the strain increment in the $x$-direction is $d\varepsilon_x = \partial u_x/\partial x$. Similarly the strain increments in the $y$ and $z$-directions are

$$d\varepsilon_x = \frac{\partial u_x}{\partial x}, \quad d\varepsilon_y = \frac{\partial u_y}{\partial y}, \quad d\varepsilon_z = \frac{\partial u_z}{\partial z}. \tag{2.9}$$

The symbols $d\varepsilon_x, d\varepsilon_y, d\varepsilon_z$ have been used here in order to underline that strain increments are small. They may be regarded as the increments of strains between the consecutive instants of the process of plastic flow. Besides the changes of the lengths of the edges, also the right angles at the corners of the parallelepiped before deformation are later subject to changes. For instance, the right angle made by the edges $dx$ and $dy$ changes by the magnitude $d\gamma_{xy} = (\partial u_x/\partial y)+(\partial u_y/\partial x)$ which represents the increment of the angle of distortion. Figure 20 shows the projection of the deformed face $dxdy$ onto the $xy$-plane. Similarly, the remaining increments of the distortion angle can be obtained. Finally we have

$$d\gamma_{xy} = \frac{\partial u_x}{\partial y} + \frac{\partial u_y}{\partial x}, \quad d\gamma_{yz} = \frac{\partial u_y}{\partial z} + \frac{\partial u_z}{\partial y}, \quad d\gamma_{zx} = \frac{\partial u_z}{\partial x} + \frac{\partial u_x}{\partial z}.$$

(2.10)

The equalities $d\gamma_{xy} = d\gamma_{yx}$, $d\gamma_{yz} = d\gamma_{zy}$, $d\gamma_{zx} = d\gamma_{xz}$ must hold.

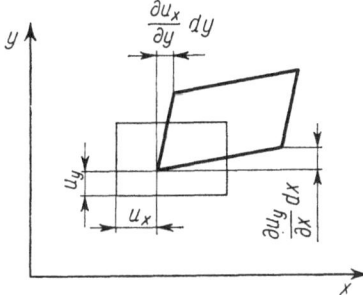

Fig. 20. Displacement and deformation of the elementary parallelepiped

Thus the strain increments at a specific point can be determined by the strain increment tensor

$$\begin{bmatrix} d\varepsilon_x & d\varepsilon_{xy} & d\varepsilon_{xz} \\ d\varepsilon_{yx} & d\varepsilon_y & d\varepsilon_{yz} \\ d\varepsilon_{zx} & d\varepsilon_{zy} & d\varepsilon_z \end{bmatrix},$$

where $d\varepsilon_{xy}$, $d\varepsilon_{yz}$, $d\varepsilon_{zx}$ are half increments of the angles of distortion. Making use of suffix notation, we can write the relations between components of the strain increment tensor and components of the small displacement vector in the form

$$d\varepsilon_{ij} = \frac{1}{2}\left(\frac{\partial u_i}{\partial x_j} + \frac{\partial u_j}{\partial x_i}\right).$$

(2.11)

The strain increment tensor has three invariants, the first of them,

$$\mathscr{I}_1 = d\varepsilon_x + d\varepsilon_y + d\varepsilon_z = 3d\varepsilon_m,$$

representing the relative volumetric increment.[1] Experimental tests show

---

[1] The volume of the deformed elementary parallelepiped is $(1+d\varepsilon_x)(1+d\varepsilon_y)(1+d\varepsilon_z) \times dxdydz$. Subtracting the initial volume $V = dxdydz$ and disregarding small terms of higher orders, we obtain $\Delta V/V = d\varepsilon_x + d\varepsilon_y + d\varepsilon_z$.

that for plastically deformed metals the change in volume is very small. Therefore one assumes almost always in the theory of plastic flow the incompressibility of the metal. Thus we have $\mathscr{I}_1 = 0$. Similarly as in the analysis of the stress state, it is advantageous to resolve the strain increment tensor into the spherical tensor with the components $d\varepsilon_m$ and the strain increment deviatoric

$$\begin{bmatrix} d\varepsilon_x - d\varepsilon_m & d\varepsilon_{xy} & d\varepsilon_{xz} \\ d\varepsilon_{yx} & d\varepsilon_y - d\varepsilon_m & d\varepsilon_{yz} \\ d\varepsilon_{zx} & d\varepsilon_{zy} & d\varepsilon_z - d\varepsilon_m \end{bmatrix}.$$

Note that, if incompressibility is assumed, we have $d\varepsilon_m = 0$ and the strain increment deviatoric coincides with the strain increment tensor.

Invariants of the deviatoric of the strain increments written in terms of the principal strain increments are

$$\mathscr{I}'_1 = 0,$$

$$\mathscr{I}'_2 = \tfrac{1}{6}[(d\varepsilon_1 - d\varepsilon_2)^2 + (d\varepsilon_2 - d\varepsilon_3)^2 + (d\varepsilon_3 - d\varepsilon_1)^2],$$

$$\mathscr{I}'_3 = (d\varepsilon_1 - d\varepsilon_m)\,(d\varepsilon_2 - d\varepsilon_m)\,(d\varepsilon_3 - d\varepsilon_m).$$

Principal strain increments can be found in the same manner as the principal stresses. The second invariant of the deviatoric of the strain increments is among the fundamental notions in the theory of plastic flow. It represents the square of a magnitude called the *increment of the distortion intensity* or *increment of the equivalent strain* $d\varepsilon_i = \sqrt{\mathscr{I}'_2}$. The notation $d\varepsilon_i$ is used in order to emphasize that this magnitude should be regarded as the increment of the distortion intensity accumulated in a material particle in a short period of time between two consecutive stages of the advancing process of plastic flow. The total intensity itself characterizing the state of strain accumulated since the beginning of the process can be obtained by summation of the increments $d\varepsilon_i$,

$$\varepsilon_i = \int d\varepsilon_i = \int \sqrt{\tfrac{1}{6}[(d\varepsilon_1 - d\varepsilon_2)^2 + (d\varepsilon_2 - d\varepsilon_3)^2 + (d\varepsilon_3 - d\varepsilon_1)^2]}. \qquad (2.12)$$

For the elementary deformation modes the distortion intensity has a simple physical meaning. For the uniaxial tension we have $d\varepsilon_1 = d\varepsilon$, and by assuming the incompressibility of the material, $d\varepsilon_2 = d\varepsilon_3 = -0.5d\varepsilon$. Hence $\varepsilon_i = \int \tfrac{1}{2}\sqrt{3}\, d\varepsilon = \tfrac{1}{2}\sqrt{3}\, \varepsilon$. For pure shear $d\varepsilon_1 = d\varepsilon$, $d\varepsilon_2 = -d\varepsilon$ and $d\varepsilon_3 = 0$. Hence the distortion intensity is $\varepsilon_i = \int d\varepsilon = \varepsilon$.

### 2.3   Strain rates

In the analysis of the mechanics of plastic forming processes it is more advantageous to use the flow velocities than displacements. This is connected with the fact that the theory of plastic flow, upon which this analysis is based, formulates relations between the rates of strain and the stresses and not between the strains themselves and stresses. Also the boundary conditions in specific problems are usually given in terms of velocities, either as velocities of the tool or as velocities of moving material, as for instance, in the case of wire drawing.

The flow velocity of the material at a given point is determined by a vector $\mathbf{v}$ with components $v_x, v_y, v_z$ parallel to the coordinate axes. If $\delta t$ is a short time increment, within which the velocities may be assumed to preserve the constant values, then the displacements can be determined as follows:

$$u_x = v_x \, \delta t, \quad u_y = v_y \, \delta t, \quad u_z = v_z \, \delta t.$$

Similarly as strain increments have been expressed by means of the displacement vector $\mathbf{u}$, the strain rates can be connected with the components of the velocity vector $\mathbf{v}$

$$\epsilon_x = \frac{\partial v_x}{\partial x}, \quad \epsilon_y = \frac{\partial v_y}{\partial y}, \quad \epsilon_z = \frac{\partial v_z}{\partial z},$$

$$\epsilon_{xy} = \frac{1}{2}\left(\frac{\partial v_x}{\partial y} + \frac{\partial v_y}{\partial x}\right), \quad \epsilon_{yz} = \frac{1}{2}\left(\frac{\partial v_y}{\partial z} + \frac{\partial v_z}{\partial y}\right), \qquad (2.13)$$

$$\epsilon_{zx} = \frac{1}{2}\left(\frac{\partial v_z}{\partial x} + \frac{\partial v_x}{\partial z}\right).$$

Therefore the strain rate at a given point is determined by the strain rate tensor

$$\epsilon_{ij} = \begin{bmatrix} \epsilon_x & \epsilon_{xy} & \epsilon_{xz} \\ \epsilon_{yx} & \epsilon_y & \epsilon_{yz} \\ \epsilon_{zx} & \epsilon_{zy} & \epsilon_z \end{bmatrix}.$$

Relations (2.13) can be written in a compact form, by using the suffix notation, as

$$\epsilon_{ij} = \frac{1}{2}\left(\frac{\partial v_i}{\partial x_j} + \frac{\partial v_j}{\partial x_i}\right). \qquad (2.14)$$

25

The condition of the incompressibility of the material takes now the form

$$\epsilon_x + \epsilon_y + \epsilon_z = 0 \tag{2.15}$$

which is very useful in practical applications.

The axially symmetrical flow, which is of great practical significance, is usually analysed in cylindrical coordinates $r, \vartheta, z$. Let $v_r$ denote the radial velocity component, $v_\vartheta$ the circumferential velocity component, and $v_z$ the axial component. Then the strain rate components are

$$\epsilon_r = \frac{\partial v_r}{\partial r}, \qquad \epsilon_\vartheta = \frac{1}{r}\left(v_r + \frac{\partial v_\vartheta}{\partial \vartheta}\right), \qquad \epsilon_z = \frac{\partial v_z}{\partial z},$$

$$\epsilon_{r\vartheta} = \frac{1}{2}\left[\frac{1}{r}\left(\frac{\partial v_r}{\partial \vartheta} - v_\vartheta\right) + \frac{\partial v_\vartheta}{\partial r}\right],$$

$$\epsilon_{\vartheta z} = \frac{1}{2}\left(\frac{\partial v_\vartheta}{\partial z} + \frac{1}{r}\frac{\partial v_z}{\partial \vartheta}\right), \tag{2.16}$$

$$\epsilon_{zr} = \frac{1}{2}\left(\frac{\partial v_z}{\partial r} + \frac{\partial v_r}{\partial z}\right).$$

### 2.4 Equations of motion and equations of equilibrium

Assume that there are prescribed at every point of the plastically deforming medium a flow velocity vector **v** and an acceleration vector $d\mathbf{v}/dt$. To obtain the differential equations of motion we have to introduce the d'Alembert inertial forces. The vector of the d'Alembert force per unit volume is $-\varrho \, d\mathbf{v}/dt$, where $\varrho$ stands for the mass density of the medium. Let us explain first of all the meaning of the symbol $d\mathbf{v}/dt$, in terms of which the vector of acceleration will be determined. The velocity vector is a function of the coordinates $x, y, z$ and also of the time $t$ if the general case of non-stationary motion is considered. Thus the velocity is a function of four independent variables $\mathbf{v} = v(x, y, z, t)$. The velocity components are

$$v_x = v_x(x, y, z, t), \qquad v_y = v_y(x, y, z, t), \qquad v_z = v_z(x, y, z, t).$$

By way of example, we shall determine the acceleration component in the $x$-axis direction. The change of the velocity component $v_x$ during a short time increment $\delta t$ is

$$\delta v_x = \frac{\partial v_x}{\partial t}\delta t + \frac{\partial v_x}{\partial x}\delta x + \frac{\partial v_x}{\partial y}\delta y + \frac{\partial v_x}{\partial z}\delta z,$$

where $\delta x$, $\delta y$, $\delta z$ are the components of displacement of a particle during the time increment $\delta t$. On dividing the two sides by $\delta t$ we get

$$\frac{\delta v_x}{\delta t} = \frac{\partial v_x}{\partial t} + \frac{\partial v_x}{\partial x}\frac{\delta x}{\delta t} + \frac{\partial v_x}{\partial y}\frac{\delta y}{\delta t} + \frac{\partial v_x}{\partial z}\frac{\delta z}{\delta t}.$$

If the time increment tends to zero, the differences may be replaced by differentials. Thus we have

$$\frac{dv_x}{dt} = \frac{\partial v_x}{\partial t} + \frac{\partial v_x}{\partial x}\frac{dx}{dt} + \frac{\partial v_x}{\partial y}\frac{dy}{dt} + \frac{\partial v_x}{\partial z}\frac{dz}{dt}.$$

The derivatives $dx/dt$, $dy/dt$, $dz/dt$ are the components of the velocity vector $v_x, v_y, v_z$ at a point considered and at a given instant. Finally we obtain the acceleration in the direction of the x-axis

$$\frac{dv_x}{dt} = \frac{\partial v_x}{\partial t} + v_x\frac{\partial v_x}{\partial x} + v_y\frac{\partial v_x}{\partial y} + v_z\frac{\partial v_x}{\partial z}. \tag{2.17a}$$

In the same manner accelerations in the y and z-axis directions are found to be

$$\frac{dv_y}{dt} = \frac{\partial v_y}{\partial t} + v_x\frac{\partial v_y}{\partial x} + v_y\frac{\partial v_y}{\partial y} + v_z\frac{\partial v_y}{\partial z}, \tag{2.17b}$$

$$\frac{dv_z}{dt} = \frac{\partial v_z}{\partial t} + v_x\frac{\partial v_z}{\partial x} + v_y\frac{\partial v_z}{\partial y} + v_z\frac{\partial v_z}{\partial z}. \tag{2.17c}$$

Passing to the conditions of equilibrium of an elementary parallelepiped shown in Fig. 13, we obtain by projecting all static and inertial forces on the coordinate directions $x, y, z$ the equations of motion

$$\frac{\partial \sigma_x}{\partial x} + \frac{\partial \tau_{xy}}{\partial y} + \frac{\partial \tau_{xz}}{\partial z} = \varrho\left(\frac{\partial v_x}{\partial t} + v_x\frac{\partial v_x}{\partial x} + v_y\frac{\partial v_x}{\partial y} + v_z\frac{\partial v_x}{\partial z}\right),$$

$$\frac{\partial \tau_{xy}}{\partial x} + \frac{\partial \sigma_y}{\partial y} + \frac{\partial \tau_{yz}}{\partial z} = \varrho\left(\frac{\partial v_y}{\partial t} + v_x\frac{\partial v_y}{\partial x} + v_y\frac{\partial v_y}{\partial y} + v_z\frac{\partial v_y}{\partial z}\right), \tag{2.18a}$$

$$\frac{\partial \tau_{xz}}{\partial x} + \frac{\partial \tau_{yz}}{\partial y} + \frac{\partial \sigma_z}{\partial z} = \varrho\left(\frac{\partial v_z}{\partial t} + v_x\frac{\partial v_z}{\partial x} + v_y\frac{\partial v_z}{\partial y} + v_z\frac{\partial v_z}{\partial z}\right),$$

which in suffix notation take the form

$$\frac{\partial \sigma_{ij}}{\partial x_j} = \varrho\left(\frac{\partial v_i}{\partial t} + v_j\frac{\partial v_i}{\partial x_j}\right). \tag{2.18b}$$

In these equations the body forces have been omitted, because their role in the mechanics of plastic forming processes in negligible.

In cylindrical coordinates $r, \vartheta, z$ (Fig. 21) the equations of motion take the form

$$\frac{\partial \sigma_r}{\partial r} + \frac{1}{r}\frac{\partial \tau_{r\vartheta}}{\partial \vartheta} + \frac{\partial \tau_{rz}}{\partial z} + \frac{\sigma_r - \sigma_\vartheta}{r}$$

$$= \varrho\left(\frac{\partial v_r}{\partial t} + v_r\frac{\partial v_r}{\partial r} + \frac{v_\vartheta}{r}\frac{\partial v_r}{\partial \vartheta} + v_z\frac{\partial v_r}{\partial z} - \frac{v_\vartheta^2}{r}\right),$$

$$\frac{\partial \tau_{r\vartheta}}{\partial r} + \frac{1}{r}\frac{\partial \sigma_\vartheta}{\partial \vartheta} + \frac{\partial \tau_{\vartheta z}}{\partial z} + \frac{2\tau_{r\vartheta}}{r}$$

$$= \varrho\left(\frac{\partial v_\vartheta}{\partial t} + v_r\frac{\partial v_\vartheta}{\partial r} + \frac{v_\vartheta}{r}\frac{\partial v_\vartheta}{\partial \vartheta} + v_z\frac{\partial v_\vartheta}{\partial z} + \frac{v_r v_\vartheta}{r}\right). \tag{2.19}$$

$$\frac{\partial \tau_{rz}}{\partial r} + \frac{1}{r}\frac{\partial \tau_{\vartheta z}}{\partial \vartheta} + \frac{\partial \sigma_z}{\partial z} + \frac{\tau_{rz}}{r}$$

$$= \varrho\left(\frac{\partial v_z}{\partial t} + v_r\frac{\partial v_z}{\partial r} + \frac{v_\vartheta}{r}\frac{\partial v_z}{\partial \vartheta} + v_z\frac{\partial v_z}{\partial z}\right).$$

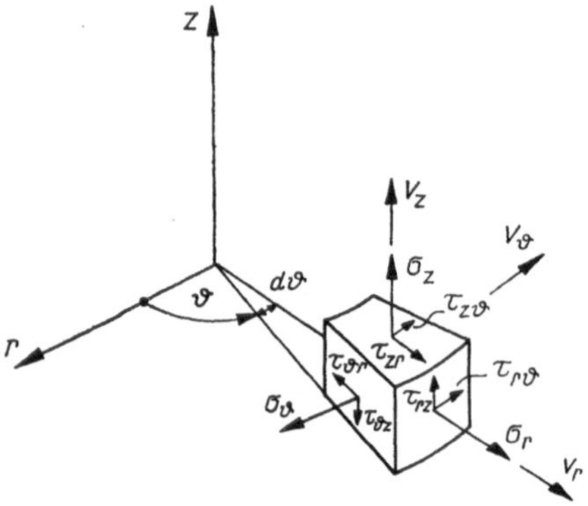

Fig. 21. Stresses and velocities in a cylindrical coordinate system

Of great practical significance is the particular case of axial symmetry, when all stress and velocity components do not depend on the coordinate $\vartheta$ and moreover $\tau_{r\vartheta} = \tau_{\vartheta z} = 0$ and $v_\vartheta = 0$. The equations of motion are then much simpler:

$$\frac{\partial \sigma_r}{\partial r} + \frac{\partial \tau_{rz}}{\partial z} + \frac{\sigma_r - \sigma_\vartheta}{r} = \varrho\left(\frac{\partial v_r}{\partial t} + v_r\frac{\partial v_r}{\partial r} + v_z\frac{\partial v_r}{\partial z}\right),$$

$$\frac{\partial \tau_{rz}}{\partial r} + \frac{\partial \sigma_z}{\partial z} + \frac{\tau_{rz}}{r} = \varrho\left(\frac{\partial v_z}{\partial t} + v_r\frac{\partial v_z}{\partial r} + v_r\frac{\partial v_z}{\partial z}\right). \tag{2.20}$$

For the spherical coordinates $r$, $\vartheta$, $\varphi$ our considerations will be limited to the case of spherical symmetry when all stress and velocity components depend on the radius $r$ only. In such case we have also $v_\vartheta = v_\varphi = 0$, $\tau_{r\vartheta} = \tau_{r\varphi} = 0$ and $\sigma_\vartheta = \sigma_\varphi$. The conditions of equilibrium reduce then to one equation

$$\frac{\partial \sigma_r}{\partial r} + \frac{2(\sigma_r - \sigma_\vartheta)}{r} = \varrho \left( \frac{\partial v_r}{\partial t} + v_r \frac{\partial v_r}{\partial r} \right). \tag{2.21}$$

If the process of plastic flow is accompanied by small accelerations of the particles, the inertial forces may be omitted in the considerations of the equilibrium conditions. The right-hand sides of the equations of motion (2.18)–(2.21) are then zero, and these equations reduce to the equations of equilibrium.

# Yield conditions and flow laws

## 3.1 General remarks

If the material is loaded by a complex stress state, the question arises when the transition from an elastic state to the plastic state takes place. For the uniaxial loading this happens when the yield point is reached. For a complex state of stress one can expect that isotropic material will yield plastically if a certain relation between the invariants of the stress tensor is satisfied. Thus the yield condition may be generally written in the form

$$F(I_1, I_2, I_3) = 0. \tag{3.1}$$

Unfortunately, there does not exist at the present moment a theoretical way of establishing this relation. The yield conditions are therefore empirical criteria but we can verify their practical significance for real metals, having at present at our disposal a large number of experimental data.

Experimental tests of metals loaded by moderate hydrostatic compression, performed first by P. W. Bridgman (see [13]), show that yielding of metals is, up to a first order approximation, unaffected by the value of the first invariant of the stress tensor. Thus the yield condition does not depend on the spherical part of the stress tensor and will be connected only with the invariants of the stress deviatoric

$$F(I_2', I_3') = 0. \tag{3.2}$$

The first invariant of the stress deviatoric $I_1'$, being zero, obviously does not appear in this relation.

Condition (3.2) may be represented as a certain surface (Fig. 22) in the stress space, because $I_2'$ and $I_3'$ are functions of stresses which are expressed by (2.5). If the state of stress is determined by a point lying inside the surface, the material is in the elastic state. The points belonging to the surface correspond to states of stress at which the material deforms plasti-

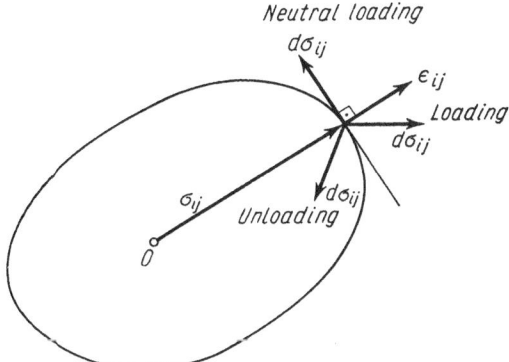

Fig. 22. Yield surface

cally. Such a surface is usually called the *yield surface*. The concept of yield surface is very advantageous in the explanation of certain notions in the theory of plasticity. For example the definition of unloading of the plastically deformed material, which is simple in the case of uniaxial loading, is much more difficult to delineate in a complex state of stress. However, using the concept of the yield surface, we may identify unloading with a stress increment which in the stress space is represented by a vector $d\sigma_{ij}$ directed inwards from the yield surface. Similarly, if the vector of stress increment is directed outwards from the yield surface, we have the loading process which can take place only in the case where the material displays strain-hardening. The discussion of this case will be given in further sections where the strain-hardening hypotheses will be considered. The stress increment $d\sigma_{ij}$ represented by a vector tangent to the yield surface corresponds to the so-called neutral loading. We shall later show that the yield surface must be convex. Certain singularities in the form of sharp edges and conical points are, however, admissible.

In practical computations of the mechanics of plastic forming processes two basic yield conditions are used:

(1) The *Huber–Mises yield condition*, which in some works is called the *condition of the constant intensity of shear stresses*.

(2) The *Tresca yield condition*, known also as the *condition of constant maximum shear stress*.

### 3.2 Huber–Mises yield condition

This condition was proposed by M. T. Huber [56] as early as 1904. Huber suggested that yielding of the metal begins when the elastic energy of distortion reaches a critical value. His idea published in Polish did not, unfortunately, attract general attention, and was independently repeated by R. von Mises in 1913, [90] in closer connection with the theory of plasticity. The Huber–Mises yield condition assumes a very simple form of the general relation (3.2), namely

$$I_2' - k^2 = 0, \tag{3.3}$$

where $k$ stands for a constant characterizing the plastic behaviour of the material.

Thus condition (3.3) does not depend on the third invariant of the stress deviatoric. In Chapter 2 we have introduced the notion of the shear stresses intensity $\sigma_i$, connected with the second invariant of the stress deviatoric by means of the simple formula $\sigma_i = \sqrt{I_2'}$. Therefore, condition (3.3) corresponds to yielding at the constant critical value of the shear stresses intensity $\sigma_i = k$, and thus it is often referred to as the *condition of the intensity of constant shear stresses*. In most works, however, it is called the *Huber–Mises yield condition*, or in some western works, the *Mises yield condition*.

Let us express the Huber–Mises condition in terms of stresses. Replacing $I_2'$ in (3.3) by the expression (2.5), we obtain the fundamental relation

$$(\sigma_1 - \sigma_2)^2 + (\sigma_2 - \sigma_3)^2 + (\sigma_3 - \sigma_1)^2 = 6k^2, \tag{3.4}$$

very convenient if problems are considered, in which the directions of the principal stresses $\sigma_1, \sigma_2, \sigma_3$ are known in advance. Various axially symmetric problems may be mentioned as typical examples. In most cases, however, principal directions are not known and should be determined. In such cases it is more advantageous to use the following expression for the yield condition

$$(\sigma_x - \sigma_y)^2 + (\sigma_y - \sigma_z)^2 + (\sigma_z - \sigma_x)^2 + 6(\tau_{xy}^2 + \tau_{yz}^2 + \tau_{zx}^2) = 6k^2, \tag{3.5}$$

obtained on substituting the general expression (2.6) for the intensity of shear stresses.

The two expressions (3.4) and (3.5) are of fundamental significance in the theory of plasticity. We shall use them in further chapters when solving particular problems.

The constant $k$ has a simple physical meaning. Assume that in the coordinate system $x$, $y$, $z$ all stress components except $\tau_{xy}$ are zero. Hence $\tau_{xy} = k$, and therefore $k$ stands for the magnitude of shear stresses in the case of pure shear at which material begins to yield. In other words, $k$ represents the yield point in a test of pure shear, which usually is obtained in experimental investigations by twisting thin-walled tubular specimens. If the thickness of the wall is small as compared with the tube diameter, one can, with the sufficient degree of accuracy, assume that the value of shear stress does not vary across the thickness of the wall. Thus there is at our disposal a simple method for the experimental evaluation of the constant $k$. However, the value of $k$ may also be obtained in a simpler way, by means of the uniaxial tension test. Let us assume that only the principal stress $\sigma_1$ having the value of the yield point $\sigma_{pl}$ is present, while the two remaining principal stresses $\sigma_2$ and $\sigma_3$ are zero. From the relation (3.4) we obtain

$$k = \frac{\sigma_{pl}}{\sqrt{3}}. \tag{3.6}$$

Thus having found the yield point $\sigma_{pl}$ from the uniaxial tension test, which is technically much simpler than any of the pure shearing tests, we may by means of the relation (3.6) compute the constant $k$.

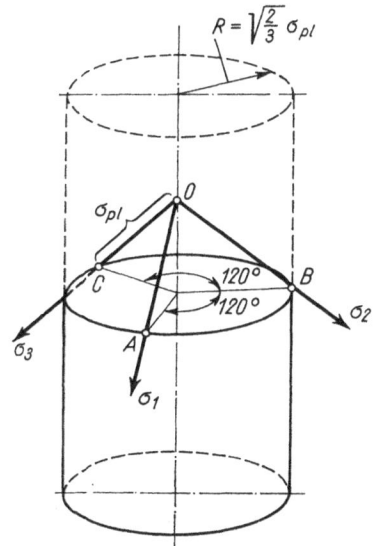

Fig. 23. Geometrical representation of the Huber–Mises yield condition in the space of principal stresses

The yield surface corresponding to the Huber–Mises yield condition takes in the principal stress space $\sigma_1$, $\sigma_2$, $\sigma_3$ the form of an infinitely long cylinder of circular cross-section, and with the axis inclined at the same angle to each of the three axes of principal stresses. In Fig. 23, by means of heavy lines, there is shown a portion of this part of the cylinder which passes through the octant of the stress space adjoining to the positive halves of the $\sigma_1$, $\sigma_2$, $\sigma_3$-axes. The remaining part of the cylinder is shown by dashed lines. Since the points $A$, $B$, $C$ at which the coordinate axes intersect the yield surface represent states of uniaxial tension in the 1, 2 and 3-directions, respectively, the segments $OA$, $OB$ and $OC$ are equal to the yield point in uniaxial tension. Hence, from simple geometrical considerations, we get the length of radius of the cylinder $R = \sqrt{\frac{2}{3}}\sigma_{\mathrm{pl}}$ or $R = \sqrt{2}\,k$.

### 3.3   Tresca yield condition

This yield condition was established much earlier than the Huber–Mises yield condition; indeed, as early as in 1864, H. Tresca published results of his experiments on the plastic flow of metals,[1] from which he inferred the hypothesis that plastic yielding starts at the moment when the largest shear stress reaches a critical value. Assuming that $\sigma_1 > \sigma_2 > \sigma_3$, we can write down this yield condition in the form

$$\sigma_1 - \sigma_3 = 2k, \tag{3.7}$$

where $k$ is a constant. Condition (3.7) may be written in terms of the invariants $I_2'$, $I_3'$ but the result is complicated and has no practical significance. The physical meaning of the constant $k$ is obtained, as in the foregoing case, from the consideration of the state of pure shear by stresses $\tau$. From condition (3.7) we obtain directly $\tau = k$. Hence $k$ is the yield locus in the pure shear test. Considering the uniaxial tension test with the tensile stress $\sigma_1 = \sigma_{\mathrm{pl}}$, ($\sigma_2 = \sigma_3 = 0$), we find the following relation between the values of yield points in pure shear and uniaxial tension tests:

$$k = \sigma_{\mathrm{pl}}/2. \tag{3.8}$$

When for a given material the yield point in uniaxial tension is determined, the value of $k$ obtained from (3.8) is approximately 14 percent smaller than

---

[1] For details of Tresca's experiments see the interesting essay in J. F. Bell's book [5].

that obtained from (3.6) for the Huber–Mises condition. This is the largest difference between both yield conditions, and in the past it was frequently exploited in experimental investigations in order to check which of the two conditions better corresponds to the real behaviour of metals.

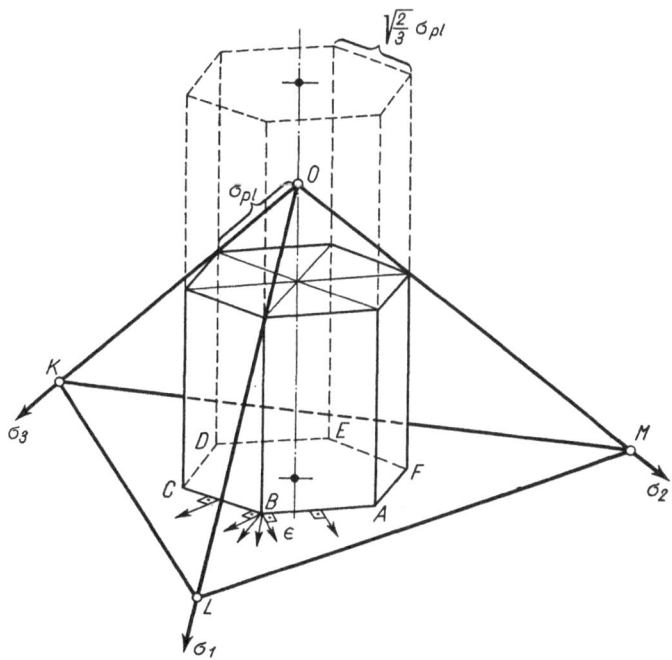

Fig. 24. Geometrical representation of the Tresca yield condition in the space of principal stresses

The yield surface is now represented by an infinitely long prism with the regular hexagon as the cross-section. In Fig. 24 only a portion of this prism is presented. It is easily seen that the Tresca prism is inscribed in the Huber–Mises cylinder. Thus its edges coincide with the generatrices of that cylinder.

### 3.4 Experimental verification of the yield condition

Experiments are performed mostly under plane stress conditions, which means that one of the principal stresses, say $\sigma_3$, is zero. If in the rectangular coordinate system $x, y, z$ the $x, y$-plane coincides with the plane of principal directions 1 and 2, then there exist only the stress components $\sigma_x$, $\sigma_y$, $\tau_{xy}$. The remaining components are zero. The state of plane stress is used in experimental studies, because it can be technically realized in a relatively simple manner. Most of the tests are performed by using thin-walled tubular specimens loaded simultaneously by an axial force, a twisting moment, and internal or external pressure. By suitably choosing the values of these loadings, any arbitrary combination of the values of the stress components $\sigma_x$, $\sigma_y$, $\tau_{xy}$ in the tube can be obtained. If the thickness of the specimen's wall is small as compared with its diameter, we may with good accuracy assume that stresses do not change across the wall-thickness.

We obtain the Huber–Mises yield condition for plane stress by putting $\sigma_z = \tau_{zx} = \tau_{yz} = 0$ in the general expression (3.5):

$$\sigma_x^2 - \sigma_x\sigma_y + \sigma_y^2 + 3\tau_{xy}^2 = 3k^2. \tag{3.9}$$

For fixed $x$ and $y$ directions this condition is geometrically represented in the stress space $\sigma_x$, $\sigma_y$, $\tau_{xy}$ by an ellipsoid (Fig. 25). One of the axes of the ellipsoid coincides with the $\tau_{xy}$-axis whereas the two others lie in the $\sigma_x$, $\sigma_y$-plane, being simultaneously bisectrices of the right angles between the axes $\sigma_x$ and $\sigma_y$. The various most frequently used ways of load-

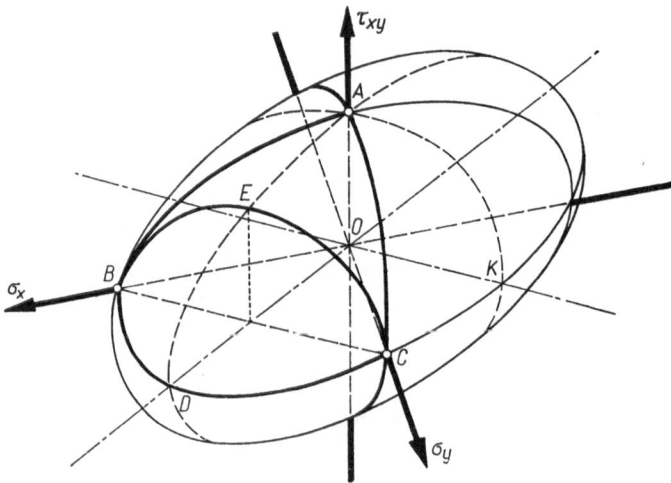

Fig. 25. Geometrical representation of the Huber–Mises yield condition for the plane state of stresses

ing in experiments with tubular specimens are represented by ellipses lying on the surface of the ellipsoid. For instance, the ellipse $AB$, for which $\sigma_y = 0$, corresponds to combined torsion and tension of the specimen by an axial force if the $x$-direction is assumed to coincide with that of the generatrices of the tube. The states of stresses determined by points lying on the ellipse $AC$ ($\sigma_x = 0$) can be obtained by a simultaneous loading of the specimen by twisting moment and internal pressure, the two ends of the tube being closed by plungers to avoid axial tension. The ellipse $BDC$ ($\tau_{xy} = 0$) corresponds to the combined loading by an axial force and internal pressure. It is readily seen that the ellipse $BEC$ represents uniaxial tensions in different directions with respect to $x$-axis. We shall discuss this problem later on in the next section.

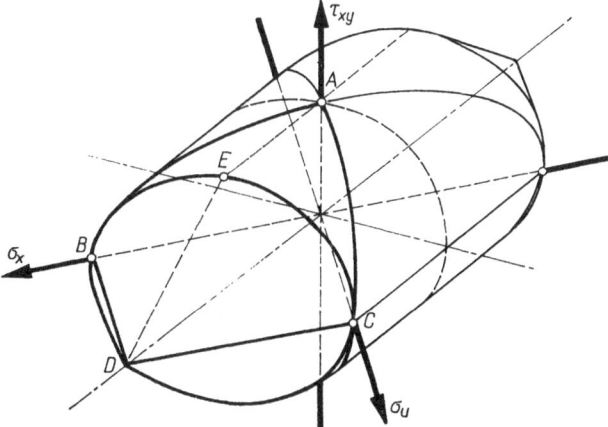

Fig. 26. Geometrical representation of the Tresca yield condition for the plane state of stresses

The yield surface for the Tresca yield condition under plane stress has a more elaborated shape (Fig. 26). Part of it is formed by an elliptic cylinder with the axis coinciding with the axis of the Huber–Mises ellipsoid. This cylinder is closed on both sides by elliptic cones. The ellipse $AB$ corresponding to combined loading by a twisting moment and axial tensile force has now a shorter vertical axis than in the case of the Huber–Mises yield condition. Instead of the ellipse $BDC$ in Fig. 25 we have now a hexagon formed by sectors of straight lines.

In most experimental studies tubular specimens are loaded by combined torsion and axial tension, thus along the ellipse $AB$. Particularly in the recent two decades there appeared a number of works presenting results of very thorough experiments on the effect of plastic prestraining on the yield surface. In each of these works, the initial yield surface of the material in the virgin state was investigated as the starting point. Figure 27 shows results for some metals taken from several papers [6, 57,

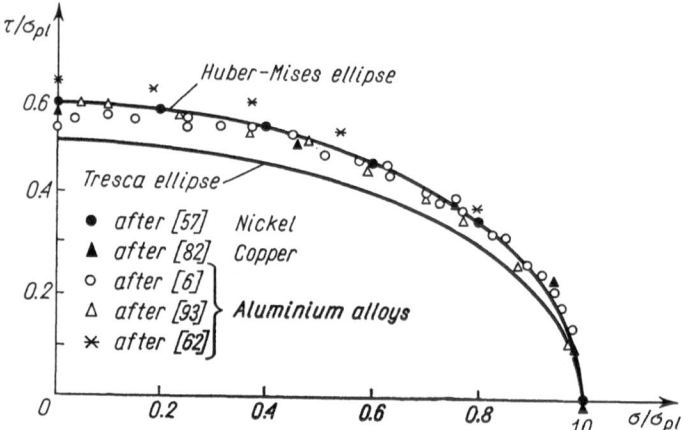

Fig. 27. Experimental verification of the Huber–Mises and Tresca yield conditions. Combined torsion and tension tests

62, 82, 93]. The majority of experimental points lie between the theoretical Huber–Mises and Tresca ellipses. It is seen, however, that they are located closer to the Huber–Mises ellipse. Some points lie even outside this ellipse, but the author suggests that this can be due to possible measuring errors since the sensitivity of the measuring device was limited in this region of loading.

In the classic paper by W. Lode [80] tubular specimens were loaded simultaneously by axial force and internal pressure, thus along the ellipse *BDC* on Fig. 25. His results and results obtained in some recent works [11, 87, 89, 136] are presented in Fig. 28, in which there are also shown the Huber–Mises ellipse and the Tresca hexagon, along which the plane $\tau_{xy} = 0$ intersects the Tresca yield surface (see Fig. 26). Also in this case the experimental points lie between both theoretical lines, being in general closer to the Huber–Mises ellipse.

All experimental works confirm the validity of the Huber–Mises yield condition for all tested metals. Only mild steel gives results in between

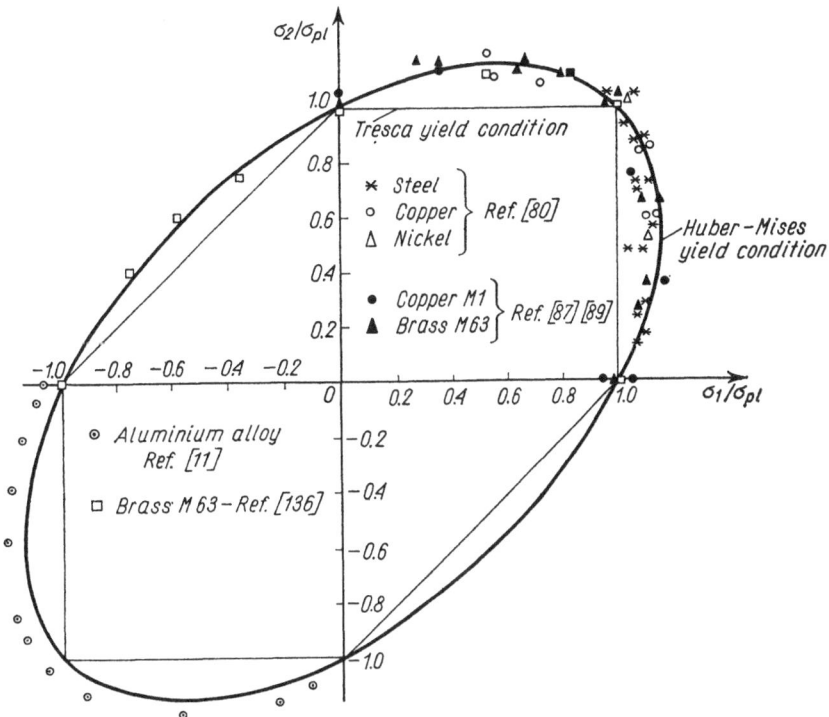

Fig. 28. Experimental verification of the Huber–Mises and Tresca yield conditions. Combined tension and internal pressure tests

the two conditions, and even frequently closer to the Tresca condition. Thus it is generally accepted that the Huber–Mises yield condition is sufficiently close to the real properties of metals, and that the Tresca yield condition constitutes a fairly good approximation. In practical applications, however, both conditions are used since the Tresca condition is very convenient for its mathematical simplicity.

## 3.5 Effect of plastic deformation on yield condition

All results referred to in the foregoing section were obtained using metals which did not undergo previous plastic cold working operations, and in most cases specimens before testing were annealed in order to remove effects of material processing as, for instance, rolling or forging. In cold

working processing the material undergoes considerable plastic deformations which strongly deform the yield surface. This problem was recently
intensively investigated. Although the number of experimental data is still
unsufficient, one can, however, use them to formulate certain general
conclusions.

In this section we discuss results of some experiments concerning the
effect of plastic deformation. In the next section theoretical concepts dealing with this problem will be presented. All experiments, except the one
described further in detail, have been performed using tubular specimens.

In the classic work by G. Taylor and H. Quinney [132] each specimen
was loaded by uniaxial tension far beyond the initial yield locus. Assuming
that the ellipsoid in Fig. 25 represents the initial yield condition of the
virgin material, such a loading is represented by a point on the $\sigma_x$-axis
located beyond $B$. Then, after partial unloading, specimens were additionally loaded by a twisting moment, the magnitude of the axial tensile stresses
being kept constant. For each specimen the diagram of elongation versus
twisting moment was plotted. The yield loci were found by extrapolating
the smooth portion of the diagram back to the intersection with the axis
of the twisting moment. This is a different approach from that used in
recent works, in which the proportional limit or various offset definitions
of yield during subsequent loading are considered as the yield locus. Using
that procedure, Taylor and Quinney obtained for steel, copper and aluminium almost perfect ellipses in the $\sigma_x$, $\tau_{xy}$-plane. This means that the
initial ellipse $AB$ in Fig. 25, after prestraining of all specimens by uniaxial
tension, expanded, the position of its central point and half-axes ratio
remaining unchanged. However, interpretation of these results in terms of
present strain-hardening concepts is rather difficult since the yield loci were
established in a peculiar way, neglecting completely the most interesting
portion of the stress-strain diagrams, where the curvature has the largest
values.

In [93] tubular specimens of aluminium alloys were plastically deformed by the twisting moment far beyond the initial yield locus, that
is, above the point $A$ in Fig. 25. Next, after unloading, the specimen was
loaded again by combined torsion and axial tension. The stress-strain
diagram was recorded and loading was stopped immediately when the
first non-linearity of the diagram appeared. As the yield locus the proportional limit was taken. Such a definition of yield is, as far as plastic
forming processes are concerned, of limited significance.

A similar mode of loading was used in [62]. The yield locus also in this work was identified with the proportional limit. J. I. Jagn and O. A. Shishmariev [57] deformed plastically tubular nickel specimens by uniaxial tension, that is, beyond the point $B$ in Fig. 25, and then investigated the yield surface by loading them by various combinations of the axial force and twisting moment. Yield surfaces were established for different offset definitions of yield locus.

In the author's work [125], contrary to the majority of other investigations, not thin-walled tubular specimens but flat specimens were used. The loading mode corresponds to the ellipse *BEC* on the yield ellipsoid (Fig. 25). This ellipse is formed by intersection of the ellipsoid with the plane $\sigma_x + \sigma_y = \sigma_{pl}$ perpendicular to the $\sigma_x$, $\sigma_y$ plane. Introducing into the yield condition (3.9) the sum $\sigma_x + \sigma_y$ instead of $\sigma_{pl}$, we obtain the relation $\sigma_x \sigma_y = \tau_{xy}^2$ which must be satisfied along our ellipse. Such a relation is possible only in the case where one of the principal stresses is equal to zero (see Fig. 29). In other words, points belonging to the ellipse *BEC*

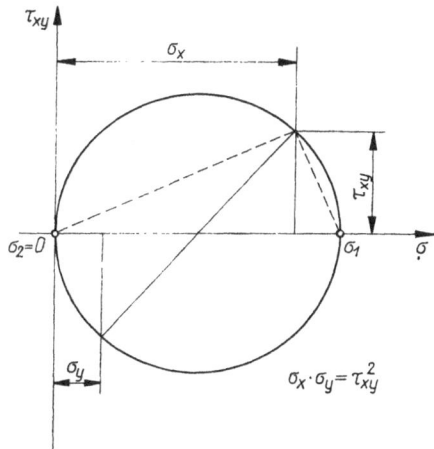

Fig. 29. Mohr circle for uniaxial tension

represent uniaxial tensions in different directions with respect to the $x$-axis, as for instance, the point $E$, for which $\sigma_x = \sigma_y = \tau_{xy}$, represents the uniaxial tension making an angle of $45°$ with the $x$-axis.

In the experiments referred to, an investigation has been made of the change of the initial ellipse *BEC* for an initially isotropic material after prestressing by uniaxial tension in the $x$-direction far beyond the yield locus. The tests were performed as follows. First a large specimen cut out from a sheet of 6.5 mm thick aluminium alloy in the $x$-direction was loaded by uniaxial tensile stress until permanent plastic deformation reached a certain prescribed value. Next, after unloading, small specimens were cut out from this prestressed specimen in various directions with respect to the $x$-axis.

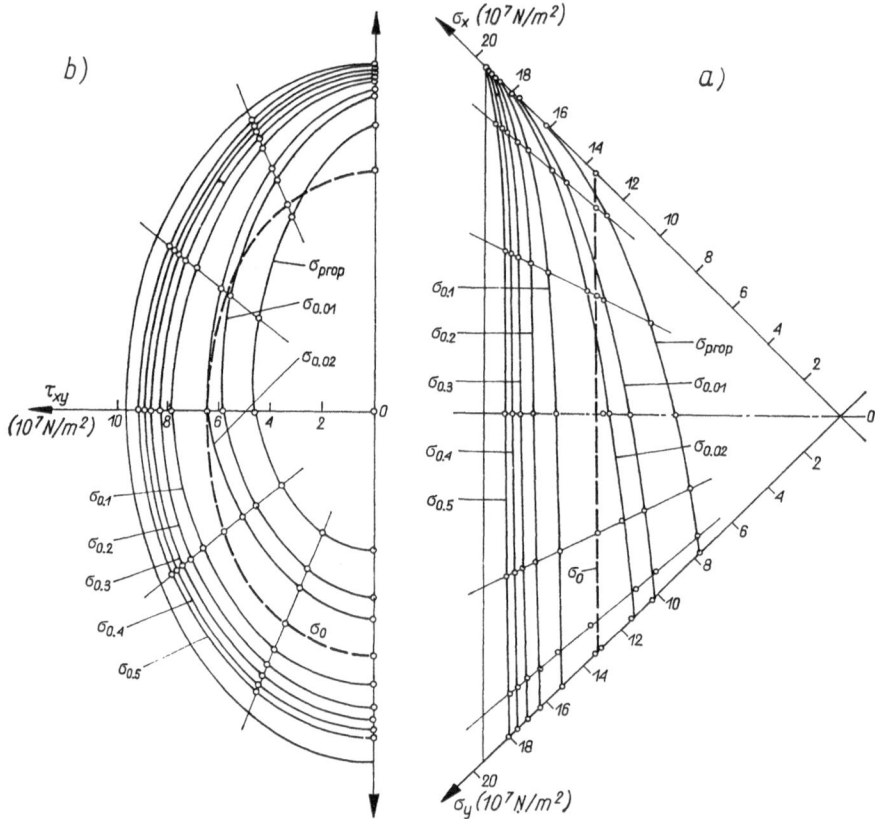

Fig. 30. The effect of plastic deformation by uniaxial tensile stress $\sigma_x$ on the variation in shape of initial ellipse *BEC* (comp. Fig. 25). The yield curve $\sigma_0$ for virgin material, corresponding to the theoretical ellipse *BEC*, is shown by broken line. Continuous lines represent yield curves for prestressed material. Each curve corresponds to a certain value of permanent deformation in the direction of subsequent tension. This deformation is equal to zero for a $\sigma_{prop}$-curve, 0.01 percent for $\sigma_{0.01}$-curve and 0.5 percent for $\sigma_{0.5}$-curve: a) projection on $\sigma_x, \sigma_y$-plane; b) projection on the plane *AOK* (comp. Fig. 25)

Each of the small specimens was loaded by uniaxial tensile stress on an hydraulic testing machine. The stress-strain diagrams were then plotted, from which the proportional limit and stresses corresponding to various small values of plastic strains were found. The final results are shown in Fig. 30 in the stress space $\sigma_x, \sigma_y, \tau_{xy}$ in two projections. The right-hand figure shows the projection on the $\sigma_x, \sigma_y$ plane, the left-hand the projection on the $AOK$ plane, perpendicular to the longer axis of the ellipsoid (Fig. 25) and passing through the $\tau_{xy}$-axis. The dashed line represents the experimental yield curve for the initial unstressed material. This curve almost ideally coincides with the theoretical ellipse $BEC$ on the ellipsoid. The continuous lines represent the states of stress corresponding to different values of permanent deformation in the direction of tension obtained by subsequent loading of previously prestressed material. It is seen that the initial elliptic yield curve is strongly deformed and shifted, but the larger the permanent strain assumed in the definition of yield locus, the more this effect changes to take the form of a uniform expansion of the initial ellipse.

An additional refinement of the technique described above and more data for an aluminium alloy were presented in the subsequent work [131]. Large specimens were prestressed far beyond the initial yield limit subsequently in two orthogonal directions.

In [89] thin-walled tubular specimens of the M-63 brass were loaded by an axial force and internal pressure. Thus principal directions coinciding with the axial and circumferential directions were fixed. Investigations were made of the change of the shape of the sector $BDC$ of the theoretical Huber–Mises ellipse obtained by intersecting the ellipsoid (Fig. 25) with the $\tau_{xy} = 0$ plane. The first set of seven annealed specimens was used to find the shape of the initial yield surface. Each of the seven specimens with no prestrain were loaded along a different radial path issuing from the origin of the coordinate system $\sigma_z, \sigma_t$, where $\sigma_z$ denotes the axial stress and $\sigma_t$ is the stress component in the circumferential direction. The deformations have been measured by means of resistance strain gauges applied on the surface of each specimen. For each specimen loaded along the prescribed loading path, the $\sigma_i(\varepsilon_i)$ diagram was plotted, where $\sigma_i$ is the shear stress intensity and $\varepsilon_i$ is the distortion intensity defined by (2.6) and (2.12), respectively. From these diagrams were obtained the stresses corresponding to the proportional limit and to plastic distortion intensities $\varepsilon_i^p = 0.01, 0.02, 0.1, 0.2, 0.3, 0.4,$ and $0.5$ percent. In this manner for each specimen seven points in the stress plane $\sigma_z, \sigma_t$ have been found. In Fig. 31

the small crosses represent experimental points corresponding to the proportional limit and to the plastic distortion intensity offset 0.5 percent. Through these points the curves were plotted corresponding to various definitions of the yield limit. Both curves, i.e. the proportional limit curve $\sigma_{prop}^{(0)}$ and the curve marked as the $\sigma_{0.5}^{(0)}$—curve for $\varepsilon_i^p = 0.5$ percent, are in good agreement with the theoretical ellipse *BDC* (Fig. 25).

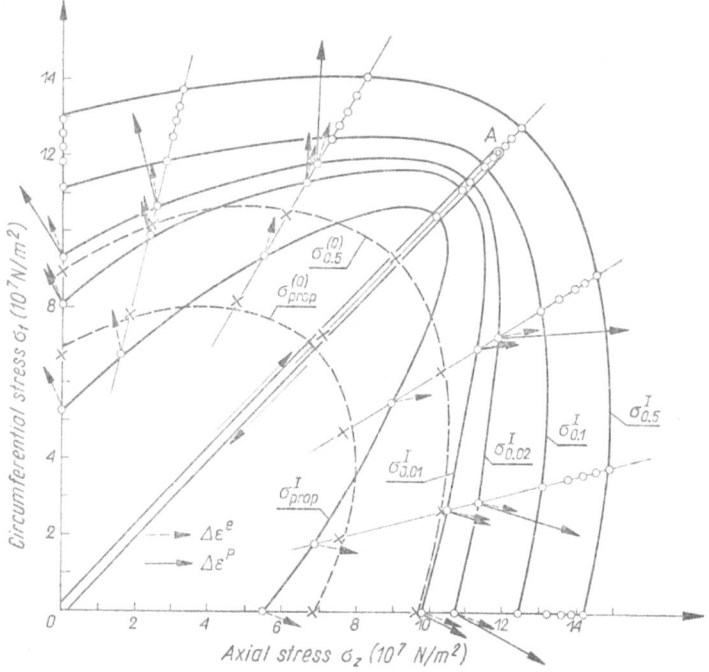

Fig. 31. The effects of plastic prestressing along the path *OA* on the shape of the yield surface of M-63 brass. Broken lines represent yield surface for virgin material, corresponding to the ellipse *BDC* (comp. Fig. 25). Plastic strain increment vectors are represented by continuous arrows. Broken arrows correspond to elastic strains

In the second series each of the seven specimens was prestrained along the path *OA* (Fig. 31) beyond the initial yield locus and then unloaded. Next each of the identically prestrained specimens was loaded along a different radial path from the origin 0. These radial loading paths coincide with those from the first series and are shown in Fig. 31. The experimental curves corresponding to $\sigma_{prop}^I$, $\sigma_{0.01}^I$, ..., $\sigma_{0.5}^I$ were found in the same manner as for the first series. Some of these curves are shown in the figure. The strong deformation of these curves, as compared with the initial ellipse, is clearly visible. This particularly holds true for the $\sigma^I_{\text{prop}}$, $\sigma^I_{0.01}$ and $\sigma^I_{0.02}$ curves. For larger offsets $\varepsilon^p_i$ the geometrical similarity with the initial Huber–Mises ellipse is approximately preserved. For the $\sigma^{(0)}_{0.3}$ and $\sigma^I_{0.5}$ curves this similarity is almost perfect.

Besides the change of the shape of yield surfaces there were also investigated in [89] the elastic and plastic strain increments. The elastic and plastic portions of the axial and circumferential components of strain increments were found by differentiating the stress-strain diagrams. These strain increments are represented in Fig. 31 as vectors, the axial strain increment being represented by the vector's component parallel to the $\sigma_z$-axis, and the circumferential strain increment by the component parallel to the $\sigma_t$-axis. The elastic strain increment vectors are shown by dashed lines, whereas the plastic strain increment vectors are represented by continuous lines. The length of plastic strain increment vectors enables us to observe the increase of the plastic part of deformation. For the proportional limit curve $\sigma^I_{\text{prop}}$ there is no plastic deformation. For the $\sigma^I_{0.01}$-curve the plastic deformation is of the same order as the elastic part and for the $\sigma^I_{0.02}$-curve its value is 2 to 3 times larger than the elastic part. For further curves this plastic part of deformation rapidly increases. One observes the important feature that plastic strain increment vectors are in general orthogonal to the corresponding yield curves.

### 3.6 Strain-hardening hypotheses

Experimental results discussed in the foregoing section show that the initial yield surface under plastic deformation changes its shape and dimensions, and moreover it undergoes a shift in the stress space. Thus the Huber–Mises yield condition (3.5) and the Tresca yield condition (3.7) may be used only in the case of a perfectly plastic material which does not display any strain-hardening effect. The stress-strain diagram for such an idealized material is shown in Fig. 12a. In such a case the realization of the stress states represented by points outside the yield surface is impossible. All the points lying inside that surface represent the states of stresses which are not accompanied by any deformations (material is rigid). The points belonging to the surface itself correspond to limit states of stress at which material

45

yields plastically. The real metals, however, display the strain-hardening phenomenon and therefore the Huber–Mises and Tresca yield conditions constitute for them the starting point determining only the incipient plastic flow. For a further analysis of the advancing process of deformation we must know at every instant the continually changing shape and position of the yield surface. In some works such an instantaneous yield surface is referred to as the loading surface [68]. The exact mathematical description of the changing yield surfaces seems to be, because of its complexity, practically impossible. Thus attempts are made to describe them by means of certain approximate relations called the *strain-hardening hypotheses*. There exist several such hypotheses, but of practical significance are at present only two of them, namely the so-called *hypotheses of isotropic and of kinematic strain-hardening*.

The isotropic strain-hardening hypothesis represents the simplest and oldest approach to the description of the strain-hardening effect. It consists in assuming that, due to the plastic deformation in the material, the initial yield surface expands uniformly, preserving the geometrical similarity of its shape and position in the stress space. If at an instant the stress state represented by a point lying outside the initial yield surface is known at a specific point (assume that no unloading took place), then the actual yield surface can be found. This is a surface geometrically similar to the initial surface and passing through the point representing the actual stress state. For the Huber–Misses yield condition this corresponds analytically to an augmentation of the constant $k$ in (3.4) and (3.5). In practical computations the stress state is usually not known and has to be determined. If the strain-hardening is taken into account, this is possible only if stresses and strains are considered together. By analogy with the stress-strain diagram for uniaxial tension or pure shear, we may relate the second invariant of the stress deviatoric $I_2'$ in (3.3) to the intensity of distortion $\varepsilon_i$. We will assume that this relation is for a given material fixed and does not depend on the loading mode. Bearing in mind that $\sigma_i = \sqrt{I_2'}$, we obtain

$$\sigma_i = \sigma_i(\varepsilon_i). \qquad (3.10)$$

This relation can be determined from the $\sigma$, $\varepsilon$ diagram for uniaxial tension, remembering that, for such loading, $\sigma_i = \sigma/\sqrt{3}$ and $\varepsilon_i = \sqrt{3}\varepsilon/2$. Thus only the scale on the coordinate axes of the diagram has to be changed. It is readily seen that relation (3.10) is equivalent to dependence of $k$ in

(3.4) and (3.5) on the distortion intensity, changing as the process advances.

For complex loading paths it is not permissible to take the final values of strains in the calculations of the $\varepsilon_i$ appearing in relation (3.10), because one can imagine loading paths for which the final strains are zero, but such that the material suffered hardening. As the simplest example, the tensile loading beyond the yield locus followed by compression reducing the deformations to zero can be mentioned. For such cases the hypothesis of isotropic hardening predicts the rise of the yield limit to the value corresponding to tensile deformation of the magnitude equal to the sum of the absolute values of the two consecutive deformations. Thus the distortion intensity should be calculated according to formula (2.12), by summing up all the increments of intensity from the beginning

$$\varepsilon_i = \int d\varepsilon_i.$$

Only for the cases of the so-called *proportional loading*, where all strain components augment proportionally, one can take the final values of strains for calculating the $\varepsilon_i$, because the result will then be the same as in the case of increments summation.

The main weakness of the hypothesis of isotropic hardening is the fact that it does not take into account the Bauschinger effect, or speaking more generally, it neglects the translation of the yield surface from its initial position, which is evidently proved by experiments. This weakness

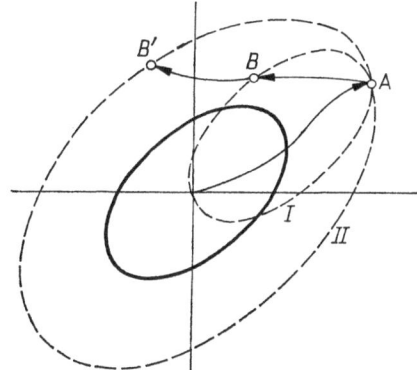

Fig. 32. Kinematical (curve *I*) and isotropic (curve *II*) strain-hardening hypotheses. Initial yield surface is represented by continuous line

47

is particularly disadvantageous in cases when during the process of deformation partial unloading takes place, for instance from a point $A$ (Fig. 32), followed by subsequent reloading until a point $B$ on the yield surface. The hypothesis of isotropic hardening predicts for such loading path that further plastic yielding will begin when the state of stress will reach the level corresponding to the point $B'$. In plastic forming operations we have usually the continued loading process without intermediate unloadings, and for that reason this weakness is of minor practical significance.

In an attempt to take into account the Bauschinger effect, A. Ishlinsky [60] and W. Prager [96] independently proposed the so-called kinematic concept of the description of strain-hardening under complex loadings. This concept will be presented here in the modified form given by R. Shield and H. Ziegler [106]. Let us imagine a simplified tension-compression diagram with the linear hardening law and the so-called idealized Bauschinger effect (Fig. 33). Let the tensile loading until the point $B$ be followed

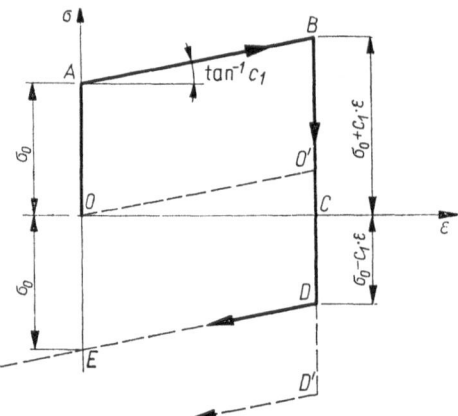

Fig. 33. Kinematical and isotropic hardening concepts applied to simplified tension-compression diagram

by unloading until $C$ and then by subsequent compression. At the point $D$ the material begins to yield again. Further, the compressive stress increases along the straight line $DE$ parallel to $AB$. Both lines intersect the $\sigma$-axis at the same distance from the origin, thus $OA = OE = \sigma_0$. The stress-strain diagram for a material prestressed until $B$ can be obtained by a shifting of the initial diagram along the $\sigma$-axis by the distance equal to $CO'$. The length of the sector $CO'$ is determined by the product $c_1 \varepsilon$, where

$c_1$ is the hardening modulus in pure tension and $\varepsilon$ denotes the preliminary tensile plastic strain.

Passing to the complex stress states, we may assume that the entire yield surface is due to the plastic deformation of the material shifted as a rigid body, without changing its shape and dimensions. Hence the notion "kinematical strain hardening". The components of the translation increment $d\alpha_{ij}$ in the six-dimensional stress space in the directions of the $\sigma_{ij}$-stress axes, are proportional to the corresponding components of the strain increment $c\,d\varepsilon_{ij}$, where $c$ is a proportionality factor. The total components of translation $\alpha_{ij}$ may be obtained by summing up these increments. Thus we have $\alpha_{ij} = c\varepsilon_{ij}$. If the Huber–Mises criterion in the form (3.5) is taken as the initial yield condition, then during the process of deformation the yield condition is expressed by the relation

$$[(\sigma_x - c\varepsilon_x) - (\sigma_y - c\varepsilon_y)]^2 + \ldots + 6[(\tau_{xy} - c\varepsilon_{xy})^2 + \ldots] = 6k^2, \tag{3.11}$$

which is obtained by substituting in (3.5) the differences $(\sigma_{ij} - c\varepsilon_{ij})$ at place of the stresses $\sigma_{ij}$. The magnitude $k$ is constant and equal to the yield locus in pure shear for the initial unstressed material; thus $k = \sigma_0/\sqrt{3}$. Let us consider the physical meaning of the proportionality factor $c$, assuming the state of uniaxial tension by the stress $\sigma_x = \sigma$. All remaining stress components are zero. The strain state is determined by the strain $\varepsilon_x = \varepsilon$. The incompressibility condition requires that $\varepsilon_y = \varepsilon_z = -0.5\varepsilon$. The remaining strain components, i.e. $\varepsilon_{xy}, \varepsilon_{yz}, \varepsilon_{zx}$, are zero. On substituting these values into (3.11) we obtain $\sigma = \sigma_0 + \frac{3}{2}c\varepsilon$, where $\sigma_0$ stands for the yield locus before deformation. Comparing this relation with Fig. 33, we get a simple relation between $c$ and the hardening modulus in simple tension $c_1$ in the form $c = \frac{2}{3}c_1$.

Let us compare now the experimental results shown in Fig. 30 with the hypothesis of kinematic hardening. For the plane stress state one should substitute into (3.11) $\sigma_z = \tau_{yz} = \tau_{zx} = 0$, $\varepsilon_{yz} = \varepsilon_{zx} = 0$ and, moreover, from the incompressibility condition it follows that $\varepsilon_z = -(\varepsilon_x + \varepsilon_y)$. [1]

After some rearrangements we obtain

$$[\sigma_x - c(2\varepsilon_x + \varepsilon_y)]^2 - [\sigma_x - c(2\varepsilon_x + \varepsilon_y)]\,[\sigma_y - c(2\varepsilon_y + \varepsilon_x)] +$$
$$+ [\sigma_y - c(2\varepsilon_y + \varepsilon_x)]^2 + 3(\tau_{xy} - c\varepsilon_{xy})^2 = 3k^2. \tag{3.12}$$

---

[1] Note that $\varepsilon_z \neq 0$. For that reason we cannot obtain the proper expression for the kinematical hardening rule under plane stress conditions by substituting the magnitudes $\sigma_x - c\varepsilon_x$, $\sigma_y - c\varepsilon_y$ and $\tau_{xy} - c\varepsilon_{xy}$ into (3.9) at place of $\sigma_x$, $\sigma_y$ and $\tau_{xy}$.

## 3   Yield conditions and flow laws

Consider now how the initial Huber–Mises ellipsoid shown in Fig. 25 will shift, after prestressing of the sheet material by an uniaxial tensile stress $\sigma$ in the $x$-direction. In such a case we have $\varepsilon_x = \varepsilon$, $\varepsilon_y = -0.5\varepsilon$ and $\varepsilon_{xy} = 0$; hence

$$(\sigma_x - \tfrac{3}{2}c\varepsilon)^2 - (\sigma_x - \tfrac{3}{2}c\varepsilon)\sigma_y + \sigma_y^2 + 3\tau_{xy}^2 = 3k^2.$$

It is readily seen that this corresponds to the translation of the ellipsoid along the $\sigma_x$-axis by the distance $\tfrac{3}{2}c\varepsilon = \sigma - \sigma_0$ (Fig. 34). The states of uni-

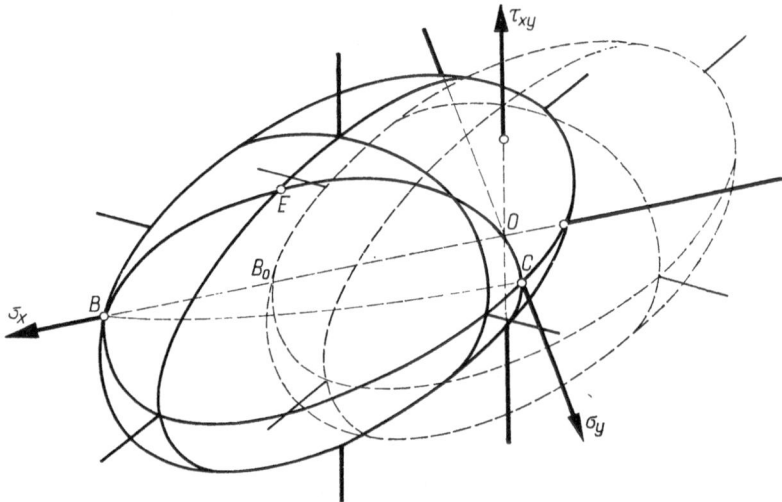

Fig. 34. Kinematical strain-hardening hypothesis for plane state of stress. Initial ellipsoid (comp. Fig. 25) is shifted along $\sigma_x$-axis after plastic deformation by uniaxial tensile stresses $\sigma_x$

axial tension in different directions with respect to the $x$-axis are now represented by the curve *BEC* lying on the shifted ellipsoid. Two projections of this curve are shown in Fig. 35 by dashed lines. The continuous lines represent experimental curves taken from Fig. 30. It is seen that the kinematic hardening hypothesis gives qualitatively good agreement with experimental results.

One of the advantages of the kinematic hardening rule is the fact that, contrary to the isotropic hardening rule, the curvature of the shifted yield surface at the loading point does not increase. This is an important point since all experimental data show that in the vicinity of that point the curvature is in any case increasing, whereas decreasing is never observed.

However, in the case shown in Fig. 31 the kinematic hardening rule is in complete disagreement with experimental results. If the $\sigma_{0.02}$ is taken as the yield surface on which plastic deformations are sufficiently large, the isotropic hardening concept is closer to the real behaviour of the metal than the kinematic one.

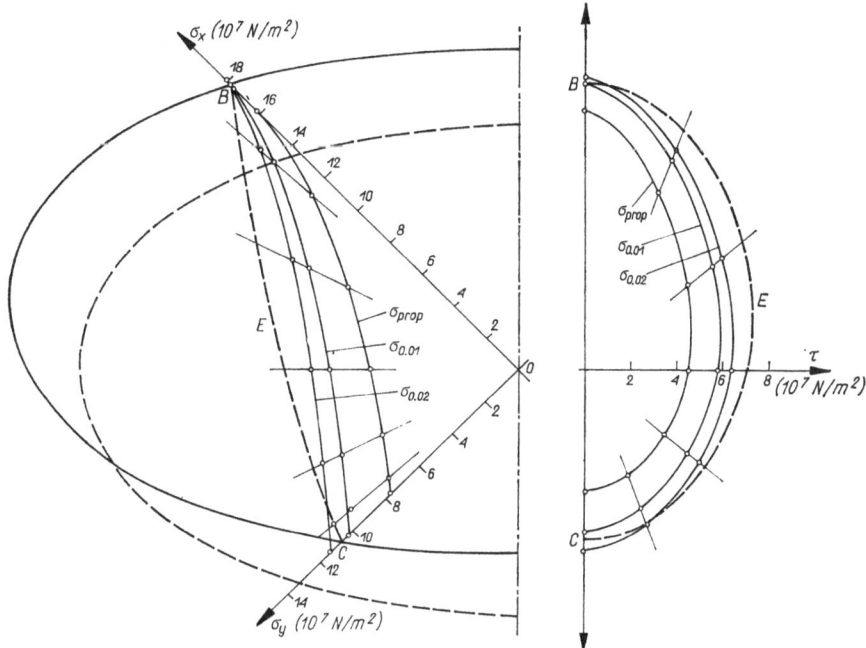

Fig. 35. Comparison of experimental curves shown in Fig. 30 with the kinematical strain-hardening hypothesis. Broken line *BEC* represents the theoretical curve shown in Fig. 34. Experimental curves are represented by continuous lines

Finally, it is interesting to note that in spite of its obvious weakness, a concept even as simple as the kinematic hardening rule is able to predict surprisingly well the response of metals undergoing complex loadings. For instance, it was shown [88, 131] that if the material was prestressed along one loading path, then unloaded and prestressed deeper into the plastic range along another path, it essentially "forgets" the first prestrain. These results were in good agreement with the kinematic rule of strain-hardening. Another type of complex loading was used in [85]. The tubular specimens of M-63 brass were loaded by alternating biaxial complex loading

51

superimposed upon a fixed biaxial isotropic state of stress. Experimental results show that the material strained cyclically beyond the initial yield locus begins to work under a fully elastic regime after a few cycles of the initial transitory period of loading. It is shown that the kinematic hardening rule adequately predicts the number of stress cycles after which the material displays no further plastic deformations.

Let us mention other concepts taking into account simultaneous expansion and translation of the initial yield surface [69, 74], or accounting for a possible rotation of the surface [3]. The fact that for some loading paths the yield surface undergoes rotation was previously observed experimentally in [89]. We will not pay here much attention to these theories since their applicability in computations of practical problems is limited. Contrary to the two strain-hardening hypotheses described above, they represent a multiparameter approach to the hardening problem. The main disadvantage of such an approach constists in the fact that if we want to use a multiparameter hardening rule in computations of any practical problem, we must have at the disposal full experimental information as to the changing configuration of the yield surface for that specific problem. Since the loading path of a material particle is usually not known in advance, the use of such theories is all the more difficult. In the one-parameter theories as in the kinematic and isotropic hardening rules the configuration of the yield surface is uniquely defined by the end-point of the loading path, and therefore no experimental data are needed.

In the plastic forming processes we have usually the so-called *active loading* which means the continued loading of each material particle without any partial unloading. Referring to Fig. 32, the stress increment is always directed outwards the actual yield surface, and not inwards as shown on the sector *AB*. If active loading takes place, the shape of the yield surface, except in the nearest neighbourhood of the loading point *A*, has no practical meaning. In the vicinity of the loading point the hypothesis of kinematic hardening describes the curvature of the actual yield surface better than the isotropic hypothesis. It should be emphasized that the two basic strain-hardening rules lead in practical problems to very close stress and strain distributions. This will be illustrated later on in Chapter 10 on an example of tube drawing.

## 3.7   Stress-strain rate relations

The theory of plastic flow originates in the works of Saint-Venant, Lévy, and Mises. It is based upon the assumption that the plastic strain increment considered as a vector in the stress space is directed along the outward normal to the yield surface at the point of loading. This corresponds to the assumption that the function (3.2),

$$F(I_2', I_3') = 0,$$

represents the so-called *plastic potential*. Strain rates are then related to stresses by means of the relations

$$\epsilon_{ij} = \lambda \frac{\partial F}{\partial \sigma_{ij}}. \tag{3.13}$$

Relations (3.13) represent the *flow law associated with the yield condition*. An associated flow law implies that the strain rate vector in the stress space is orthogonal to the yield surface, which is shown schematically in Fig. 22. Also the vector of the strain increment is orthogonal to this surface, provided the strain increment computed for a sufficiently short time increment is small enough. As we have mentioned in the previous section, plastic strain increment vectors have been experimentally determined in [89]. These vectors are represented in Fig. 31 by continuous lines. They are generally orthogonal to the respective yield surfaces. Also for other loading paths the normality criterion was experimentally confirmed.

The scalar multiplier $\lambda$ is constant at a given instant and at a specific point of the body, which follows directly from (3.13), but generally its value changes with time and can assume different values at each point of the deforming region. This is, therefore, not a material constant, and stands only for the factor of proportionality between the strain rate components and the respective derivatives of the plastic potential. Of course, these derivatives are functions of components of the stress tensor.

It may easily be shown that $\lambda$ must be positive when the material is plastically deformed. Such a conclusion follows from the condition that the work done on the plastic irreversible deformations must be positive. This is obvious, since to carry on the process of plastic deformation, we must supply the energy from outside. This condition can be expressed as follows:

$$D_p = \sigma_{ij} \epsilon_{ij} = \sigma_{ij} \lambda \frac{\partial F}{\partial \sigma_{ij}} > 0. \tag{3.14}$$

The product $\sigma_{ij}\,\partial F/\partial\sigma_{ij}$ is positive, for, as will be shown in the next section, the yield surface is convex and therefore the angle between the stress vector $\sigma_{ij}$ and the normal to the yield surface in the stress space is always acute. Thus $\lambda > 0$ is necessary for condition (3.14) to be satisfied.

For the Huber–Mises yield condition and for perfectly plastic non-hardening material we obtain from (3.13)[1]

$$\frac{\epsilon_x}{\sigma_x - \sigma_m} = \frac{\epsilon_y}{\sigma_y - \sigma_m} = \cdots = \frac{\epsilon_{zx}}{\tau_{zx}}, \qquad (3.15\text{a})$$

where $\sigma_m$ is the mean stress, or in suffix notation,

$$\epsilon_{ij} = \lambda(\sigma_{ij} - \tfrac{1}{3}\delta_{ij}\sigma_{kk}). \qquad (3.15\text{b})$$

Relations (3.15) are known as the *Lévy–Mises stress-strain rate relations*. When strain rates are expressed in terms of the velocities $v_x, v_y, v_z$ by means of (2.14), relations (3.15) may be written in the form

$$\frac{1}{\sigma_x - \sigma_m}\frac{\partial v_x}{\partial x} = \frac{1}{\sigma_y - \sigma_m}\frac{\partial v_y}{\partial y} = \cdots = \frac{1}{2\tau_{xy}}\left(\frac{\partial v_x}{\partial y} + \frac{\partial v_y}{\partial x}\right) = \cdots, \quad (3.16\text{a})$$

or in suffix notation,

$$\frac{1}{2}\left(\frac{\partial v_i}{\partial x_j} + \frac{\partial v_j}{\partial x_i}\right) = \lambda(\sigma_{ij} - \tfrac{1}{3}\delta_{ij}\sigma_{kk}). \qquad (3.16\text{b})$$

Relations (3.15) and (3.16) hold valid also in the case when isotropic hardening is assumed.

In the case of kinematic hardening one should replace in (3.15) and (3.16) the stress components $\sigma_{ij}$ by the differences $(\sigma_{ij} - c\varepsilon_{ij})$, respectively.

In numerous works the flow law (3.13) is used jointly with different approximate expressions for the function $F(\sigma_{ij})$. In the case of the Tresca yield criterion this leads to very simple relations, allowing to solve numerous practical problems in the elementary way. Also for various linearized approximations of the Huber–Mises criterion the associated flow law gives simple stress-strain rate relations. From the mathematical point of view such associated flow laws are correct. However, as shown for instance by V. V. Sokolovsky [112–114] on a number of examples, they give in some

---

[1] Differentiating the function $F$ with respect to shear stresses, e.g. $\tau_{xy}$, we must remember that the term $\tau_{xy}^2$ in (3.5) represents half the sum $\tau_{xy}^2 + \tau_{yx}^2$, which due to the symmetry condition $\tau_{xy} = \tau_{yx}$ is equal to $2\tau_{xy}^2$. In a formal way one should differentiate separately with respect to $\tau_{xy}$ and $\tau_{yx}$. In practice we divide the derivative of the function (3.5) by two. Similarly, for $\tau_{yz}$ and $\tau_{zx}$.

cases unrealistic results, completely different from those obtained for the flow law (3.15)–(3.16). For instance, in the case of tube drawing they may predict that the drawn tube does not change its thickness, which is not observed in real processes.

In spite of this the flow law associated with the Tresca yield condition is frequently used, first of all in the cases where it leads to a substantial mathematical simplification of the problem, thus making it possible to obtain a solution or radically simplifying computations. For instance, in the case of axially-symmetric plastic flow and in certain problems of plane stress, we can obtain theoretical solutions only by using such an approach. With the Tresca yield condition there is connected a certain singularity resulting from the existence of sharp edges in the yield surface. Let us consider a plane *KLM* perpendicular to the axis of the Tresca hexahedral prism (Fig. 24). This plane intersects the prism along a regular hexagon *ABCDEF*. If the state of stress corresponds to a point on one of the faces of the prism, i.e. on the edge of the hexagon, the strain rate vector is, according to the normality rule, orthogonal to this face. If, however, the stress state is represented by a point on the edge of the prism, i.e. by a vertex of the hexagon, say by the point *B*, then the direction of the strain rate vector is not uniquely defined. Its direction may lie within the angle formed by normals to the two adjacent faces *AB* and *BC*.

In some works [120, 50] the Tresca yield condition is used together with the flow law (3.15) associated with the Huber–Mises yield condition. This violates the normality criterion, but such an approach is of a practical significance since in certain cases it gives results close to those obtained with the Huber–Mises yield condition, and it is mathematically simpler.

## 3.8 Drucker's postulate; convexity of the yield surface

In the foregoing sections we have mentioned that the yield surface is convex and that the strain rate represented as a vector in the stress space should be orthogonal to the yield surface. These important criteria may be based upon a more fundamental approach to the plastic stress-strain relations.

In Fig. 36 are shown schematically two types of the stress-strain diagrams in pure tension. In the case (a) the curve is monotonically increasing, and any stress increment $d\sigma > 0$ causes a positive strain increment $d\varepsilon > 0$. Thus for such a material, which is called a *stable material*, the product of the increments is positive, $d\sigma\,d\varepsilon > 0$. The idealized rigid-perfectly plastic material with no hardening is also reckoned to belong to the class of stable materials. In this case we have obviously $d\sigma = 0$, and therefore $d\sigma\,d\varepsilon = 0$. Thus, generally for stable materials, we can write $d\sigma\,d\varepsilon \geqslant 0$.

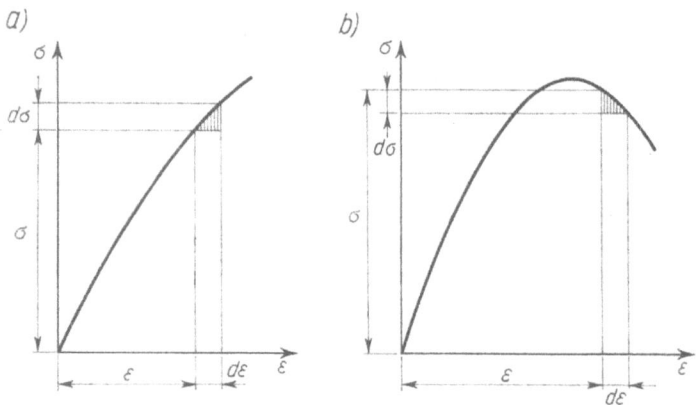

Fig. 36. Two types of stress-strain diagrams in uniaxial tension: a) stable material, b) unstable material

For an unstable material (b) a portion of the stress-strain diagram is descending, and the negative stress increments $d\sigma < 0$ are accompanied there by positive increments of strain $d\varepsilon > 0$. Thus for unstable materials we have $d\sigma\,d\varepsilon < 0$.

The theory of plastic flow is concerned with the class of stable materials only, to which belong real metals, with the exception of the local instability which is displayed by just a few metals (see Fig. 3). D. C. Drucker in his fundamental work [26] generalized the concept of stable behaviour of materials to complex loadings.

Since throughout this book the elastic strains are neglected, our considerations will be limited here to the rigid-plastic materials, although in Drucker's original paper the elastic-plastic materials are considered. According to Drucker's postulate, the simple condition $d\sigma\,d\varepsilon \geqslant 0$ for a stable material under pure tension takes in the case of complex loading the form

$$d\sigma_{ij}\,d\varepsilon_{ij} \geqslant 0. \tag{3.17}$$

The stress increment $d\sigma_{ij}$ causing the plastic strain increment $d\varepsilon_{ij}$ must be directed outward of the actual yield surface (Fig. 37). To ensure the scalar product $d\sigma_{ij}d\varepsilon_{ij}$ being positive, the angle formed by the two vectors $d\sigma_{ij}$

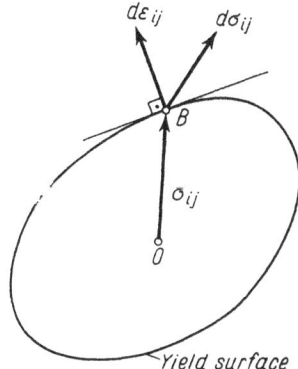

Fig. 37. Orthogonality of the strain increment vector

and $d\varepsilon_{ij}$ in the stress space must be acute. This will be satisfied in any case only if the vector $d\varepsilon_{ij}$ is orthogonal to the yield surface. Thus the normality criterion has been obtained now as the direct consequence of the stability condition (3.17).

Let us now return to the case of pure tension of a rigid-plastic material (Fig. 38). If the material loaded by the stress $\sigma$ suffers an additional strain increment $d\varepsilon$, the work done by the stress $\sigma$ must be positive ($\sigma d\varepsilon > 0$).

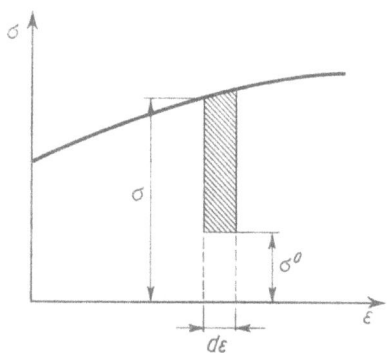

Fig. 38. Reloading in uniaxial tension from $\sigma^0$ to $\sigma$

It is readily seen that if the reloading starts from any stress level $\sigma^\circ$, the inequality $(\sigma - \sigma^\circ)d\varepsilon \geqslant 0$ must be satisfied. Drucker's generalization of this inequality to complex loading has the form

$$(\sigma_{ij} - \sigma_{ij}^\circ)d\varepsilon_{ij} \geqslant 0, \tag{3.18}$$

which implies that the yield surface must be convex. If the yield surface is convex, the angle contained between the vector $(\sigma_{ij} - \sigma_{ij}^\circ)$ and the strain increment vector $d\varepsilon_{ij}$ orthogonal to the surface, is in any case acute (Fig. 39a). Thus the scalar product of the two vectors satisfies inequality (3.18).

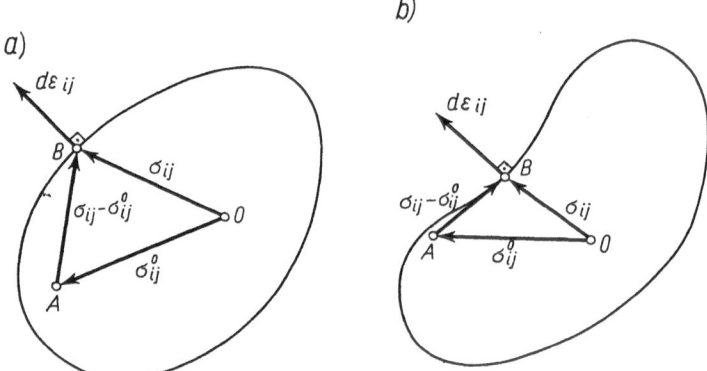

Fig. 39. Convex (a) and concave (b) yield surfaces. For concave yield surface condition (3.18) will not be satisfied for certain loading modes

If the yield surface is not convex (Fig. 39b), then we can always choose the point $A$ so that the scalar product of the two vectors is negative.

It is readily seen that both conclusions on the normality of the strain increment vector and the convexity of the yield surface hold valid for a perfectly plastic material with no hardening. The stress increment $d\sigma_{ij}$ is tangent to the yield surface and therefore we have $d\sigma_{ij}d\varepsilon_{ij} = 0$. Thus inequality (3.17) is satisfied. Also in the case where the yield surface is formed by a number of planes, as for the Tresca yield condition, inequalities (3.17) and (3.18) are satisfied provided that the associated flow rule is assumed.

### 3.9  Extremum principles of plasticity

In numerous problems on plastic forming of metals solutions satisfying all static and kinematic conditions are still not yet available due to their mathematical complexity. In such cases, however, the upper and lower bounds on the unknown exact value of the forces required to cause plastic yielding, may be found by using the extremum principles of plasticity [27, 28]. We shall now outline these principles in the form required in applications. A more detailed treatment of the general theorems of the theory of plasticity can be found in Koiter's work [71], and also in specialist monographs devoted to the mathematical theory of plasticity, e.g. in books of W. Prager and P. Hodge [97] and of L. Kachanov [68].

Consider a perfectly plastic non-hardening body of volume $V$ undergoing plastic deformations. Inside the body both the plastic and rigid regions may exist, the strain rates being zero in the latter. Thus the energy equations can be written with respect to the whole volume including rigid regions. Let $X_i$ denote the components of the external forces acting on the surface of the body, $du_i$ increments of displacement of the specific points of the surface, $\sigma_{ij}$ the stress tensor at an arbitrary point of the body, and let $d\varepsilon_{ij}$ be the strain increment tensor at that point. Thus the virtual work principle can be written in the form

$$\int_V \sigma_{ij} d\varepsilon_{ij} dV = \int_S X_i du_i dS, \tag{3.19}$$

where $S$ stands for the area of the external surface of the body.

Denoting by $v_i$ the components of the velocity of a surface point, and by $\epsilon_{ij}$ the strain rate tensor, we can write equation (3.19) in a different form as

$$\int_V \sigma_{ij} \epsilon_{ij} dV = \int_S X_i v_i dS. \tag{3.20}$$

Equation (3.20) simply states that the rate of energy dissipation inside the deforming body equals the rate at which the tractions applied on $S$ do work on the velocities of their points of application.

(a) *Lower bound theorem*

Let there be given on a portion $S_F$ of the surface of the body the forces $X_i$ and on the remaining portion $S_v$ let there be prescribed the velocities of displacement $v_i$. Assume that due to these external factors there comes into existence the limit state, determinned by the actual stress field $\sigma_{ij}$ and the actual strain rates $\epsilon_{ij}$.

Let us consider a certain arbitrarily chosen stress field $\sigma_{ij}^*$ in the body, satisfying all static conditions, that is, the equations of internal equilibrium, boundary conditions on the portion $S_F$ of the surface, ensuring the tractions there to be equal to the given loads $X_i$, and the condition that the yield criterion is not violated anywhere. This stress field may not satisfy the kinematic conditions of the problem, i.e. the relations (3.13), the compatibility condition and the kinematic boundary conditions on $S_v$.

The stress field $\sigma_{ij}^*$ will be referred to as the *statically admissible stress field*, as opposed to the *actual stress field* $\sigma_{ij}$, which satisfies all static and kinematic conditions.

The energy equation (3.20) assumes for the statically admissible stress field $\sigma_{ij}^*$ and the actual strain rate field $\epsilon_{ij}$ the form

$$\int_V \sigma_{ij}^* \epsilon_{ij} dV = \int_S X_i^* v_i dS, \tag{3.21}$$

where $X_i^*$ are the external forces connected with the stress field $\sigma_{ij}^*$. On the portion $S_F$ of the surface we have obviously $X_i^* = X_i$.

By subtracting (3.21) from (3.20) we obtain

$$\int_V (\sigma_{ij} - \sigma_{ij}^*) \epsilon_{ij} dV = \int_{S_v} (X_i - X_i^*) v_i dS. \tag{3.22}$$

Now we shall show that the left-hand side of equation (3.22) is always positive. To prove this the stresses will be represented as vectors in the six-dimensional stress space. The actual stress field satisfies the yield condition, thus the end-points of the vectors $\sigma_{ij}$ lie on a convex yield surface in this space. Similarly, a vector $\epsilon_{ij}$, normal to the yield surface at the end-point $P$ (Fig. 40) of the actual stress vector $OP$, will represent the

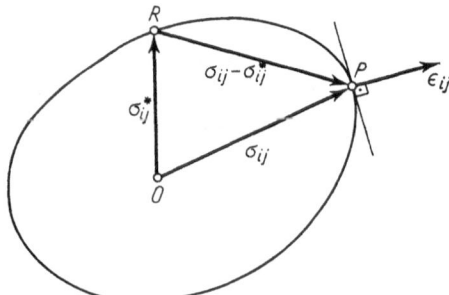

Fig. 40. Statically admissible stress $\sigma_{ij}^*$ and actual stress $\sigma_{ij}$

strain rate tensor. The end-point $R$ of the vector $\sigma_{ij}^*$ lies on the yield surface or inside it. It is seen that the vector $\sigma_{ij} - \sigma_{ij}^*$ makes an acute angle with the vector $\epsilon_{ij}$. Thus their scalar product on the left-hand side of (3.20) is non-negative,

$$(\sigma_{ij} - \sigma_{ij}^*)\epsilon_{ij} \geqslant 0. \tag{3.23}$$

The equality sign occurs when the stress field $\sigma_{ij}^*$ coincides with the actual stress field $\sigma_{ij}$. Now it follows from (3.22) that also the right-hand side must be positive, and therefore, we finally obtain

$$\int_{S_v} X_i v_i \, \mathrm{d}S \geqslant \int_{S_v} X_i^* v_i \, \mathrm{d}S. \tag{3.24}$$

Inequality (3.24) states that when plastic flow takes place, the rate of work at which the actual external forces do work on the given velocities is greater than the rate of work done by forces resulting from the arbitrarily assumed statically admissible stress field on the same velocities. From this statement follows an important conclusion, permitting to estimate a lower bound on the actual values of external forces, needed to cause plastic yielding, in the cases where the exact values cannot be found.

THEOREM 1: *A lower bound on the limit load may be found from any statically admissible stress field assumed in the body. This stress field must satisfy equations of internal equilibrium, static boundary conditions and ensure the yield criterion to be nowhere violated. Kinematic boundary conditions and compatibility conditions may not be satisfied.*

(b) *Upper bound theorem*

Similarly as in the previous case let us consider a body of volume $V$. On the portion $S_F$ of its surface the external forces $X_i$ are given, whereas on the remaining portion $S_v$ the velocities $v_i$ are prescribed. The limit state of the body is determined by stresses $\sigma_{ij}$ and strain rates $\epsilon_{ij}$.

Assume now a certain arbitrarily chosen deformation mechanism $\epsilon_{ij}^*$, which satisfies all kinematic conditions of the problem, i.e. conditions of compatibility and incompressibility, and which is corresponding to the given velocities on the surface $S_v$. The equations of equilibrium and stress boundary conditions may not be satisfied. It is also not necessary to verify whether the yield condition is not anywhere violated. With a deformation mechanism satisfying these assumptions there is connected the existence of the velocities $v_i^*$ on the surface $S_F$ of the body.

The field $\epsilon_{ij}^*$ will be called the *kinematically admissible deformation mechanism,* in contrast to the *actual strain rate field* $\epsilon_{ij}$ which satisfies all kinematic and static conditions of the problem.

Now we can write condition (3.20) for the kinematically admissible strain rate field $\epsilon_{ij}^*$ and the actual stress field $\sigma_{ij}$ as

$$\int_V \sigma_{ij}\epsilon_{ij}^* dV = \int_S X_i v_i^* dS. \tag{3.25}$$

Consider now the stress field $\sigma_{ij}^*$ related by means of the flow law (3.13) to the arbitrarily chosen strain rate field $\epsilon_{ij}^*$. This stress field does not have to be statically admissible. Similarly as in the case of the lower bound theorem, we can write (see Fig. 41)

$$(\sigma_{ij}^* - \sigma_{ij})\epsilon_{ij}^* \geqslant 0. \tag{3.26}$$

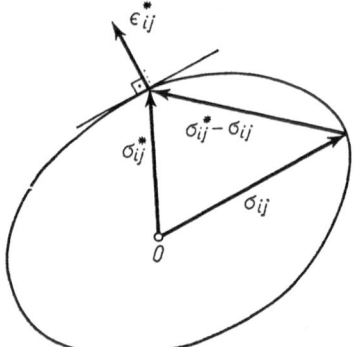

Fig. 41. Kinematically admissible rate of strain $\epsilon_{ij}^*$ and actual stress $\sigma_{ij}$

Indeed, the two vectors $(\sigma_{ij}^* - \sigma_{ij})$ and $\epsilon_{ij}^*$ make an acute angle in the stress space and therefore their scalar product is positive.

From (3.25) and (3.26) we immediately obtain

$$\int_S X_i v_i^* dS \leqslant \int_V \sigma_{ij}^* \epsilon_{ij}^* dV. \tag{3.27}$$

Bearing in mind the obvious equality $v_i = v_i^*$, which must be satisfied on the portion $S_v$ of the surface, we arrive at the inequality

$$\int_{S_v} X_i v_i dS \leqslant \int_V \sigma_{ij}^* \epsilon_{ij}^* dV - \int_{S_F} X_i v_i^* dS. \tag{3.28}$$

The left-hand side represents the rate at which the external forces do work on the given velocities. It is readily seen that in the case where the arbitrarily chosen field $\epsilon_{ij}^*$ coincides with the actual field $\epsilon_{ij}$, both sides of

(3.28) are equal. For other kinematically admissible fields $\epsilon_{ij}^*$ which are different than the actual field, the left-hand side is smaller than the right-hand side. This means that the rate of work of the external forces on the given velocities attains its minimum for the actual deformation mechanism. This leads to the following conclusion which permits to find an upper bound on the actual limit load in the case where its exact value cannot be determined.

THEOREM 2: *An upper bound on the limit load may be found from any kinematically admissible deformation mechanism. This mechanism must satisfy conditions of compatibility and incompressibility and the kinematic boundary conditions.*

In practical problems an upper bound on the limit load may be found by equating the rate of work done by the forces $X_i$ on the portion $S_v$ of the external surface to the rate of the internal dissipated energy.

The extremum principles presented above hold valid also in the cases where the deforming region contains discontinuities in the tangential velocity across certain surfaces. The proofs for this more general case may be found in specialist works (see, for instance, [68, 97]).

### 3.10 Brief summary; equations of the three-dimensional plastic flow

In the general case of three-dimensional plastic flow nine unknowns, 6 components of the stress tensor $\sigma_{ij}$ and three components of the velocity vector $v_i$ must be determined. The set of ten equations in ten unknowns is composed of three equations of motion [see (2.18)]

$$\frac{\partial \sigma_{ij}}{\partial x_j} = \varrho \left( \frac{\partial v_i}{\partial t} + v_j \frac{\partial v_i}{\partial x_j} \right), \tag{3.29}$$

of the yield condition

$$F(\sigma_{ij}) = 0, \tag{3.30}$$

and of the six relations between the velocity components $v_i$ and the components of the stress tensor

$$\frac{1}{2} \left( \frac{\partial v_i}{\partial x_j} + \frac{\partial v_j}{\partial x_i} \right) = \lambda \frac{\partial F}{\partial \sigma_{ij}}, \tag{3.31}$$

obtained by substituting (2.14) in the associated flow law (3.13). In this set of equations appears the tenth unknown function $\lambda$, which can easily be eliminated from (3.31).

Equations (3.29)-(3.31) describe the *dynamic* problem of the plastic flow in the general three-dimensional case. Unfortunately as yet we do not have at our disposal any effective method of solving dynamic problems, even in less general cases of the plane plastic flow or axially symmetric flow. Except for the few particular problems (see, for instance, [94, 95]), the dynamic solutions are limited to relatively simple axially symmetric problems when the directions of the principal stresses are known in advance. We shall discuss examples of such dynamic solutions in Chapter 10. In other cases discussed in this book the *quasi-static* approach is used when instead of the equations of motion (3.29) we assume the equations of equilibrium

$$\frac{\partial \sigma_{ij}}{\partial x_j} = 0. \tag{3.32}$$

This is a rather drastic simplification of the problem caused by mathematical difficulties. It must be, however, emphasized that quasi-static solutions are of great practical significance, because in most operations of plastic forming the accelerations are small, and the quasi-static approach is therefore justified.

Any attempt to solve the general three-dimensional quasi-static plastic flow problem is also connected with serious mathematical difficulties. Effective methods of solution exist for certain particular classes of problems such as the plane plastic flow discussed in Chapter 4, the axially symmetric flow (Chapter 8) and the plane stress problems (Chapter 9).

Having found the solution to any particular problem, we should check whether the condition of positive rate of dissipated energy (3.14) is satisfied everywhere in the deforming region. Solutions violating this condition are incorrect.

# 4

# *The theory of plane plastic flow*

## 4.1 Basic relations

The state of plane flow is defined by the fundamental property that the displacements of all particles of the body are parallel to a given plane, which will be chosen as the $x, y$-plane of the Cartesian coordinate system $x, y, z$, and that these displacements are independent of the coordinate $z$ of the particle. The velocity field is determined by the components

$$v_x = v_x(x, y), \quad v_y = v_y(x, y), \quad v_z = v_z(x, y).$$

Thus the velocity and stress distribution is the same in all planes perpendicular to the $z$-axis.

At every point of the body four components of stress $\sigma_x, \sigma_y, \sigma_z, \tau_{xy}$ (Fig. 13) are to be found. Since the remaining components $\tau_{xz}$ and $\tau_{yz}$ are zero, by symmetry, $\sigma_z$ is a principal stress. The two velocity components $v_x$ and $v_y$ are also to be found. We shall solve the problem of plane flow only for the rigid-plastic material displaying no strain-hardening, since we are compelled by mathematical difficulties to disregard both the elastic part of deformation and the strain-hardening effect.

Let us pass to the equations relating the six unknowns sought for. The equations of motion (2.18) reduce now to the form

$$
\begin{aligned}
\frac{\partial \sigma_x}{\partial x} + \frac{\partial \tau_{xy}}{\partial y} &= \varrho \left( \frac{\partial v_x}{\partial t} + v_x \frac{\partial v_x}{\partial x} + v_y \frac{\partial v_x}{\partial y} \right), \\
\frac{\partial \tau_{xy}}{\partial x} + \frac{\partial \sigma_y}{\partial y} &= \varrho \left( \frac{\partial v_y}{\partial t} + v_x \frac{\partial v_y}{\partial x} + v_y \frac{\partial v_y}{\partial y} \right).
\end{aligned}
\tag{4.1}
$$

Unfortunately, the general solution of problems of plane flow accounting for the inertia forces on the right-hand side of equations (4.1) is still not available. On the other hand, in most plastic forming operations accelerations of the material particles are small and therefore the influence of inertial forces is negligible. This refers to almost all processes discussed in Chapters 5–7. An estimation of the influence of inertial forces will be given in Chapter 6. Thus in our present considerations the inertial forces in equations (4.1) will be neglected; the problem will be considered as a quasi-static one. The equations of motion reduce to the well-known equations of equilibrium

$$\frac{\partial \sigma_x}{\partial x} + \frac{\partial \tau_{xy}}{\partial y} = 0,$$

$$\frac{\partial \tau_{xy}}{\partial x} + \frac{\partial \sigma_y}{\partial y} = 0. \tag{4.2}$$

The next equations are the yield condition and the associated flow rule written in the general form (3.13). Assume first the flow rule (3.15) associated to the Huber–Mises yield condition. In the present case, taking into account that obviously for the plane flow there must be $\epsilon_z = \epsilon_{yz} = \epsilon_{xz} = 0$, we obtain

$$\frac{\epsilon_x}{\sigma_x - \sigma_m} = \frac{\epsilon_y}{\sigma_y - \sigma_m} = \frac{\epsilon_{xy}}{\tau_{xy}}, \tag{4.3}$$

where as before $\sigma_m = \frac{1}{3}(\sigma_x + \sigma_y + \sigma_z)$.

Since $\epsilon_z = 0$ for the plane flow, we immediately obtain from the condition of incompressibility (2.15) that $\epsilon_x = -\epsilon_y$. By substituting this equality in (4.3) we get the fundamental relation

$$\sigma_z = \frac{1}{2}(\sigma_x + \sigma_y). \tag{4.4}$$

Thus the stress component $\sigma_z$ is the arithmetic mean of the normal stresses $\sigma_x$ and $\sigma_y$. Moreover, the component $\sigma_z$ is a principal stress, i.e. $\sigma_z = \sigma_3$. Let us denote the two remaining principal stresses in the $x, y$-plane by $\sigma_1$ and $\sigma_2$, and assume for definiteness that $\sigma_1 > \sigma_2$. Now, instead of the relation (4.4), we can write its equivalent form

$$\sigma_3 = \frac{1}{2}(\sigma_1 + \sigma_2). \tag{4.5}$$

Substituting (4.5) in the general form of the Huber–Mises yield condition (3.4), we obtain for the plane flow the following simple expression:

$$\sigma_1 - \sigma_2 = 2k. \tag{4.6}$$

It is readily seen that (4.6) holds valid also for the Tresca yield condition since $(\sigma_1 - \sigma_2)$ represents the double value of the maximum shear stress. Note that, according to (4.5), $\sigma_3$ lies between $\sigma_1$ and $\sigma_2$. Let us recall that, assuming the Huber–Mises yield condition, we have to put $k = \sigma_{pl}/\sqrt{3}$, whereas for the Tresca yield condition $k = \sigma_{pl}/2$, where $\sigma_{pl}$ is the yield locus in pure tension.

Thanks to the fact that (4.6) holds for the two basic yield conditions, we have for the Huber–Mises and Tresca conditions the same relation written in terms of stress components $\sigma_x$, $\sigma_y$, $\tau_{xy}$,

$$(\sigma_x - \sigma_y)^2 + 4\tau_{xy}^2 = 4k^2, \tag{4.7}$$

obtained directly from the general relation (3.5) by substituting (4.4) and $\tau_{yz} = \tau_{zx} = 0$. For the Tresca criterion the relation (4.7) may be obtained directly from Fig. 42. The radius of the Mohr circle must be equal to $k$, as required by (4.6).

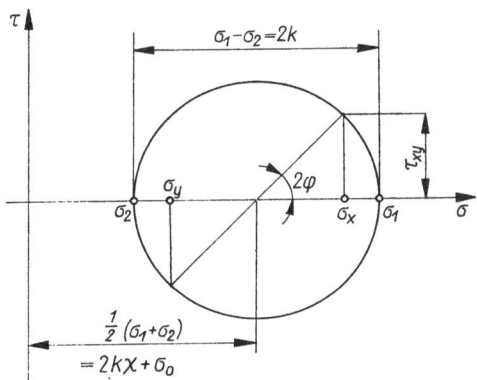

Fig. 42. Auxiliary functions $\chi$ and $\varphi$ defining the state of stress in the theory of plane flow

Condition (4.7) is geometrically represented in the stress space $\sigma_x$, $\sigma_y$, $\tau_{xy}$ by an infinitely long cylinder with the axis lying in the $(\sigma_x, \sigma_y)$-plane and making the same angles with the $\sigma_x$ and $\sigma_y$-axes. In Fig. 43 there is presented a portion of this cylinder. The perpendicular cross-section of the cylinder has a form of an ellipse with the half axes $k$ and $k/\sqrt{2}$, respectively. Since the values of $k$, as related to the yield point in pure tension, are different for the two yield conditions, the dimensions of the cylinder corresponding to Huber–Mises yield condition are slightly larger than those for the Tresca condition. Nevertheless the cylinders for the two conditions preserve geometrical similarity.

Let us pass to the stress-strain rate relations. Since condition (4.7) holds for both yield criteria, the associated flow rule (3.13) leads to the conclusion that relations (4.3) correspond to flow rules associated with the Huber–Mises and the Tresca yield conditions.

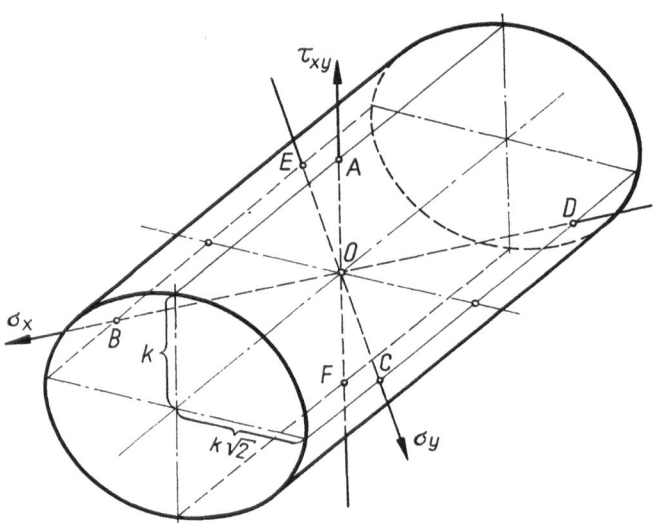

Fig. 43. Geometrical representation of yield condition (4.7) in the space of stresses $\sigma_x$, $\sigma_y$, $\tau_{xy}$

By substituting (4.4) and (2.13) into (4.3) we obtain finally the relations between flow velocities and stresses

$$\frac{1}{\sigma_x - \sigma_y}\frac{\partial v_x}{\partial x} = \frac{1}{\sigma_y - \sigma_x}\frac{\partial v_y}{\partial y} = \frac{1}{4\tau_{xy}}\left(\frac{\partial v_x}{\partial y} + \frac{\partial v_y}{\partial x}\right). \tag{4.8}$$

Due to equation (4.4), the number of unknowns reduces to five, namely: $\sigma_x$, $\sigma_y$, $\tau_{xy}$ and the velocities $v_x$, $v_y$. The equations of equilibrium (4.2) and the yield condition (4.7) together with relations (4.8) constitute a set of five equations in five unknowns. An important property of this set is the fact that the velocities appear neither in the equations of equilibrium nor in the yield condition. These three equations constitute the system in three unknowns $\sigma_x$, $\sigma_y$, $\tau_{xy}$. The problem is in this sense statically determinate, although in practical applications usually the number of boundary conditions involving stresses is not sufficient to permit a determination of stresses without considering velocities. If, however, the number of stress boundary conditions is sufficient, the stress field can be found independently, and afterwards the solution ought to be checked for all the kinematical conditions.[1]

## 4.2   Determination of the stress field

By the above, the determination of the state of stress at every point reduces to the solution of the system of the three equations (4.2) and (4.7). The method of solution, originally due to Lévy [76] consists in introducing two new auxiliary unknowns, the function $\chi$, depending on the sum of the principal stresses $\sigma_1$ and $\sigma_2$, and on the angle $\varphi$ between the direction of

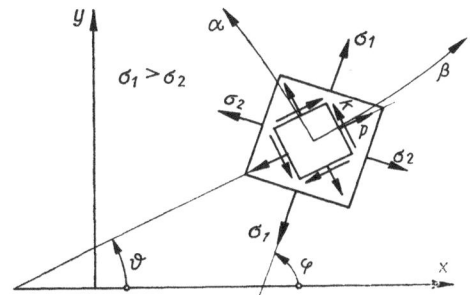

Fig. 44. Directions of principal stresses and of slip-lines

the greater principal stress $\sigma_1$ and the $x$-axis (Fig. 44). The difference of principal stresses is constant as required by the yield condition (4.6). Thus we have

$$\tfrac{1}{2}(\sigma_1 + \sigma_2) = \sigma_0 + 2k\chi, \qquad \tfrac{1}{2}(\sigma_1 - \sigma_2) = k. \qquad (4.9)$$

$\sigma_0$ is a constant, which should be chosen before the procedure of solving is started. Usually one takes $\sigma_0 = 0$, but in some cases it is more advantageous to take a non-zero value, as it is for instance in the case of the sheet drawing operation discussed in Chapter 6.

---

[1] For a more detailed discussion of this problem see Hill's book [50].

Looking at the Mohr circle whose radius is, according to the yield condition (4.6), equal to $k$ (Fig. 42), one can see that

$$\sigma_x = \sigma_0 + 2k\chi + k\cos 2\varphi,$$
$$\sigma_y = \sigma_0 + 2k\chi - k\cos 2\varphi, \tag{4.10}$$
$$\tau_{xy} = k\sin 2\varphi.$$

These expressions satisfy identically the yield condition (4.7). Substituting (4.10) into the equations of equilibrium (4.2), one obtains the basic set of the quasi-linear partial differential equations in the two unknown functions $\chi$ and $\varphi$

$$\frac{\partial \chi}{\partial x} - \sin 2\varphi \frac{\partial \varphi}{\partial x} + \cos 2\varphi \frac{\partial \varphi}{\partial y} = 0,$$

$$\frac{\partial \chi}{\partial y} + \cos 2\varphi \frac{\partial \varphi}{\partial x} + \sin 2\varphi \frac{\partial \varphi}{\partial y} = 0. \tag{4.11}$$

This system is of the hyperbolic type for all possible values of $\chi$ and $\varphi$. Hence it can be solved by means of the method of characteristics. Solutions of particular cases may be obtained by solving the appropriate boundary value problems, when the values of the two functions $\chi$ and $\varphi$, or some relations between these functions, are given along certain lines. These conditions are usually sufficient to define the values of $\chi$ and $\varphi$ uniquely within regions adjacent to those lines. We shall discuss this problem later on in Section 4.5 devoted to the boundary value problems. At present let us consider the so-called *Cauchy boundary value problem* when along a certain line $y = y(x)$ all three stress components, and consequently the auxiliary functions $\chi$ and $\varphi$, are known. We shall see later on that in problems of plastic flow the knowledge of the sought for functions along a certain line not always suffices to find their values in the adjacent region. It is generally proved in the theory of hyperbolic equations that such a solution is possible provided the partial derivatives of the two functions with respect to the coordinates can be determined along the starting line. In our case this can be done by supplementing equations (4.11) with the following obvious equalities

$$d\chi = \frac{\partial \chi}{\partial x} dx + \frac{\partial \chi}{\partial y} dy,$$

$$d\varphi = \frac{\partial \varphi}{\partial x} dx + \frac{\partial \varphi}{\partial y} dy, \tag{a}$$

which must be satisfied along the line $y = y(x)$.

We shall find the values of the derivatives $\partial \chi / \partial x, \ldots, \partial \varphi / \partial y$ by solving the above set of four equations with respect to them. This can be done provided the characteristic determinant, formed by the coefficients at the derivatives, is not equal to zero. In the other case, when this determinant is equal to zero, the derivatives along the line $y = y(x)$ are not defined. The lines displaying such a property are called the *characteristics* of the system (4.11). A simple and advantageous method of determining the differential equations determining the characteristics is given in Appendix 2 at the end of this book. To make the presentation of the theory of plane flow more self-contained, we shall derive here these equations directly from the determinants of the system (4.11), although this method is more laborious. Note, however, that the two methods are fully equivalent. We obtain the differential equations of characteristics by solving with respect to $\mathrm{d}y/\mathrm{d}x$ the quadratic equation

$$\begin{vmatrix} 1 & 0 & -\sin 2\varphi & \cos 2\varphi \\ 0 & 1 & \cos 2\varphi & \sin 2\varphi \\ \mathrm{d}x & \mathrm{d}y & 0 & 0 \\ 0 & 0 & \mathrm{d}x & \mathrm{d}y \end{vmatrix} = 0. \tag{b}$$

Along the characteristics there must be satisfied certain differential relations between the functions $\chi$ and $\varphi$. It is known from the theory of hyperbolic equations that characteristics are defined as the lines along which not only the characteristic determinant must be equal to zero, but also any determinant obtained by replacing in the characteristic one an arbitrary column by the column formed from these terms which do not contain derivatives. Those free terms are zero in equations (4.11) and are equal to $\mathrm{d}\chi$ and $\mathrm{d}\varphi$ in the supplementary expressions (a).

Equation (b) can be written in another form as

$$\cos 2\varphi \left(\frac{\mathrm{d}y}{\mathrm{d}x}\right)^2 - 2\sin 2\varphi \frac{\mathrm{d}y}{\mathrm{d}x} - \cos 2\varphi = 0. \tag{c}$$

Hence we obtain two solutions for the direction coefficients of the characteristics

$$\frac{\mathrm{d}y}{\mathrm{d}x} = \frac{\sin 2\varphi + 1}{\cos 2\varphi} = \tan\left(\varphi + \frac{\pi}{4}\right),$$

$$\frac{\mathrm{d}y}{\mathrm{d}x} = \frac{\sin 2\varphi - 1}{\cos 2\varphi} = \tan\left(\varphi + \frac{\pi}{4}\right). \tag{4.12}$$

The differential relations which must be satisfied along the characteristics, are

$$d\chi + d\varphi = 0 \text{ for } \frac{dy}{dx} = \tan\left(\varphi + \frac{\pi}{4}\right),$$

$$d\chi - d\varphi = 0 \text{ for } \frac{dy}{dx} = \tan\left(\varphi - \frac{\pi}{4}\right). \tag{4.13}$$

Finally the equations of characteristics take the form

$$\frac{dy}{dx} = \tan\left(\varphi + \frac{\pi}{4}\right), \quad \chi + \varphi = \xi = \text{const}, \tag{4.14a}$$

for the first family, called the α-*lines family*, and

$$\frac{dy}{dx} = \tan\left(\varphi - \frac{\pi}{4}\right), \quad \chi - \varphi = \eta = \text{const}, \tag{4.14b}$$

for the second family, referred to as the β-*lines family*.

In equations (4.14) one introduces the auxiliary functions $\xi$ and $\eta$ defined by

$$\xi = \chi + \varphi, \quad \eta = \chi - \varphi, \tag{4.15a}$$

or in other way, by

$$\varphi = \tfrac{1}{2}(\xi - \eta), \quad \chi = \tfrac{1}{2}(\xi + \eta). \tag{4.15b}$$

In certain particular problems it is more advantageous to introduce the angle $\vartheta$, at which a β-characteristic is inclined to the x-axis (Fig. 44), instead of the angle $\varphi$ between the larger principal stress $\sigma_1$ and the x-axis. Bearing in mind that $\varphi = \vartheta + \pi/4$, equations (4.14) may be written in another form as

$$\frac{dy}{dx} = -\cot\vartheta, \; \chi + \vartheta = \text{const for an α-line};$$

$$\frac{dy}{dx} = \tan\vartheta, \quad \chi - \vartheta = \text{const for a β-line}. \tag{4.16}$$

Thus the integration of the basic system of partial differential equations (4.11) reduces to the integration of the ordinary differential equations (4.14) or (4.16) since the characteristics lie on the integral surface $S(\chi, \varphi, x, y)$ in the four-dimensional space $\chi, \varphi, x, y$. Thus to find them is equivalent to a solution of the initial set of equations (4.11).

The role of characteristics is fundamental in the theory of plane plastic flow, not only because they allow us to obtain effective solutions to many problems, but also for their physical meaning. Equations (4.12) directly indicate that they make the angles of $+\pi/4$ and $-\pi/4$ with the principal directions defined by the angle $\varphi$. Thus characteristics represent the lines of maximum shear, or in other words, *slip-lines*. Slip-lines may be observed on the etched surfaces of certain plastically deformed metals. They have the form of dark strips on the lighter surface of the metal. Numerous observations have shown that experimentally obtained slip-lines astonishingly well coincide with characteristics obtained from theoretical solutions. We shall see later that the characteristics (4.14) of the system (4.11) coincide with the characteristics of equations (4.8) determining the flow velocities. This important property of the stress and velocity fields remarkably simplifies the determination of the velocity distribution.

### 4.3 Properties of slip-lines

The determination of the net of slip-lines is of fundamental significance, for when this net is known, then so is the value of the angle $\varphi$ of the principal direction at every point. Suppose now that at a specific point $A$ there is given the value $\chi_A$ connected with stresses by means of (4.9). This suffices

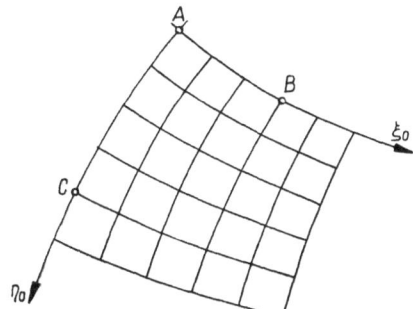

Fig. 45. Slip-line net

to find the values of $\chi$ in the entire field of slip-lines (Fig. 45). First we find $\chi$ along the slip-lines $\xi_0 = \text{const}$ and $\eta_0 = \text{const}$ passing through $A$. At any arbitrary point $B$ of the line $\xi_0$ we have according to (4.14a)

$$\chi_B = \chi_A + \varphi_A - \varphi_B.$$

Similarly at any arbitrary point $C$ of the slip-line $\eta_0$

$$\chi_C = \chi_A - \varphi_A + \varphi_C.$$

Now, starting from the points lying on the slip-lines $\xi_0$ and $\eta_0$, one can calculate the value of $\chi$ at an arbitrary point of the remaining slip-lines. Having found $\varphi$ and $\chi$, we can determine stresses by means of (4.10).

In a general case the net of slip-lines can be obtained by numerical integration of equations (4.14). In numerous cases, however, the net can be determined in a simple way employing geometrical properties of these lines. These properties enable us to determine the slip-line nets graphically, which is of great practical importance because of the simplicity of graphical methods. The most important properties of slip-lines are listed below.

1.   *The slip-lines form an orthogonal net.* This property is a direct consequence of equations (4.12).

2.   *The angle between two slip-lines of one family, where they are cut by a slip-line of the other family, is constant independently of the choice of the latter.* This property is known as *Hencky's first theorem.*

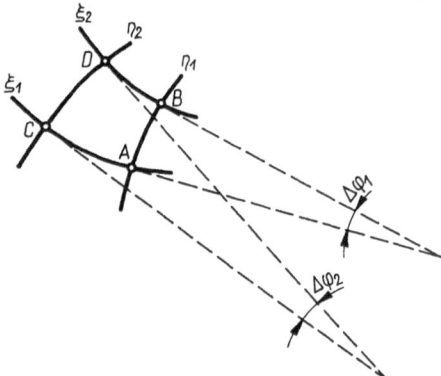

Fig. 46. Hencky's first theorem: $\varDelta\varphi_1 = \varDelta\varphi_2$

To prove this theorem let us consider two pairs of slip-lines (Fig. 46), two of which $\xi_1$ and $\xi_2$ belong to the first family, and two others $\eta_1$, $\eta_2$ to the second family. From (4.15b) we obtain that at the point $A$, where the slip-lines $\xi_1$ and $\eta_1$ intersect each other, $\varphi_A = \frac{1}{2}(\xi_1 - \eta_1)$. Similarly, at $B$, $\varphi_B = \frac{1}{2}(\xi_2 - \eta_1)$. Thus the angle $\varDelta\varphi_1 = \varphi_A - \varphi_B$ has the value $\varDelta\varphi_1 = \frac{1}{2}(\xi_2 - \xi_1)$. In the same way we get $\varDelta\varphi_2 = \frac{1}{2}(\xi_2 - \xi_1)$. Hence $\varDelta\varphi_1 = \varDelta\varphi_2$. This equality has been found for arbitrarily chosen slip-lines, thus Hencky's first theorem is generally proved.

3.  *If a segment of a slip-line is straight, the stresses along it do not change.*
This is obvious, because at every point of the straight segment the angle $\varphi$
has the same value. Hence from (4.14) we immediately obtain that also $\chi$
must be constant along that segment.

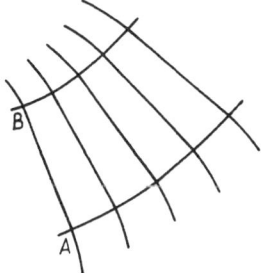

Fig. 47. If segment *AB* of one of the slip-lines is straight, then corresponding
segments of all slip-lines of the same family are also straight

4.  *If a segment AB of one of the slip-lines is straight* (Fig. 47), *then so
are the segments of all slip-lines of the same family, cut off by the two slip-
lines of the other family through A and B.* This property is a consequence
of Property 2 (*Hencky's first theorem*). The state of stress is constant along
each of the straight segments, but it varies from segment to segment.

5.  *Straight segments of slip-lines of one family, cut off by slip-lines of the
other family, have the same length.*

## 4.4   Elementary nets of slip-lines

The simplest state of stress is the one to which there corresponds a net of
slip-lines consisting of two families of straight lines. According to Property
3, along all slip-lines of the first family $\xi = \xi_0$ must hold, and along the
lines of the second family $\eta = \eta_0$, where $\xi_0$ and $\eta_0$ are the same for all
lines. From (4.15a) we immediately obtain that at an arbitrary point $M$
of the slip-line field

$$\varphi_M = \tfrac{1}{2}(\xi_0 - \eta_0) \quad \text{and} \quad \chi_M = \tfrac{1}{2}(\xi_0 + \eta_0).$$

Thus the direction of the principal stresses, defined by the angle $\varphi_M$, and
the mean stress connected with $\chi_M$, are the same at all points. In other
words, the state of stress is homogeneous throughout the field.

Consider next the net in which the slip-lines of one family are straight. The particular very important case of such nets is the so-called *centred fan*, composed of homocentric arcs of circles and straight radii from a central point $O$ (Fig. 48). Let the straight slip-lines belong to the first

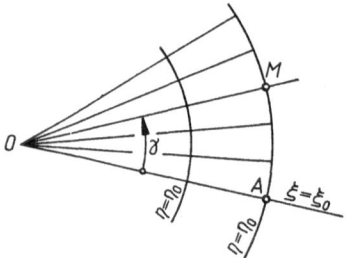

Fig. 48. Centred fan of slip-lines

family $\alpha$ and the curvilinear ones to the second family $\beta$. The value of $\eta$ must be the same on all slip-lines of the curvilinear family, since, according to Property 3, along $OA$ $\chi$ and $\varphi$ must be constant. Consequently the magnitude $\eta = \chi - \varphi$ is constant along $OA$. Let us now determine the state of stress at an arbitrary point $M$ of the radius $OM$ inclined at the angle $\gamma$ to the radius $OA$, if $\varphi_A$ and $\chi_A$ at $A$ are known. From (4.14b) we get

$$\chi_M = \chi_A + \gamma, \tag{4.17}$$

bearing in mind that $\varphi_M = \varphi_A + \gamma$. Relation (4.17) determines the stress state on any straight slip-line if the stresses are given on one of them. The magnitudes $\chi_M$ and $\varphi_M$ are obviously constant along $OM$. Thus along each radius the state of stress is different, and consequently the point $A$ is a singular one. Relation (4.17) holds valid also in the case where the straight slip-lines do not have a common point. In such a case the other family of slip-lines is not composed of the arcs of circles.

In Fig. 49 is presented a net of slip-lines, a portion of which is composed of the elementary nets described above. In the squares $ABCD$ and $DEFG$ there exist homogeneous, but different, states of stress. The two stress states must be different as required by relation (4.17). Substituting the value of the angle $CDE$ at place of $\gamma$, we obtain the relation between the stress states on the radii $CD$ and $ED$, and consequently between those in the two squares. If we know the state of stress on the segment $EF$, we

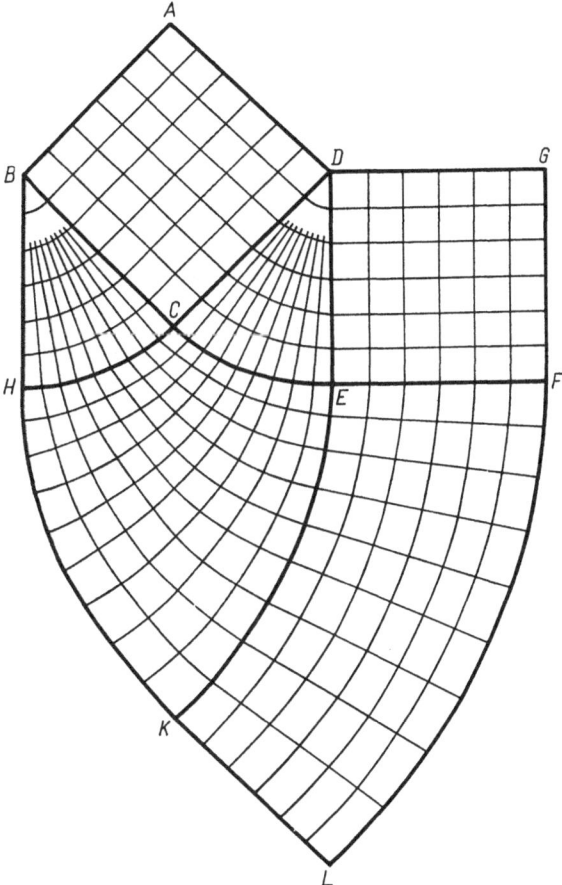

Fig. 49. A net of slip-lines composed of elementary nets

can find by means of (4.17) the stresses in the region *EFLK* in which the slip-lines of one family are straight, according to Property 4.

The slip-lines in the region *HCEK* cannot be obtained in an elementary way. The state of stress within this region is not an elementary one. The method of solution of such more advanced problems will be explained in the next section. For practical reasons we shall give here the coordinates of the nodal points of this net (Fig. 50), for it is this type of the net, built on two circular arcs *AB* and *AC*, which frequently appears, as we shall see later on, in solutions to numerous plastic forming operations. Often

77

Table 1. Coordinates $x$ and $y$

| β \ α | | 0 | 1 | 2 | 3 | 4 | 5 |
|---|---|---|---|---|---|---|---|
| 0 | $x$ | 0.000 | 0.059 | 0.112 | 0.159 | 0.199 | 0.232 |
| | $y$ | 0.000 | 0.064 | 0.133 | 0.207 | 0.284 | 0.365 |
| 1 | $x$ | | 0.129 | 0.194 | 0.253 | 0.305 | 0.350 |
| | $y$ | | 0.000 | 0.071 | 0.147 | 0.229 | 0.316 |
| 2 | $x$ | | | 0.271 | 0.343 | 0.408 | 0.466 |
| | $y$ | | | 0.000 | 0.078 | 0.163 | 0.255 |
| 3 | $x$ | | | | 0.428 | 0.507 | 0.581 |
| | $y$ | | | | 0.000 | 0.087 | 0.183 |
| 4 | $x$ | | | | | 0.602 | 0.692 |
| | $y$ | | | | | 0.000 | 0.097 |
| 5 | $x$ | | | | | | 0.798 |
| | $y$ | | | | | | 0.000 |
| 6 | $x$ | | | | | | |
| | $y$ | | | | | | |
| 7 | $x$ | | | | | | |
| | $y$ | | | | | | |
| 8 | $x$ | | | | | | |
| | $y$ | | | | | | |
| 9 | $x$ | | | | | | |
| | $y$ | | | | | | |
| 10 | $x$ | | | | | | |
| | $y$ | | | | | | |
| 11 | $x$ | | | | | | |
| | $y$ | | | | | | |
| 12 | $x$ | | | | | | |
| | $y$ | | | | | | |

the slip-lines fields in these solutions are composed entirely of elements shown in Fig. 50. In Table 1 are collated the coordinates of the nodal points, obtained by assuming that the radius of both circular fans, on which the net is founded, is equal to unity, i.e. $O_1 A = O_2 A = 1$. The angular interval between the two consecutive radii in the fans is $5°$. Each for the net of slip-lines in Fig. 50

| 6 | 7 | 8 | 9 | 10 | 11 | 12 |
|---|---|---|---|---|---|---|
| 0.259 | 0.278 | 0.289 | 0.293 | 0.289 | 0.278 | 0.259 |
| 0.448 | 0.533 | 0.620 | 0.707 | 0.794 | 0.880 | 0.966 |
| 0.388 | 0.418 | 0.439 | 0.452 | 0.457 | 0.452 | 0.439 |
| 0.407 | 0.502 | 0.600 | 0.700 | 0.801 | 0.903 | 1.005 |
| 0.518 | 0.560 | 0.594 | 0.619 | 0.635 | 0.640 | 0.635 |
| 0.353 | 0.457 | 0.565 | 0.678 | 0.793 | 0.911 | 1.031 |
| 0.647 | 0.704 | 0.753 | 0.792 | 0.821 | 0.839 | 0.845 |
| 0.286 | 0.397 | 0.515 | 0.639 | 0.768 | 0.902 | 1.040 |
| 0.774 | 0.849 | 0.914 | 0.970 | 1.015 | 1.048 | 1.068 |
| 0.205 | 0.322 | 0.448 | 0.583 | 0.725 | 0.874 | 1.030 |
| 0.898 | 0.991 | 1.076 | 1.151 | 1.214 | 1.266 | 1.304 |
| 0.110 | 0.231 | 0.364 | 0.508 | 0.662 | 0.826 | 0.998 |
| 1.017 | 1.131 | 1.236 | 1.333 | 1.418 | 1.491 | 1.551 |
| 0.000 | 0.124 | 0.261 | 0.413 | 0.577 | 0.754 | 0.943 |
| | 1.265 | 1.394 | 1.514 | 1.624 | 1.722 | 1.807 |
| | 0.000 | 0.143 | 0.297 | 0.470 | 0.659 | 0.862 |
| | | 1.547 | 1.693 | 1.830 | 1.956 | 2.069 |
| | | 0.000 | 0.160 | 0.338 | 0.537 | 0.753 |
| | | | 1.867 | 2.033 | 2.190 | 2.335 |
| | | | 0.000 | 0.182 | 0.387 | 0.614 |
| | | | | 2.232 | 2.422 | 2.603 |
| | | | | 0.000 | 0.208 | 0.443 |
| | | | | | 2.649 | 2.868 |
| | | | | | 0.000 | 0.239 |
| | | | | | | 3.129 |
| | | | | | | 0.000 |

nodal point is denoted by two numbers, the first of which being the number of the slip-line of the first family $\alpha$, and the second—the number of the line of the second family $\beta$, intersecting each other at the node. One can also obtain from the table the relation between the angle $\delta$ of the fan and the length $b$ of the net. A more detailed table with data for the net up

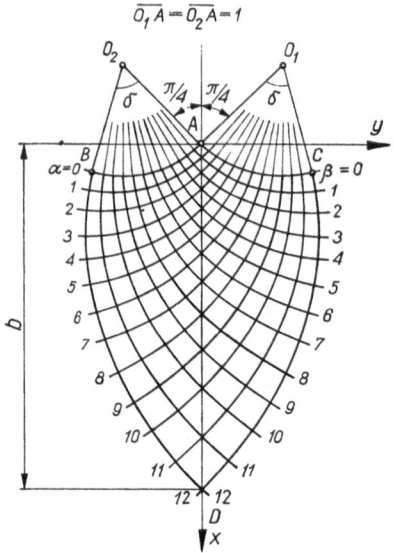

$$\overline{O_1A} = \overline{O_2A} = 1$$

Fig. 50. A net of slip-lines built on two centered fans. Coordinates of its nodal points are collated in Table 1 (p. 78–79)

to $\delta = 135°$, and coordinate values up to four decimal points may be found in [135].

### 4.5 Basic boundary value problems

As we have mentioned in Section 2, the solution of various particular problems consists in the successive solving of appropriate boundary value problems for the equations of characteristics (4.14). Now we shall discuss the main features of basic boundary value problems.

### 1 *The Cauchy boundary value problem*

Let the normal stress $\sigma_n$ and the shear stress $\tau_{nt}$ be given on the edge $AB$ of a region under consideration (Fig. 51). Both stress components are continuous along $AB$ and their first derivatives exist. If the material is to be in the plastic state, the third stress component $\sigma_t$ on $AB$ can be determined from the yield condition (4.7), by assuming that at a specific point of the edge the $x$ and $y$-directions coincide with the directions $n$ and $t$. Hence we get

$$\sigma_t = \sigma_n \pm 2\sqrt{k^2 - \tau_{nt}^2}.$$

Note that a positive or negative sign on the right-hand side must by chosen in accordance with the physical sense of the problem. In most cases the shear stress $\tau_{nt}$ vanishes on the edge; hence $\sigma_t = \sigma_n \pm 2k$. If we know the values of $\sigma_n$ and $\tau_{nt}$ on the edge, we can also determine the values and directions of the principal stresses, and consequently the auxiliary functions $\chi$ and $\varphi$.

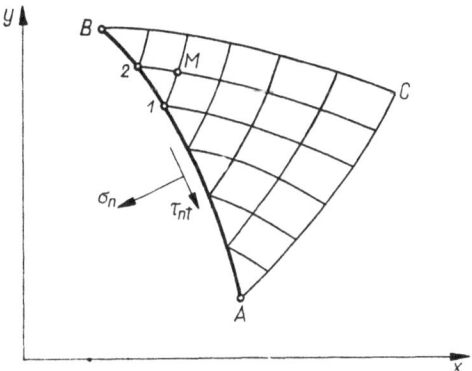

Fig. 51. The Cauchy boundary value problem

These data suffice to find the solution in the whole triangular domain $ABC$, bounded by the edge $AB$ and by the arcs $AC$ and $BC$ of two characteristics of different families. An essential property of the line $AB$ is the fact that it intersects only once each of the characteristics. At no point can $AB$ coincide with a characteristic.

In the particular case, where the edge $AB$ is straight and loaded by uniformly distributed normal stresses or is stress-free, the state of stress in the region $ABC$ is homogeneous, and the mesh of characteristics is formed by two families of straight lines inclined at 45° to the edge (Fig. 52).

In a general case the solution is carried out numerically by means of the method of finite differences (*Massau's method*—see for instance [50, 110]). Divide now the edge $AB$ by a number of points into arbitrarily small segments. At each of the points the values of $\sigma_n$ and $\tau_{nt}$ are given as everywhere on $AB$. Let us show how to find the coordinates of the nodal point $M$ at which the characteristics of different families through the points $1$ and $2$ lying on $AB$ intersect each other, and how to calculate the values $\chi_M$ and $\varphi_M$ at $M$. Assume for definiteness that the characteristic through $1$

81

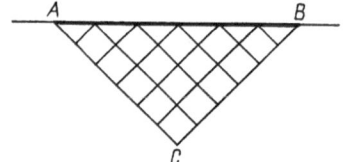

Fig. 52. An elementary case of the Cauchy boundary value problem

and $M$ belongs to the first family of $\alpha$-lines, and the characteristic through 2 and $M$ to the second family $\beta$.

We shall determine the values of $\chi_M$ and $\varphi_M$, determining the state of stress at $M$ from equations (4.14) which lead to the equations

$$\chi_M + \varphi_M = \chi_1 + \varphi_1,$$
$$\chi_M - \varphi_M = \chi_2 - \varphi_2. \qquad (4.18)$$

The values which are taken by the functions $\chi$ and $\varphi$ at the points *1* and *2* are denoted by the suffixes 1 and 2, respectively.

Having found the angle $\varphi_M$ from (4.18), we shall find the coordinates $x_M$ and $y_M$ by replacing differentials in (4.14) by finite differences. In the system of equations

$$y_M - y_1 = \tfrac{1}{2}[\tan(\varphi_1 + \tfrac{1}{4}\pi) + \tan(\varphi_M + \tfrac{1}{4}\pi)]\,(x_M - x_1),$$
$$y_M - y_2 = \tfrac{1}{2}[\tan(\varphi_2 - \tfrac{1}{4}\pi) + \tan(\varphi_M - \tfrac{1}{4}\pi)]\,(x_M - x_2), \qquad (4.19)$$

so obtained only the coordinates $x_M$ and $y_M$ are not known, the coordinates of the points *1* and *2* being obviously known.[1]

Making use of formulae (4.18) and (4.19), we can find the solution in the whole region *ABC*. Denoting by *1* and *2* the just determined new points of the net of characteristics, we may find the coordinates of the

---

[1] Other approaches are also possible. In Hill's book [50] the following formulae are recommended

$$y_M - y_1 = [\tan(\tfrac{1}{2}(\varphi_1 + \varphi_M) + \tfrac{1}{4}\pi)]\,(x_M - x_1),$$
$$y_M - y_2 = [\tan(\tfrac{1}{2}(\varphi_2 + \varphi_M) - \tfrac{1}{4}\pi)]\,(x_M - x_2).$$

The simplest are the formulae (see V. V. Sokolovsky [110])

$$y_M - y_1 = (x_M - x_1)\tan(\varphi_1 + \tfrac{1}{4}\pi),$$
$$y_M - y_2 = (x_M - x_2)\tan(\varphi_2 - \tfrac{1}{4}\pi)$$

in which the arc is replaced by a chord whose slope has the initial value. The latter formulae give good results when a fine mesh of characteristics is used, i.e. when solutions are obtained by programming the problem on a digital computer.

new point *M* and the associated state of stress. Thus (4.18) and (4.19) are used as recurrence formulae.

There exists a number of relatively simple graphical methods of construction of slip-line (characteristic) nets. Roughly speaking, they are based upon the Hencky first theorem (see Property 2 in Section 4.3). These methods will be described in the next section.

### 2 *The characteristic boundary value problem*

Along the two characteristics *AB* and *AC* (Fig. 53) belonging to different families the values of the angle $\varphi$ and of the function $\chi$ are given. Note that the angle $\varphi$ is determined by the direction of the characteristic. The

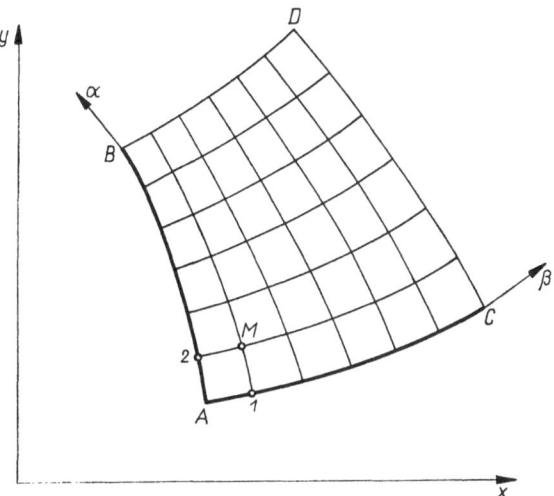

Fig. 53. The characteristic boundary value problem

values of the function $\chi$ along each of the two characteristics can be immediately found if $\chi$ is given even at a single point only; indeed, the relations $\chi+\varphi=$ const and $\chi-\varphi=$ const determine its value at the remaining points of the characteristic of the first and of the second family, respectively. In practice, the situation forming a characteristic problem usually results from previous calculations of other boundary value problems such as the Cauchy problem. These data suffice to find the solution in the region *ABDC*, bounded by the arcs *AB*, *AC* and by the characteristics *BD* and *CD* of different families through the points *B* and *C*. An example represents the net shown in Fig. 50, which was obtained by solving the characteristic problem defined by the data along *AB* and *AC*.

The numerical procedure is the same as in the case of the Cauchy problem. The recurrence equations (4.18) and (4.19) hold valid also in this case. Let *AB* be a segment of the characteristic of the first family $\alpha$, and *AC* of the second family $\beta$. Partitioning both segments by sets of points, we can find the values of $\varphi$ and $\chi$ corresponding to these points. It is seen that the points *1* and *2* in the vicinity of the corner *A* form the same arrangement as that considered in the Cauchy problem. Thus we can find the values $\chi_M$ and $\varphi_M$ at the point *M*, and the coordinates $x_M$, $y_M$ of this point. Having found all the magnitudes at *M*, we may extend the solution since these magnitudes along with the data at the next points on *AB* and *AC* form the new three-point configuration, and so on. Thus formulae (4.19) and (4.18) allow us to compute all required magnitudes at every nodal point of the network of the characteristics.

### 3    The characteristic boundary value problem with a singular point

This problem is a very important particular case of the characteristic problem. We obtain it when the length of one of the segments of an initial characteristic, say *AB* (Fig. 53), tends to zero whereas the increment of

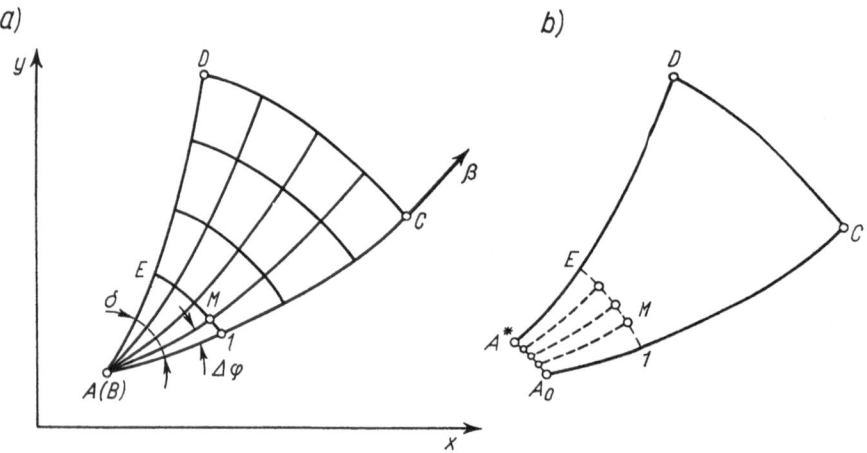

Fig. 54. The characteristic boundary value problem with a singular point *A*

the angle $\varphi$ on the segment $AB$, shrunk into a single point $A$, has a finite value $\Delta\varphi$. Thus we know the values of $\chi$ and $\varphi$ along $AC$, and the change in the angle $\varphi$ at the point $A$ (Fig. 54a). An elementary case of such a problem is shown in Fig. 48.

To begin the numerical calculations, the total increment $\Delta\varphi$ of the angle $\varphi$ at $A$ must be divided into a number $n$ of small increments $\delta\varphi = \Delta\varphi/n$. We shall show how to determine the values of $\chi_M$ and $\varphi_M$ at a point $M$ adjacent to the singular point $A$, and how to find the coordinates of $M$. Let $AC$ be a characteristic of the second family. To make the explanation of the numerical procedure more clear, let us imagine that at $A$ a number of points overlap from $A_0$ to $A^*$ (Fig. 54b). All of them have the same coordinates but the value of the angle $\varphi$ consecutively increases by $\delta\varphi$ from one point to the other. Thus $\varphi_{A_0} = \varphi_A$ at $A_0$, and at the end-point $A$ we have $\varphi_{A^*} = \varphi_A + \Delta\varphi$.

From the first of Hencky theorems (Property 2) we obtain immediately

$$\varphi_M = \varphi_1 + \delta\varphi.$$

From (4.14a) we get

$$\chi_M = \chi_1 + \varphi_1 - \varphi_M,$$

since $1{-}M$ is a segment of the characteristic of the first family.

We obtain both coordinates $x_M$ and $y_M$ from (4.19), remembering that point $2$ overlaps the point $A$, and therefore $x_2 = x_A, y_2 = y_A$. The angle $\varphi$ at point $2$ is $\varphi_2 = \varphi_A + \delta\varphi$.

Now the next point can be computed. The point $M$ just found will now be denoted as the new point $1$. All the relations are unchanged. Calculating the coordinates $x_M$ and $y_M$ of the new point $M$, we must bear in mind that now $\varphi_2 = \varphi_A + 2\delta\varphi$. Similarly, calculating the third point on the arc $1{-}E$, we have $\varphi_2 = \varphi_A + 3\delta\varphi$, and so on. In this way all points on $1{-}E$ can be obtained. The function $\chi$ is not constant at $A$; it increases through overlapping points by the value $\delta\varphi$, because the point $A$ represents a peculiar form of the characteristic of zero length, along which $\chi + \varphi$ = const. The change in the angle $\varphi$ is $\delta\varphi$, and therefore the same change must occur in $\chi$. Thus the stresses suffer a jump at $A$. Hence the point $A$ is a singular one.

The data just found along the arc $1{-}E$ of the characteristic of the first family, and the data given along the arc $AC$ of the characteristic of the second family, form a situation which is identical with that on Fig. 53.

Thus further calculations are conducted according to the scheme of the standard characteristic problem, described in the foregoing section. In this way the solution can be found within the whole domain $ACD$.

In the same way the problem can be solved in the other case when the length of a characteristic of the second family tends to zero.

## 4    Mixed boundary value problems

There exists a number of mixed boundary value problems. Generally speaking in a mixed problem boundary values are given along a characteristic line, and moreover along a non-characteristic one. The most important variants only will be discussed here.

Let along a segment $AC$ of a characteristic the values of $\chi$ and $\varphi$ be given. Along a segment $AB$ of a line $y = y(x)$, which does not coincide with any characteristic, the angle $\varphi$ is known (Fig. 55). These data are sufficient to solve the problem within the whole region $ABC$.

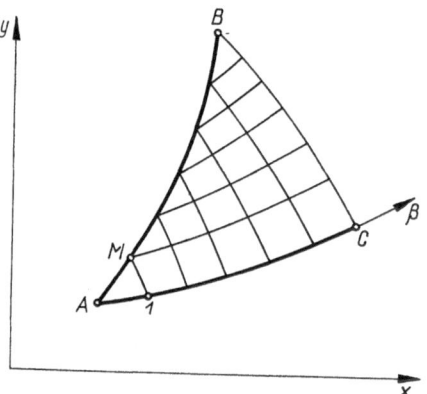

Fig. 55. A mixed boundary value problem

Numerical calculations can be performed by means of the recurrence formulae (4.18) and (4.19), except for the points lying on the line $AB$. Let the characteristic $AC$ belong to the second family. The coordinates $x_M$ and $y_M$ of the point $M$, at which the characteristic of the first family through the point $1$ intersects $AB$, can be obtained from the equation

$$y_M - y_1 = \tfrac{1}{2}[\tan(\varphi_1 + \tfrac{1}{4}\pi) + \tan(\varphi_M + \tfrac{1}{4}\pi)]\,(x_M - x_1),$$

and from the equation $y = y(x)$ of the line $AB$, which obviously must be satisfied by these coordinates. The angle $\varphi_M$ is known by assumption. The value $\chi_M$ can be determined from the relation $\chi_M + \varphi_M = \chi_1 + \varphi_1$. In the same way one can solve the mixed problem in the case where $AC$ is a characteristic of the first family $\alpha$.

The other important variant of the mixed problem is when along a non-characteristic line $AB$ there is given a certain relation between $\chi$ and $\varphi$ in a general form $f(\varphi, \chi) = 0$. These data also suffice to find the solution within the region $ABC$. Problems of such kind appear in the cases where for instance $AB$ represents the contact surface between the deforming plastic material and the rigid tool. In the presence of Coulomb friction we have $t = \mu p$, where $p$ is the contact pressure, $t$ is the friction force, and $\mu$ stands for the friction coefficient.

## 4.6 Graphical construction of slip-line nets

Although numerical calculations of slip-line nets with the use of digital computers form at present the routine method used in practical applications, in some cases it is advantageous to construct these nets graphically. Simple geometrical properties of slip-line nets permit their graphical construction with satisfactory accuracy.

There exists a number of graphical methods of constructing slip-line nets. The simplest of them will be briefly described below. Other methods can be found in works of W. Prager and P. G. Hodge [97], W. Prager [96], L. D. Tomlenov [133], and E. G. Thomsen, C. T. Yang and S. Kobayashi [135]. In the latter a simple and interesting method is presented. In this method proposed by E. G. Thomsen [134] the slip-lines are constructed by unrolling a wedge cut of a cardboard.

The standard method presented here is based on Hencky's first theorem (Fig. 46). The accuracy of the method is in most cases satisfactory.

In many applications of the slip-line theory, which are discussed in Chapters 5, 6, and 7, the essential portion of the slip-line net reduces to the particular characteristic problem shown in Fig. 50. In this problem the two circular slip-lines $AB$ and $AC$ are known, and it is desired to establish the slip-line net within the region $ABDC$. On account of symmetry the right-hand half of the net only will be considered.

Let us plot a centred fan of straight lines through the point $O_1$ in

Fig. 56. It is convenient to assume a constant angular interval $\Delta\gamma = \delta/n$ between these lines. Then let the circular arc $AC$ be replaced by a number of straight segments $A$–0.1, 0.1–0.2, ..., 0.5–$C$. The slip-line 0.1–1.1 must intersect the axis of symmetry $AD$ at the angle of 45°. This sector, however,

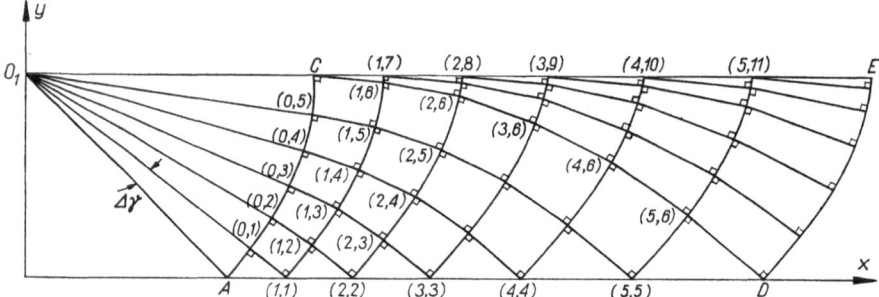

Fig. 56. Graphical construction of slip-line net

when replaced by a straight line will be inclined to $AD$ at an angle of the mean value. To determine this angle we divide the difference $\Delta\gamma$ of the slope angles at 0.1 and 1.1 by two, by plotting a chord 0.1–1.1 making the angle $\Delta\gamma/2$ with the radius $O_1$–0.1. This chord is perpendicular to the segment $A$–0.1. Similarly, the segment 0.2–1.2 is perpendicular to 0.1–0.2, and so on. The segment 1.1–1.2 is plotted parallel to $A$–0.1 since according to Hencky's theorem the angle $\Delta\varphi$ between 1.1–1.2 and 0.1–0.2 must preserve a constant value. Similarly, 1.2–1.3 is parallel to 0.1–0.2, and so on. In this way we construct the slip-line net in the whole region $ACD$. As an example of an application of this procedure, the construction of the net in the adjacent domain $CDE$ is also shown in the figure. The boundary condition $\varphi = \pi/4$ is assumed along $CE$. Thus the slip-lines of one family are perpendicular to $CE$, and the lines of the other family are tangent to it. This condition, together with just the found slip-line $CD$, constitutes a typical mixed boundary value problem. The figure in which the right angles are clearly shown is self-explanatory. The net on Fig. 56 constitutes a part of the slip-line net for the problem of compression of a plastic block discussed in the next chapter (Fig. 83).

The smaller the angular interval $\Delta\gamma$ between the slip-lines, better the accuracy of the graphical method. In Fig. 56 the interval $\Delta\gamma = 7° 30'$ was assumed. This value gives satisfactory results in practical applications, but the use of the smaller value $\Delta\gamma = 5°$ is recommended.

The method outlined above is particularly suitable when a drawing-board is used, but using simple triangles we may also easily construct even very complex slip-line nets.

## 4.7 Determination of the velocity field

Determination of the velocity field is necessary in each particular case since in some cases the slip-line net is kinematically inadmissible, which means that the kinematic conditions of the problem are violated. Examples of processes for which the slip-line nets are kinematically inadmissible will be discussed in Chapters 5 and 6. Such solutions are not complete and they give only an approximate estimate of the load. In most cases there is available a velocity field compatible with the slip-line field.

When determining the velocities, we start from the relations (4.8) which can be written in the form

$$\left(\frac{\partial v_x}{\partial y} + \frac{\partial v_y}{\partial x}\right)(\sigma_x - \sigma_y) - 4\frac{\partial v_x}{\partial x}\tau_{xy} = 0,$$

$$\frac{\partial v_x}{\partial x} + \frac{\partial v_y}{\partial y} = 0. \tag{4.20}$$

These equations contain two unknowns $v_x$ and $v_y$ since the stresses are already found by solving the appropriate boundary value problems for the stress field. The system (4.20) is of the hyperbolic type and it can be solved by the method of characteristics. Supplementing (4.20) by the expressions

$$\frac{\partial v_x}{\partial x}dx + \frac{\partial v_x}{\partial y}dy = dv_x,$$

$$\frac{\partial v_y}{\partial x}dx + \frac{\partial v_y}{\partial y}dy = dv_y,$$

and equating to zero the characteristic determinant formed by the coefficients associated with the derivatives, we obtain the differential equations of the characteristics

$$\frac{dy}{dx} = \tan(\varphi + \tfrac{1}{4}\pi),$$

$$\frac{dy}{dx} = \tan(\varphi - \tfrac{1}{4}\pi).$$

The stresses have been replaced here by the angle $\varphi$, by means of (4.10). Thus the equations of characteristics for velocities and equations (4.12) for stresses are identical. This leads to the important conclusion of great practical significance that the characteristics for velocities coincide with those for stresses.[1]

Having found the net of characteristics (slip-lines) from the stress solution, we may determine the velocity components by integrating certain differential relations which must be satisfied along the characteristics. These relations may be obtained by equating to zero any determinant obtained from the characteristic determinant by replacing one column by the column of terms containing no derivatives. The method explained in Appendix 2 may also be used. Finally we obtain

$$dv_x + dv_y \tan(\varphi + \tfrac{1}{4}\pi) = 0,$$
$$dv_x + dv_y \tan(\varphi - \tfrac{1}{4}\pi) = 0, \tag{4.21}$$

for the characteristics of the first and the second family, respectively. We shall now transform relations (4.21), replacing the velocity components $v_x$ and $v_y$ by the components $v_\alpha$ and $v_\beta$ directed at a specific point along the slip-lines of the $\alpha$-family and the $\beta$-family, respectively (Fig. 57). Differentiating the geometrical relations

$$v_\alpha = -v_x \sin(\varphi - \tfrac{1}{4}\pi) + v_y \cos(\varphi - \tfrac{1}{4}\pi),$$
$$v_\beta = v_x \cos(\varphi - \tfrac{1}{4}\pi) + v_y \sin(\varphi - \tfrac{1}{4}\pi), \tag{4.22}$$

and then introducing (4.21) we obtain the fundamental relations

$$dv_\alpha - v_\beta \, d\varphi = 0 \text{ along } \alpha\text{-lines,}$$
$$dv_\beta - v_\alpha \, d\varphi = 0 \text{ along } \beta\text{-lines,} \tag{4.23}$$

which must be satisfied along slip-lines. These relations, originally due to H. Geiringer [38], have a simple geometrical meaning. Consider two adjacent points $A$ and $B$ of a slip-line of the family $\alpha$ (Fig. 58). Suppose that the angle of inclination of the line changed from $A$ to $B$ by $d\varphi$, and the velocity components increased by $dv_\alpha$ and $dv_\beta$, respectively. The increment of the velocity component directed along $AB$ is $dv_\alpha + v_\beta d\varphi$, the small magnitudes of the higher order being neglected. Hence it follows that, according to (4.23), the slip-lines are the lines of zero elongation.

---

[1] This is a general property of the theory of plastic flow when the flow rule associated with the yield condition is assumed. We shall see later on that stress and velocity characteristics coincide also in problems of axially symmetric flow (Chapter 8) and in those of plane stress (Chapters 9 and 11).

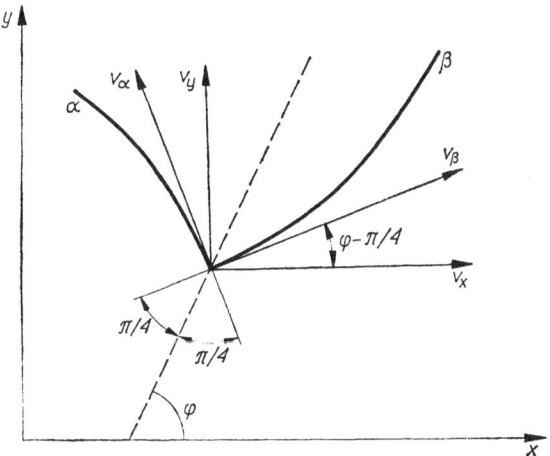

Fig. 57. Components of the velocity vector

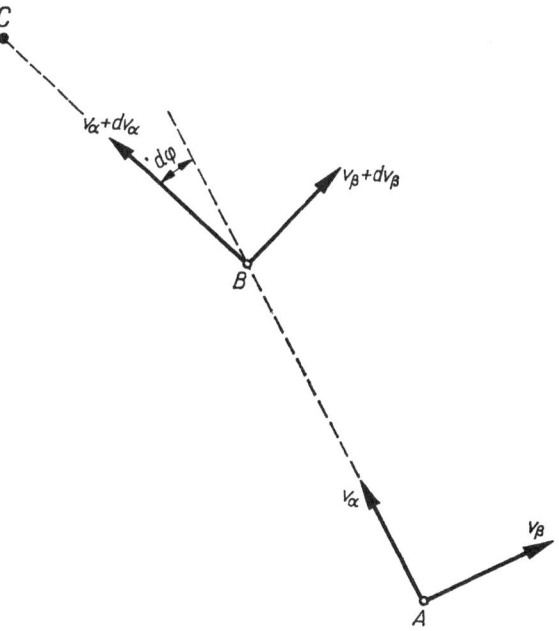

Fig. 58. Geometrical interpretation of the Geiringer equations (4.23)

This property of slip-lines can also be derived directly from equations (4.20). Consider a certain line $L$, assuming a local system of coordinates $x, y$ at an arbitrary point on it, with the $x$-axis directed tangentially to $L$. Let us assume that the velocity components $v_x$ and $v_y$ along $L$ are known, and therefore the derivatives $\partial v_x/\partial x$ and $\partial v_y/\partial x$ along $L$ are also known. When the line $L$ is not a characteristic, also the two remaining derivatives $\partial v_x/\partial y$ and $\partial v_y/\partial y$ should be determined along it. This is impossible when simultaneously the two conditions

$$\sigma_x - \sigma_y = 0 \quad \text{and} \quad \partial v_x/\partial x = 0$$

are satisfied on $L$. The first of them, $\sigma_x = \sigma_y$, indicates that in such a case the line $L$ is a slip-line, or in other words a velocity characteristic. The second condition, $\partial v_x/\partial x = 0$, indicates that the slip-line is a line of zero extension.

In order to determine the velocity components in a specific problem one has to solve the boundary value problems for equations (4.21) or (4.23) (compare [50]). Similarly as in the case of stress analysis, there may be distinguished the Cauchy problem, the characteristic and the mixed problems.

## 1 *The Cauchy boundary value problem*

Let along an arbitrary line $AB$, which does not coincide with a characteristic, the velocity components $v_\alpha$ and $v_\beta$ be given (compare Fig. 51). The net of slip-lines is assumed to be known from the static solution. Thus the $\alpha$ and $\beta$-directions are known everywhere. If other velocity components, say $v_x$, $v_y$ or $v_n, v_t$ ($n$ is the normal and $t$ is the tangential direction along $AB$), are given on $AB$, then the required components $v_\alpha$ and $v_\beta$ can easily be found. These data suffice to determine the velocities in the region $ABC$. The velocity components can be computed numerically by means of finite difference equivalents of Geiringer's equations (4.23). The velocity components $v_{\alpha M}$ and $v_{\beta M}$ at the point $M$, at which slip-lines of different families passing through the points $1$ and $2$ on $AB$ intersect each other, can be found from the equations

$$v_{\alpha M} - v_{\alpha 1} + \tfrac{1}{2}(v_{\beta M} + v_{\beta 1})\,(\varphi_M - \varphi_1) = 0,$$
$$v_{\beta M} - v_{\beta 2} - \tfrac{1}{2}(v_{\alpha M} + v_{\alpha 2})\,(\varphi_M - \varphi_2) = 0. \tag{4.24}$$

Having found a row of points adjacent to $AB$, we repeat the computations for the other row, and so on.

## 2 The characteristic boundary value problem

Let along two intersecting slip-lines $AB$ and $AC$ (see Fig. 53) of different families the normal velocity components be given. Thus if $AB$ is the slip-line of the $\alpha$-family, the component $v_\beta$ is given along it. Similarly along $AC$ the component $v_\alpha$ is known. The tangential components cannot be given arbitrarily since they must satisfy equations (4.23). These data determine uniquely the velocity distribution within the whole region $ABDC$. Calculations for consecutive nodal points of the slip-lines net are performed by means of the finite difference equations (4.24), beginning from the point adjacent to $A$.

## 3 Mixed boundary value problems

Let along a slip-line $AC$ the normal velocity component and along a non-characteristic line $AB$ a certain relation $f(v_\alpha, v_\beta) = 0$ be given (see Fig. 55). The tangential velocity component along $AC$ can be found from (4.23). Such boundary conditions suffice to find the solution in the region $ABC$. Let the slip-line $AC$ belong to the $\beta$-family. The normal velocity component $v_\alpha$ is known along it, and the $v_\beta$-component can be found from (4.23). We shall show how to find the velocities at the point $M$ at which the characteristic of the first family through $1$ intersects $AB$. One equation is identical with the first equation (4.24), and the other is determined by the given relation between the velocity components on $AB$. Thus we have

$$v_{\alpha M} - v_{\alpha 1} + \tfrac{1}{2}(v_{\beta M} + v_{\beta 1})(\varphi_M - \varphi_1) = 0,$$
$$f(v_{\alpha M}, v_{\beta M}) = 0.$$

From these equations the components $v_{\alpha M}$ and $v_{\beta M}$ can be calculated. Calculations for other nodal points of the slip-line net are performed according to equations (4.24). In the same way the problem can be solved in the case where $AC$ represents a slip-line of the first family.

## 4 Particular cases

Equations (4.20) are obviously satisfied in the particular case when both velocity components have constant values $v_x = $ const, $v_y = $ const within certain domain. Thus this domain moves as a rigid whole without any deformation within it. This kind of motion frequently appears in theoretical solutions to plastic forming operations.

Consider now the case where slip-lines of one family are straight. When an $\alpha$-line is straight, then the first equation (4.23) will be satisfied only if along this line $v_\alpha = $ const, since the increment of the angle $\varphi$ is zero along a straight line. Similarly, along a straight $\beta$-line we have $v_\beta = $ const.

In Fig. 52 there was presented an elementary slip-line net composed of two families of straight lines. Such a net corresponds to a homogeneous state of stress. The velocity field, however, does not have to be necessarily uniform. The velocity component directed along a slip-line is constant, but can vary from one line to another according to the given boundary conditions. Suppose that on the edge $AB$ in Fig. 52 both velocity components are uniformly distributed. Thus the solution for velocities reduces to the elementary case of the Cauchy problem. The velocities preserve their values along the straight slip-lines, and the whole region $ABC$ moves as a rigid block. If, however, the velocity distribution on $AB$ is non-uniform, then the velocity field in the region $ABC$ is non-homogeneous.

### 4.8   Stress and velocity discontinuities

In most practical cases we cannot obtain a solution with a continuous stress or velocity distribution. On the other hand, in the theory of plane plastic flow, described by the equations of hyperbolic type, the jumps of stress and velocity across some lines, called the *lines of discontinuity*, are admissible. Solutions with the lines of discontinuity are correct, provided they satisfy the boundary conditions, conditions of equilibrium and compatibility. Consider the basic types of velocity and stress discontinuity.

### 1   *Velocity discontinuities*

Let a line $L$ be the line of velocity discontinuity. Assume that at a specific point of $L$ the $x$-direction is tangent and $y$-direction is normal to it. Any jump of the $v_y$ velocity component across $L$ is inadmissible since in such a case the compatibility condition would be violated, whereas the $v_x$ velocity component directed along $L$ can be different on both sides of the line of discontinuity. The derivative $\partial v_x / \partial y$ is then infinite on the line $L$. Then the relations (4.8) between stresses and velocities imply that $\sigma_x - \sigma_y = 0$ on $L$. Thus from the yield condition (4.7) we directly get $\tau_{xy} = k$. The line of velocity discontinuity coincides therefore with a slip-line, or with the envelope of slip-lines. Geiringer's equations (4.23) indicate that when the tangential velocity component suffers a jump across the line of discontinuity, then the magnitude of the jump remains constant along that line. Each material particle crossing the line of discontinuity changes rapidly the direction of motion and suffers finite distortion. The appearance of lines of velocity discontinuity in particular solutions is connected with the assumption of the rigid-perfectly plastic model of the material, displaying no strain-hardening effect. In real strain-hardening materials narrow transitory strips, where the distortion is particularly large, are observed between the less distorted zones. The more distinct is the strain-hardening effect, the wider are transitory strips.

In the theory of plastic forming operations particularly important is the case where a line of velocity discontinuity intersects the contour of a rigid tool. The component of the vector of velocity discontinuity normal to the tool surface must obviously be zero. Let an $\alpha$-slip-line be the line of discontinuity, along which the velocities suffer a jump $\Delta v'$ (Fig. 59). At

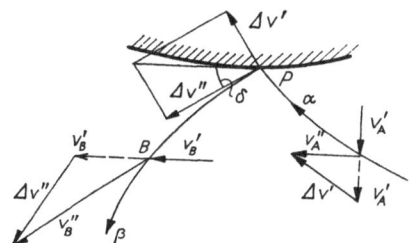

Fig. 59. At point $P$, where the line of velocity discontinuity $L$ intersects the contour of the tool, the velocity jump is reflected and propagates along the slip-line $\beta$

the point $P$, where the line of discontinuity intersects the contour of the tool, the velocity jump is reflected and propagates as $\Delta v''$ along the slip-line of the other family $\beta$. From the condition that the resultant vector of the two velocity discontinuities at $P$ must be parallel to the contour of the tool we obtain

$$\Delta v'/\Delta v'' = \tan \delta.$$

Thus the reflected velocity jump $\Delta v''$ is different from the original jump $\Delta v'$. In the particular case where $\delta = \pi/4$, which means that there is no friction between the tool and the material, we have $\Delta v' = \Delta v''$.

## 2    *Stress discontinuities*

In the solutions to problems of plastic flow of an ideally plastic material there are permitted certain lines across which the stresses suffer a jump. Let $L$ be a line of stress discontinuity. Consider the equilibrium of a small element on the line $L$ (Fig. 60). It is seen that the normal stress $\sigma_n$ acting

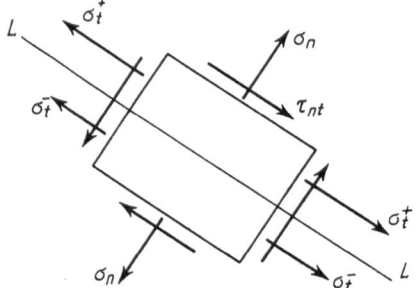

Fig. 60. The line of stress discontinuity. The stress component $\sigma_t$ assumes a different values on each side of this line

on the edges parallel to $L$ must have the same value on both sides of $L$. Also the tangential stress component $\tau_{nt}$ must be the same on both sides. However, the values of the normal stress component $\sigma_t$ applied on the edges perpendicular to $L$ can be different on each side. Any elementary force $\sigma_t\,dA$ acting on some elementary surface $dA$ is in equilibrium with an identical force acting in the opposite direction on a symmetrical elementary surface of the opposite face. If the material on both sides of the line $L$ is to be in the plastic state, we obtain from (4.7) that

$$\sigma_t = \sigma_n \pm 2\sqrt{k^2 - \tau_{nt}^2}\,.$$

The upper and lower signs before the square root correspond to the values of the $\sigma_t$ stress on both sides of $L$, respectively. The jump suffered by the stress $\sigma_t$ across the line $L$ is

$$[\sigma_t] = 4\sqrt{k^2 - \tau_{nt}^2}\,.$$

Note that the line of stress discontinuity cannot coincide with any of the slip-lines. On the slip-line the tangential stress $\tau_{nt}$ is $k$ and, therefore, the jump of $\sigma_t$ must be zero. Thus we have the important property of plane plastic flow stating that the line of stress discontinuity cannot coincide with any line of velocity discontinuity.

The simplest example of a stress discontinuity is the case of bending of a beam (Fig. 61). Let the beam be made of an elastic-plastic material with no-hardening effect. In the elastic state the stress distribution is linear (Fig. 61a). With increasing bending moment the outer zones begin to yield

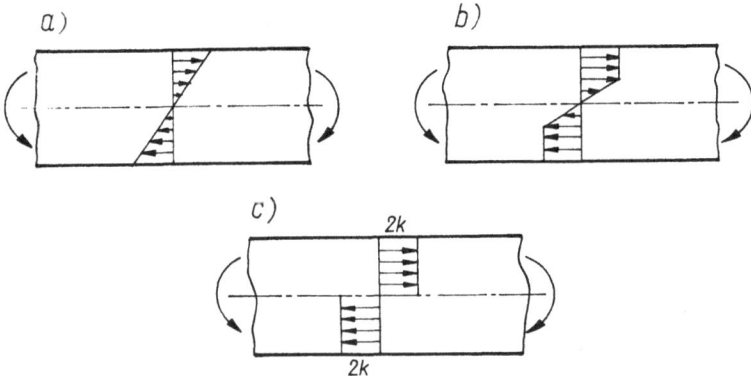

Fig. 61. The simplest example of the stress discontinuity. Bending of a beam

plastically (Fig. 61b). When the moment reaches the limit value the plastic zones meet at the axis of the beam (Fig. 61c). Hence this axis is the line of stress discontinuity. The stresses above it are tensile, while below it, compressive.

Thus it is seen that the lines of stress discontinuity represent the remnants of the elastic zones which exist in the real elastic-plastic materials. These zones diminish with increasing load, and finally under assumption of a rigid-plastic model of the material, turn into the lines of discontinuity. A more rigorous proof of this property may be found in [97].

The line of stress discontinuity (Fig. 60) is formed by intersection of a certain surface of stress discontinuity with the $x$, $y$-plane. When the plane flow problem is considered as a three-dimensional one, the discontinuities of stress can be analysed in a more general way. Assume a local coordinate system $x, y, z$ with the $x$-axis coinciding with the line of discontinuity $L$, and the $z$-axis oriented in the same manner as in the plane flow theory (Fig. 62). The surface of stress discontinuity $P$ intersects the $x$, $y$-plane along the line $L$. On both sides of the surface $P$ the state of stress is different. The normal stresses $\sigma_z^+$ and $\sigma_z^-$ can be determined from (4.4):

$$\sigma_z^+ = \tfrac{1}{2}(\sigma_n + \sigma_t^+),$$
$$\sigma_z^- = \tfrac{1}{2}(\sigma_n + \sigma_t^-).$$

Thus not only $\sigma_t$ but also the $\sigma_z$ component is different on either side. This type of stress discontinuity is statically admissible since the conditions of local equilibrium are satisfied.

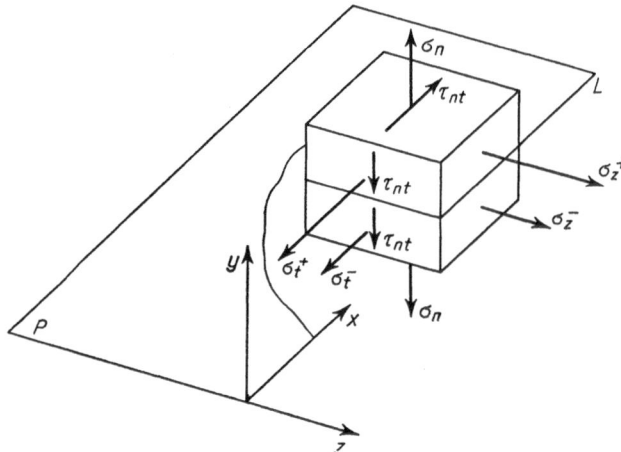

Fig. 62. The surface of stress discontinuity in the case of plane flow

A more general case of stress discontinuity is presented in Fig. 63. The stress components $\sigma_y$, $\tau_{yx}$, and $\tau_{yz}$ must be continuous. The conditions of equilibrium will not be violated if the remaining components $\sigma_x$, $\sigma_z$, and $\tau_{xz}$ have different values below and above the discontinuity surface.

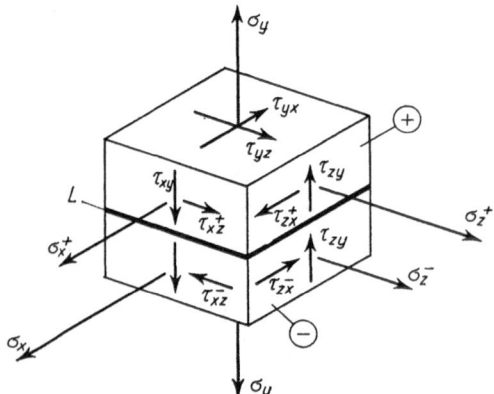

Fig. 63. A general case of stress discontinuity

## 4.9   The velocity hodograph

In Section 4.7 we described the standard numerical method by means of which the velocity field can be found. This method, however, is usually less advantageous in practical applications than the graphical method proposed by A. P. Green [39] and later developed in [41] and [96].

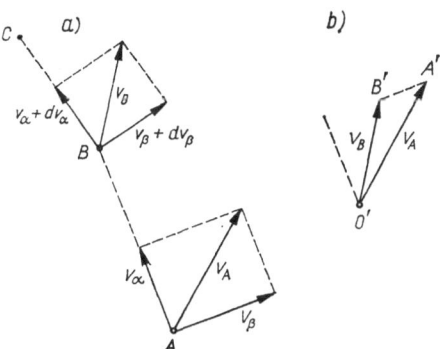

Fig. 64. Construction of the velocity hodograph: a) physical plane, b) velocity plane. Segment $A'B'$ of the hodograph is perpendicular to segment $AB$ of the corresponding slip-line

Consider two adjacent segments $AB$ and $AC$ of a slip-line (Fig. 64a). The flow velocity at the point $A$ is defined by the vector $v_A$, and at $B$ by the vector $v_B$. According to Geiringer's equations (4.23) the rate of elongation of the segment $AB$ is zero. Thus the projections of both velocity vectors $v_A$ and $v_B$ on $AB$ must be of the same magnitude.

Let in the velocity plane (Fig. 64b) the vectors of velocities $v_A$ and $v_B$ be represented by the segments $O'A'$ and $O'B'$ issuing from the origin $O'$. Since the projections of the two vectors on $AB$ are of the same length, the segment $A'B'$ in the velocity plane must be perpendicular to the segment $AB$ in the physical plane. This leads to the important conclusion that an element of a slip-line and its mapping onto the velocity plane are orthogonal. Therefore the velocity hodograph displays the same geometrical properties as the net of slip-lines since the changes in angles between corresponding points are the same in both nets.

Let us now examine how a line of velocity discontinuity is represented in the velocity plane. Let $L$ in Fig. 65a be a line of velocity discontinuity.

99

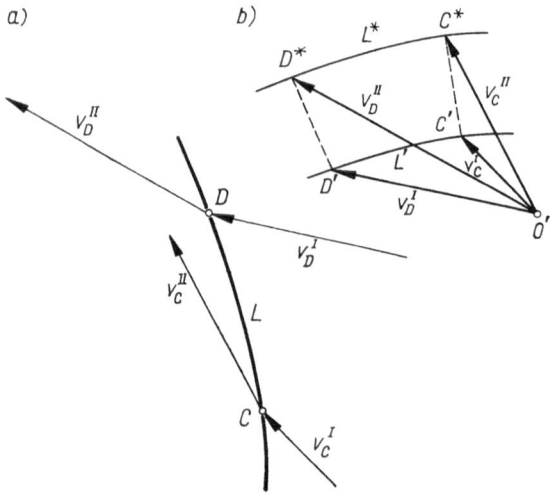

Fig. 65. The line of velocity discontinuity $L$ in the physical plane (a) is mapped onto two equidistant lines $L'$ and $L^*$ in the velocity plane (b)

In the velocity plane both sides of this line must be represented by two different lines $L'$ and $L^*$, for the velocities are different on each side of $L$. Let $C$ and $D$ be arbitrarily chosen points on $L$. The point $C$ is mapped onto two points $C'$ and $C^*$ lying in the velocity plane on the lines $L'$ and $L^*$, respectively. Similarly, the point $D$ is mapped onto $D'$ and $D^*$. As we already know, the line of velocity discontinuity $L$ must coincide with a slip-line. Hence, according to the fundamental property of orthogonality of the hodograph and slip-line nets, the tangents to $L'$ and $L^*$ at $C'$ and $C^*$, respectively, must be perpendicular to the line $L$ at $C$ in the physical plane. Similarly, the tangents to $L'$ and $L^*$ at $D'$ and $D^*$ are orthogonal to $L$ at $D$. The segment $C'C^*$ of the hodograph represents the velocity jump at $C$, and the segment $D'D^*$, the velocity jump at $D$. Both segments $C'C^*$ and $D'D^*$ are orthogonal to the two lines $L'$ and $L^*$. On the other hand, the velocity jump is constant along the discontinuity line. Thus the lengths of the segments $C'C^*$ and $D'D^*$ are the same. Therefore the curves $L'$ and $L^*$ are equidistant.

We have, therefore, the general property that a *line of velocity discontinuity is mapped in the hodograph plane onto two equidistant lines*.

In the particular case when the region on one side of the line of velocity discontinuity $L$ moves as a rigid whole with constant velocity, this side of $L$ is obviously represented in the velocity plane by a single point $l'$. If $L$ is a curve, its other side is represented by an arc $l*$ of a circle with the centre at $l'$. If, however, the line of discontinuity is straight, then also its other side is represented by a single point $l*$, which means that on both sides the material moves as rigid blocks with different velocities.

Examples of hodographs will be discussed later on, when considering particular plastic forming operations.

### 4.10   The condition of non-negative rate of internal energy dissipation

The methods of solution presented in the foregoing sections do not take into account the important condition (3.14) which states that the rate of internal energy dissipation cannot be negative at any point of the body. In the case of a negative rate of energy dissipation, energy would be produced in the deformed material, which is physically inadmissible. If a theoretical velocity field does not satisfy this physical requirement, it cannot be regarded as a correct one, even though all the kinematic conditions are satisfied. Thus in each particular solution we must check whether this condition holds everywhere. Unfortunately in most cases the procedure is time-consuming, since we must check the sign of the energy dissipation at sufficiently large number of points. There exists a number of methods proposed for instance by A. P. Green [40], by W. Prager [96], and by O. D. Grigoriev [43]. For practical purposes, however, the method proposed in H. Ford's book [37] seems to be the most suitable and it will be used throughout this book. More details concerning this method can be found in [72].

In plane plastic flow the general condition (3.14) can be expressed in terms of the principal stresses $\sigma_1$ and $\sigma_2$ as

$$D = \sigma_1 \epsilon_1 + \sigma_2 \epsilon_2 \geqslant 0, \tag{4.25}$$

since we have $\epsilon_3 = 0$.

If the slip-line net and the hodograph are found, condition (4.25) can be checked graphically. Consider two pairs of slip-lines forming the element $ABCD$ of the slip-line net (Fig. 66a) and the corresponding element $A'B'C'D'$ of the net of the velocity hodograph (Fig. 66b). The relative velocity of the point $A$ with respect to the point $C$ is represented in the velocity plane by the vector $\overrightarrow{C'A'}$. Similarly, the point $D$ moves with respect

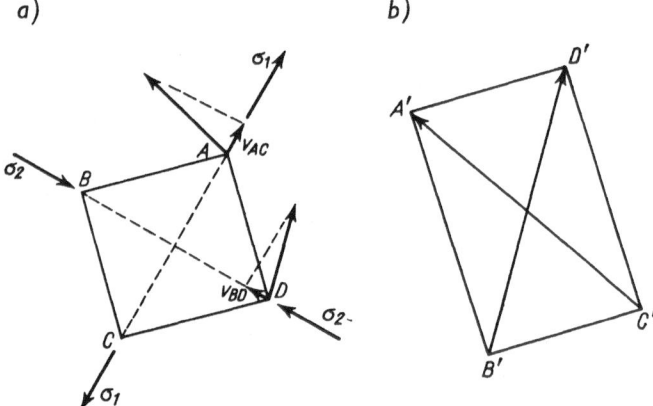

Fig. 66. An element of the slip-line net *ABCD* and corresponding element *A'B'C'D'* of the hodograph

to the point $B$ with the relative velocity defined by the vector $\overrightarrow{B'D'}$. Projections of these vectors on the diagonals of the element *ABCD* are represented by the vectors $v_{AC}$ and $v_{BD}$. Hence the rates of strain of the two diagonals are

$$\epsilon_{AC} = \frac{v_{AC}}{\overline{AC}} \quad \text{and} \quad \epsilon_{BD} = \frac{v_{BD}}{\overline{BD}}.$$

Note that both diagonals coincide with the directions of principal stresses $\sigma_1$ and $\sigma_2$, and that due to the incompressibility of the material, assumed in the theory, the moduli of the two vectors $v_{AC}$ and $v_{BD}$ are the same. When one of the diagonals, say $BD$, elongates by the increment $\Delta l$, the other diagonal shortens by the same magnitude.

Thanks to these properties of the slip-line and hodograph nets, condition (4.25) will be satisfied if on this diagonal, along which the larger principal stress $\sigma_1$ is acting, the rate of extension is of the same sign as the stress $\sigma_1$.

# 5

# *Indentation and compression operations*

## 5.1 Introduction

In this chapter and in the following two chapters we shall consider certain practically important forming operations taking place under conditions of plane flow. To solve the problems, we shall use the slip-line theory given in the foregoing chapter. We shall see later that most problems are statically indeterminate, because very seldom we are able to define in advance the magnitudes of stresses acting on the boundaries of the plastically deformed material. In most cases these boundaries are in contact with the rigid surface of the tool. The boundary conditions are therefore of the kinematic type. If for instance the tool is fixed and does not move, the normal velocity component at the contact surface must be zero. In the cases where the tool is moving, the region of plastically deformed material adjacent to the tool must have a normal velocity compatible with its motion. In most cases we solve such problems assuming a slip-line net, and then checking whether it satisfies all kinematic conditions of the process. We shall see that such a method proves to be very efficient, allowing to obtain solutions to a large number of practical problems.

All the problems are solved for the rigid-perfectly plastic material with no strain-hardening. In the solutions there appear the lines of velocity discontinuity and the rigid regions, where no deformation takes place. The rigid zones correspond to the appearance of the elastic regions in real materials. In theoretical solutions the rigid regions are frequently adjacent to the surface of the tool, and move together with it as a whole. The real weakness of these solutions consists in the fact that they do not take into account the strain-hardening effect. The serious mathematical difficulties, arising when this effect is included into the theory of plane flow, cause that at present we do not know any strain-hardening solution to the practical plastic forming operations. One should, however, emphasize that the solutions obtained for the perfectly plastic material are of great practical significance, furnishing information as to the mechanism of deformation and the influence of various factors on the course of the process. For numerous metals the stress-strain curve is for large deformations rather flat, causing that the differences between the solution with the strain-hardening and without it would be small. It is worth mentioning that the theoretical deformation patterns obtained for the non-hardening material usually are in satisfactory agreement with the experimentally investigated deformation modes of real metals, displaying the strain-hardening effect.

In the following sections we shall discuss problems of indentation of a flat punch, compression of a plastic block between rigid platens, indentation of a wedge, cutting of a plastic block by a flat and wedge-shaped tool and of the compression of a plastic wedge by a flat rigid platen.

### 5.2  Indentation of a flat punch into a half-space

The problem of indentation under conditions of plane flow has been solved as early as in 1920 by L. Prandtl [98] and later treated in a more rigorous way by R. Hill [49] and others. We shall begin with the solution of Prandtl. Later other more recent results will be discussed.

Let a rigid punch with a flat bottom be pushed downwards into a plastic half-space (Fig. 67). We shall show later on that the solution holds valid also in the case where the plastic body is bounded by two lines $x = \pm c$, provided $c$ is sufficiently large. The punch is assumed to be infinitely long in the $z$-axis direction to ensure conditions of plane

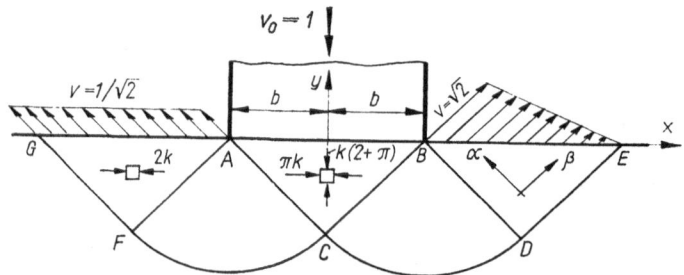

Fig. 67. Indentation of a flat punch into a half-space. Prandtl's solution. Two of the possible velocity distributions are shown on each side of the symmetry axis

plastic flow. Assume moreover that there is no friction on the contact surface. It is well known from the classic elasticity solution to the indentation problem that the stresses tend to infinity at the points $A$ and $B$. This means that in real elastic-plastic material the plastic zones begin to form in the vicinity of the two points, even at small values of the pushing force. These plastic zones grow gradually with the increasing pushing force, and finally merge together at symmetry axis. From this instant begins the process of large plastic flow, connected with the outflow of the material on both sides of the punch. Thus the boundary of the plastic material, being initially straight begins to bend due to the flow of the material. The theoretical solution to such an advanced process is connected with serious mathematical difficulties,[1] and at present only the solution to the incipient plastic flow, when the boundary is straight, is known. This incipient stage will be discussed below.

We are dealing with a typical problem, in which both the static and kinematic boundary conditions are given. Along the contact line $AB$, where the vertical velocity component must be equal to the velocity of the punch, the distribution of stresses is not known. On the two free edges $AG$ and $BE$ the velocities are not known, but some stress conditions are given since the stress component normal to the edge is zero. None of these conditions is sufficient to solve the problem directly, and, therefore, we must use an inverse method assuming in advance that the contact pressure $p$ on $AB$ is uniformly distributed.[2]

We shall show later that this assumption leads to the correct solution in which all static and kinematic conditions are satisfied. On account of symmetry only the right-hand half of the plastic domain will be considered. Now, when the contact pressure $p$ is assumed to be distributed uniformly, we may plot the net of slip-lines below the punch. It consists of two families of straight lines intersecting the contact line $AB$ at the angle of $45°$, since we have assumed that there is no friction on the interface, and therefore $AB$ is a line of principal direction. In the triangle $ABC$ the state of stress is homogeneous. $A$ and $B$ are singular points, because the stresses on the boundary of the half-space are different on both sides

---

[1] For the approximate solutions see, for instance, [67].

[2] We may also assume, instead of the uniform distribution of $p$, that the material is in the plastic state on both free edges $AG$ and $BE$. Both assumptions lead to the same solution.

of these points. Thus $B$ is a pole of the centred fan $CBD$, composed of the straight radii and of arcs of circles. The state of stress in the triangle $BED$ is homogeneous and is determined by the conditions on the free edge $BE$. The principal stress $\sigma_1$ orthogonal to $BE$ is zero. If the material is to be in the plastic state, the other principal stress $\sigma_2$ must have the value $+2k$ or $-2k$. In our case the stress $\sigma_2$ must obviously be a compressive one, and therefore we have $\sigma_2 = -2k$. The angle $\varphi$ between the direction of the greater principal stress $\sigma_1$ and the $x$-axis is $\pi/2$. The slip-lines parallel to $DE$ are inclined at the angle $(\varphi - \pi/4)$ to the $x$-axis. Thus, according to the terminology introduced in Chapter 4, they belong to the second family, or in other words to the $\beta$-lines family. Thus along them the relation $\chi - \varphi = -0.5 - \pi/2$ must be satisfied, because in the region $BDE$ we have $\varphi = \pi/2$ and $\chi = -0.5$. When these slip-lines reach the triangle $ABC$, the angle $\varphi$ changes its value to 0. Therefore we have $\chi = -0.5 - \pi/2$ in this region, and the normal stress component on the edge $AB$ is $\sigma_2 = -k(2+\pi)$. The other principal stress in the triangle $ABC$ has the value $\sigma_1 = -\pi k$. The contact pressure $p$ on the interface $AB$ is obviously equal to the stress $\sigma_2$. Hence we obtain the value of the pressure corresponding to the incipient plastic flow

$$p = 2k(1+\tfrac{1}{2}\pi). \tag{5.1}$$

Let us pass now to the determination of the velocity field, in order to check whether the slip-line field satisfies the kinematical conditions of the problem. The boundary condition to be satisfied is that the normal component of the flow velocity on $AB$ must be equal to the velocity of the punch $v_0$. We may assume without loss of generality that $v_0 = -1$. Thus we have the condition $v_y = -1$ on $AB$. Along the slip-line $CDE$ of the $\beta$-family, we have $v_\alpha = 0$, because the material below it is assumed to be in the rigid state. The component $v_\beta$ may have different values. We shall present the most general solution proposed by G. I. Bykovtsev [14].

In the triangle $ABC$ the characteristics are rectilinear. Thus from Geiringer's equations (4.23) we obtain that $v_\alpha = $ const along each of the $\alpha$-characteristics, and along each of the $\beta$-characteristics we must have $v_\beta = $ const. However, on different characteristics of the same family the velocities can be of different magnitude. Thus in the region $ABC$ we have $v_\alpha = v_\alpha(\beta)$ and $v_\beta = v_\beta(\alpha)$.

In the region $BCDE$ we have $v_\alpha = 0$ everywhere, and the velocity component is constant along each $\beta$-line. Thus equations (4.23) are satisfied identically. In the same way the velocities can be found in the symmetrically situated region $ACFG$.

Assume on the contact line $AB$ an arbitrary distribution of one of the velocity components, say $v_\alpha(x)$. The motion of the punch implies that $v_y = -1$ on $AB$. This condition determines uniquely the second velocity component $v_\beta$ on $AB$. All conditions will be satisfied if we assume, for instance, the following distribution of the velocities

$$v_\alpha = \frac{\sqrt{2}(b-x)}{2b}, \qquad v_\beta = \frac{\sqrt{2}(b+x)}{2b}. \tag{a}$$

Note that together with the component $v_y = -1$, there exists on $AB$ also the horizontal velocity component $v_x = x/b$. Thus there is taking place a sliding of the material along the contact line. Since we have assumed that there is no friction on the foot of the punch, such a sliding is admissible in our solution because the slip-lines intersect $AB$ at the angle $\pi/4$. On the right-hand side of Fig. 67 there are shown velocities of the material particles located on the stress free edge $BE$.

Another particular kind of velocity field has been proposed by L. Prandtl [98]. The velocity components on $AB$ are assumed to be

$$v_\alpha = 1/\sqrt{2}, \qquad v_\beta = 1/\sqrt{2}. \tag{b}$$

Thus in the region $ABC$ we have everywhere $v_y = -1$ and $v_x = 0$. Therefore this triangle moves vertically downwards together with the punch, as a rigid whole. The triangle $BDE$ moves also as a rigid block in the direction parallel to $DE$ with the velocity $1/\sqrt{2}$. Similarly, the triangle $AFG$ is shifted along $FG$. On the left-hand side of Fig. 67 are shown the velocities of the points on the boundary $AG$. In contrast to the solution determined by the relations (a), where the velocities change continuously, we have now a velocity jump across the lines $ACDE$ and $BCFG$. Prandtl's velocity solution can be assumed to match well the real conditions in the case where the punch has a rough surface, preventing the sliding of the material along the contact line.

In the solution proposed by R. Hill [49] the deforming region is smaller and the velocity field is different than that in the foregoing solution. The slip-line field is shown in Fig. 68. The triangle $BEF$ is stressed by an uniaxial compressive stress $-2k$. In the triangle $CBD$ we have the same state of stress as in the triangle $ABC$ in Fig. 67. Hence the contact pressure is the same in the two solutions, and has the value $p = 2k(1+\frac{1}{2}\pi)$.

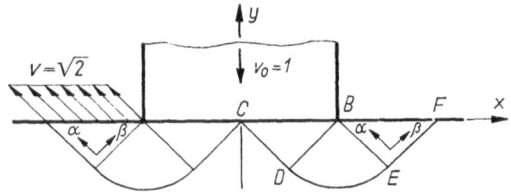

Fig. 68. Indentation of a smooth punch into half-space; Hill's solution

In Hill's solution the velocity field is determined uniquelly. The conditions $v_y = -1$ on the contact line $CB$ and $v_\alpha = 0$ along the outer slip-line $CDEF$ and along the symmetrical line on the left imply the solution

$$v_\alpha = 0, \qquad v_\beta = \sqrt{2} \quad \text{for} \quad x > 0,$$
$$v_\alpha = \sqrt{2}, \qquad v_\beta = 0 \quad \text{for} \quad x < 0.$$

The triangle $CBD$ is shifted with the velocity equal to $\sqrt{2}$ along $CD$, whereas the triangle $BEF$ moves with the same velocity along $EF$. The slip-line $CDEF$ respresents the line of velocity discontinuity, below which the material particles remain at rest.

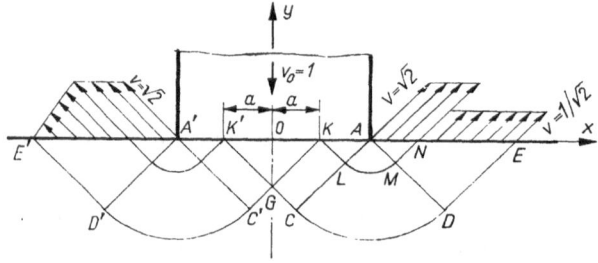

Fig. 69. Indentation of a smooth punch into half-space. Prager and Hodge solution. Two of the possible velocity distributions are shown on each side of the symmetry axis

W. Prager and P. G. Hodge [97] have given another type of velocity solution (Fig. 69). The stress distribution is the same as before, and also the same is the magnitude of the contact pressure $p$. The flow velocities just under the contact line $AOA'$ are assumed to be

$$v_\beta = 1/\sqrt{2}, \qquad v_\alpha = 1/\sqrt{2} \quad \text{for} \quad -a \leqslant x \leqslant a,$$
$$v_\beta = \sqrt{2}, \qquad v_\alpha = 0 \quad \text{for} \quad x < -a,$$
$$v_\beta = 0, \qquad v_\alpha = \sqrt{2} \quad \text{for} \quad x > a.$$

The velocity distribution in the region *KLMNAK* is the same as that in Hill's solution, and in the region *OGCDENMLKO* it is identical with that in Prandtl's solution. The slip-line *K'GCDE* is the line of velocity discontinuity since below it we have $v_\beta = 0$, and above it $v_\beta = 1/\sqrt{2}$. Also *KLMN* is the line of discontinuity; below it $v_\beta = 1/\sqrt{2}$ and above it $v_\beta = \sqrt{2}$. The outflow velocities at the boundary *AE* are shown on the right-hand side of the figure.

It is evident that we are dealing with a typical problem of the theory of plastic flow for which the number of kinematically admissible velocity solutions is infinite. For instance G. I. Bykovtsev [14] proposed still another type of solution, where on *A'OA* (Fig. 69) we have

$$v_\beta = \frac{\sqrt{2}(a-x)}{2a}, \qquad v_\alpha = \frac{\sqrt{2}(a+x)}{2a} \qquad \text{for} \quad -a \leqslant x \leqslant a,$$

$$v_\beta = \sqrt{2}, \qquad\qquad v_\alpha = 0 \qquad\qquad \text{for} \quad x < -a,$$

$$v_\beta = 0, \qquad\qquad v_\alpha = \sqrt{2} \qquad\qquad \text{for} \quad x > a.$$

We can easily verify that along the segment *AA'* the vertical velocity component $v_y$ equals $-1$, as required by the motion of the punch. On the left-hand side of Fig. 69 are presented the outflow velocities on the boundary. The velocity field does not contain any lines of velocity discontinuity. Hill's solution represents the limit case ($a = 0$) of that solution, when the region of the non-uniqueness of the velocity field transforms into a line of velocity jump. J. Rychlewski [101] pointed out that also non-symmetric variants of velocity solutions are possible.

To make certain that the solutions described above are complete, it is necessary to verify whether the rigid zone below the deforming region is able to carry the required load without violating the yield condition. This can be done by extending the slip-line field into the rigid region following J. F. W. Bishop's [10] procedure (Fig. 70). This procedure is based on the lower bound theorem of the theory of plasticity (see Section 3.9). This theorem states that the estimate of the load obtained from any arbitrary statically admissible stress field is not larger that the actual carrying capacity.

We begin the construction of the extension of the slip-line field from the outer slip-line *BCD*, by solving the mixed boundary value problem.

109

This problem is determined by the known values of the functions $\varphi$ and $\chi$ along *BCD* and by the symmetry condition ($\varphi = 0$) on *OR*. Note that for the solutions in Figs. 68, 69 and for the non-symmetric solutions

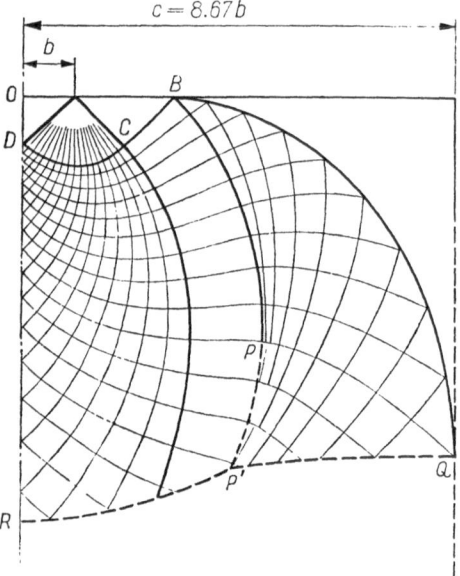

Fig. 70. Bishop's extension of the stress field into rigid region for a punch indentation problem

discussed in [101] the first stage of that procedure leads to the slip-line net in Fig. 67. Thus we begin our considerations from this net. The mixed boundary value problem formulated above determines uniquely the state of stress within the region *BCDRP'PB*. Having found the slip-line *BP*, we may solve the boundary value problem inverse to the Cauchy problem. This problem consists in finding the shape of a hypothetical free edge *BQ*, outside which the stress state is zero. Such an edge is uniquely determined by the slip-line *BP*. In the extended slip-line net there is appearing a statically admissible line of stress discontinuity *PP'*. Now, starting from the point *Q*, where the width of the slip-line field is the largest, we construct the line of stress discontinuity *QP'R* in such a manner that the stress state below it reduces to uniaxial tensile or compressive stresses directed vertically. It may be shown that these stresses do not exceed anywhere the yield locus. The stress discontinuity across $QP'R$ is of the general type as that in Fig. 63. Thus the column of width equal to the doubled distance of the point $Q$ from the axis $OR$ is able to carry the load applied to the punch. Having found the extension of the slip-line field, we are sure that, according to the lower bound theorem, the bearing capacity of rigid region below the slip-line $BCD$ is not smaller than the load required to deform plastically the material below the punch.

The extension of the slip-line field indicates that the width of the block must satisfy the condition $c \geq 8.67b$ since in the region outside the contour $BQ$ the zero stress state is assumed. For $c < 8.67b$ the mechanisms of deformation can be different than those discussed above, and the mean pressure required to deform the material plastically can be smaller than $p = 2k(1 + \frac{1}{2}\pi)$.

We have seen that there is an infinite number of kinematically admissible solutions to the velocity field in the deforming region. This non-uniqueness of the velocity solution is a consequence of the rigid-perfectly plastic model of the material assumed in the analysis. Therefore the question arises which of these solutions have a real practical meaning. Mathematically, the most rigorous is Hill's solution (Fig. 68). In other solutions (see Figs. 67, 69) the state of stress in some region below the punch is determined by boundary conditions on the stress-free edges of the material simultaneously existing on both sides of the punch. Thus even small geometrical imperfections, connected with the appearance of slight local slopes of the edges, lead to different estimates of the stress state at a specific point, depending on the edge from which the solution is started. Such a non-uniqueness of the stress solution may occur in the region $ABC$ in Fig. 67 and in the region $KGK'$ in Fig. 69. A similar conclusion is obtained if there is considered the possibility of constructing a variational solution for any arbitrary variation of the yield locus of the material and of the boundary conditions [101].

However, such considerations are only of theoretical significance for the ideal model of a rigid-plastic material with no hardening. In real metals, displaying strain-hardening effect, the incipient flow according to Hill's scheme will instantly cause local strain-hardening along the lines of velocity discontinuity. Therefore the deforming region will be forced to spread outside the initial discontinuity lines, where the material is weaker, and so on. The same concerns the non-symmetric velocity solutions. All of them tend to the velocity field shown in Fig. 67.

### 5.3   Indentation of a plastic block by two opposite narrow punches

Consider a block of a plastic material, having the thickness $2h_0$, compressed by two opposite flat punches of width $2a$, where $a < h_0$ (Fig. 71). Such a process corresponds to certain types of cutting and forging operations. Experimental evidence shows that the plastic zones begin to form

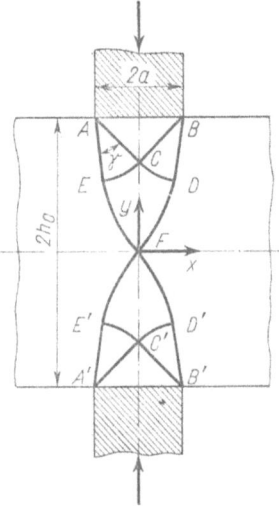

Fig. 71. Indentation of a plastic block by two opposite narrow punches. Prandtl's slip-line field

at the corners of the punches, and they grow gradually with increasing load. Finally the plastic zones meet at the vertical axis of symmetry. The regions located at a larger distance from the punches remain rigid and move outwards when the punches approach one to the other. As in the foregoing case, we begin the construction of the slip-line net by assuming that the pressure $p$ is uniformly distributed along the contact line between the punch and the indented material. Thus in the triangle $ABC$ below the punch we have a homogeneous state of stress. The points $A$ and $B$ are the poles of centred fans $CAE$ and $CBD$, respectively. The slip-line net in the curvilinear quadrangle $ECDF$ is determined by the circular arcs $EC$ and $CD$ which form a characteristic boundary value problem. A net of this type has been shown in Chapter 4 in Fig. 50 and its geometrical parameters are given in Table 1 (pp. 78–79). The angle $\gamma$ is uniquely determined by the ratio of dimensions $h/a$ since on account of symmetry, the outer slip-lines must meet at the point $F$. The stress field can be found by assuming that the unknown value of the pressure $p$ is a parameter. We begin from the data in the triangle $ABC$, where $\varphi = 0$ and $\chi = 0.5$. The latter value is obtained by taking into account that $\sigma_1 = -p + 2k$ and $\sigma_2 = -p$. The constant $\sigma_0$ in (4.9) is assumed to be $\sigma_0 = -p$. One performs the computations of the values of $\chi$ at nodal points of the slip-line net in the elementary way, by means of relations (4.14), $\chi + \varphi = \text{const}$ and $\chi - \varphi = \text{const}$, which must be satisfied along the slip-lines of the first and of the second family, respectively. Now, having found the stresses, we may find the value of the pressure $p$ from the condition of equilibrium of a half of the strip to one side of the vertical axis of symmetry. The resultant force acting on such a cross-section must be zero. We obtain the stress component $\sigma_x$ on the vertical symmetry axis from the first of the equations (4.10), taking into account that along this axis $\varphi = 0$. The pressure $p$ can be found from the integral condition of equilibrium $\int_{-h}^{h} \sigma_x \mathrm{d}y = 0$. Since the stress state in the region $ECDF$ has been computed numerically, the relation between the pressure $p$ and the ratio $h/a$ cannot be given in a closed form. It is represented in Fig. 72 by the curve $AB$ taken from [50]. Note that for $h/a = 8.713$ [1] the contact pressure reaches the value given

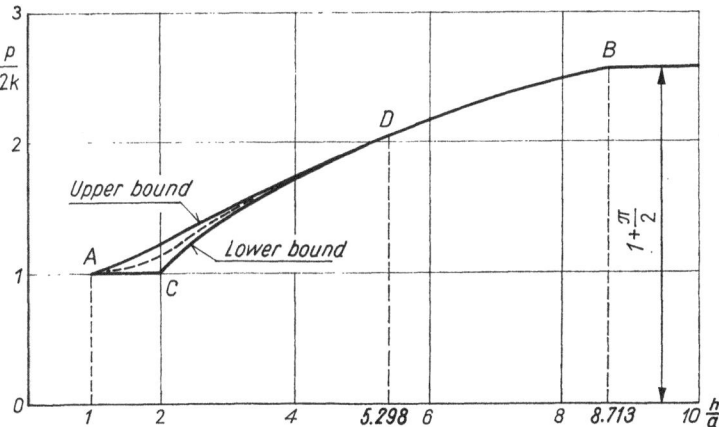

Fig. 72. The relationship between the indentation pressure $p$ and the ratio $h/a$ for a block indented by two narrow punches

---

[1] This value has been communicated to the author by Dr J. Salençon in a private letter. In earlier works (see [50]) the value $h/a = 8.74$ was given.

by (5.1). This means that in thicker blocks the material begins to flow plastically according to the scheme discussed in the foregoing section.

Consider now the velocity field. It must be such that the normal velocity component on the contact line $AB$ has the constant value $v_0$, equal to the speed of the punch and the normal velocity component on the lines $BDF$ and $AEF$ is compatible with the motion of the outer regions constituting rigid wholes. The velocity of these rigid regions is $u = v_0 a/h$ since the volume of the material does not change. The normal velocity

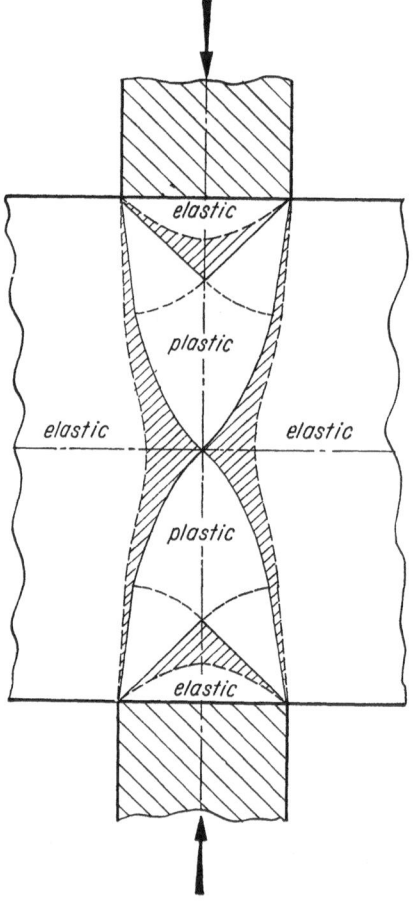

Fig. 73. In real elastic-plastic material there appear transitory regions (hatched in the figure) between elastic regions and regions of large plastic flow

component on the segments $AE$ and $BD$ has a constant value, and consequently is constant also along $AC$ and $CB$. Thus we have a homogeneous velocity field in the triangle $ABC$. This triangle moves as a rigid whole with the punch and no deformation takes place within it. The slip-line field is therefore compatible with the kinematic conditions of the problem. The velocities can easily be computed numerically by solving a sequence of boundary value problems. First we solve a characteristic problem in the curvilinear quadrangle $ECDF$, determined by the known values of normal velocities on the arcs $EF$ and $DF$, which must be compatible with the motion of the rigid zones. Then the velocities within the centred fans $CBD$ and $EAC$ can be found. The outer slip-lines $AEF$ and $BDF$ represent thé lines of velocity discontinuity. The velocity jump propagates along these lines from the central point $F$ up to the points $A$ and $B$. Therefore the velocity field is kinematically admissible. Detailed analysis of the velocity field will be presented at the end of the section, where we shall apply the method of graphical construction of the velocity hodograph.

In real elastic-plastic materials there appear certain transitory narrow regions between the regions of large plastic flow and the fully elastic zones. The exact extent of these transitory zones is not known; their boundaries are schematically shown in Fig. 73 by broken lines. The dashed regions are in the plastic state. However, the plastic deformations are small within these regions.

In order to verify whether the solution is complete, we shall construct the extension of the slip-line field into the rigid regions. Figure 74 shows such an extension for $h/a = 5.40$ [128].[1] It has been constructed following Bishop's procedure described in the foregoing section (see Fig. 70). Starting from the slip-line $BDF$, the inverse Cauchy problem is solved and the hypothetical free boundary $BLQ$ obtained. The material outside $BLQ$ is assumed to be stress free. The line of stress discontinuity $QPF$ through the point $Q$ where the hypothetical free boundary $BLQ$ is parallel to the vertical axis of symmetry, and the analogous line $Q'P'F$ for the lower part of the block, merge at the point $F$. The material in the region $QPFP'Q'$ is loaded by uniaxial compressive stresses parallel to the vertical axis. The numerical analysis shows that these stresses are below the yield locus, except at the point $Q$, where the yield locus is reached. Thus it has been shown that the slip-line solution in Fig. 71, originally due to L. Prandtl

---

[1] This extension has been independently proposed by J. Salençon [102].

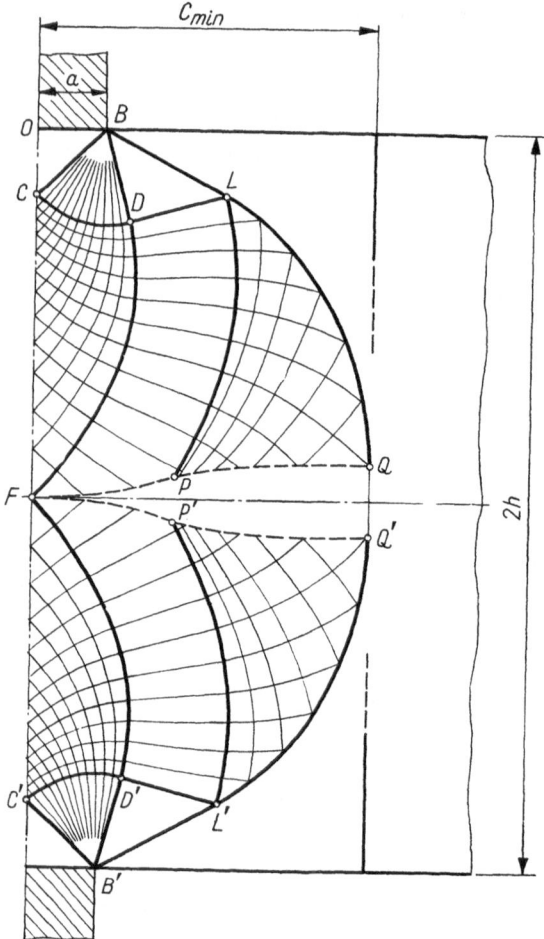

Fig. 74. Extension of the slip-line field into rigid regions for a block indented by two punches

[99], is complete for $h/a = 5.40$, satisfying all static and kinematic conditions. Such an extension can be constructed for any ratio $h/a > 5.298$.[1] In the extended slip-line field the angle of the fan $CBD$ will be larger than that in the deforming region. This angle is defined by the condition that

---

[1] This value has been communicated to the author by Dr J. Salençon in a private letter. In the author's work [128] a less accurate value $h/a = 5.40$ has been established by the graphical procedure and the field shown in Fig. 74 was found to be a limiting one.

the discontinuity line $QPF$ must pass through the central point $F$. For $h/a = 8.713$ the sector $BL$ is horizontal and the extended slip-line field consists of the two fields in Fig. 70. In this case two kinematically admissible deformation modes are possible, namely the separation of both parts of the block or a local forcing out the material at the surface. Thus the ratio $h/a = 8.713$ is a critical one. On the other hand, the ratio $h/a = 5.298$ is also critical, because, for $h/a < 5.298$, the extension of the slip-line field is not known. Thus, for $5.298 \leqslant h/a \leqslant 8.713$, Prandtl's solution is complete, and for $h/a < 5.298$ it is only kinematically admissible. Hence in this range it provides the upper estimate of the unknown value of the required force.

However, for $h/a \lesssim 5.298$, a good lower bound on the indentation force can be found from an appropriate statically admissible stress field. Figure 75 shows such a field for a particular ratio $h/a = 3.24$. The angle of the slip-line fan at the point $B$ is assumed to have a value $\gamma$ such that the line of discontinuity $QK$ forms the configuration shown in the figure. The line $BLQ$ represents a hypothetical stress-free boundary obtained by solving the inverse Cauchy boundary value problem. To the right of $BLQ$ the material is assumed to be stress-free. As previously, at the point $Q$ the free boundary is parallel to the vertical axis $OF$. In the region $QKQ'$ the material is compressed uniaxially, the stresses not exceeding at any point the yield locus. For the slip-line field under consideration the associated velocity field cannot be found and consequently only the lower estimate of the indentation force is obtained. If $BL$ represents a segment of the stress-free boundary, we have in the region $BLG$ $\chi = -0.5$ and $\varphi = \gamma$. Passing to the region $OBC$ along the slip-line $LGC$, belonging to the $\beta$-family, we obtain from (4.14b) the value $\chi = -0.5 - \gamma$ in $OBC$, where $\varphi = 0$. Thus the contact pressure resulting from the assumed stress field is $p = 2k(1 + \gamma)$. Having found the values of $\gamma$ for various ratios $h/a$, the lower bound for the contact pressure $p$ can be computed in the range $2 \leqslant h/a < 5.298$. It is represented by the curve $DC$ in Fig. 72.

In the range $1 \leqslant h/a < 2$, the lower bound may be obtained from the elementary statically admissible stress field in which the strip bounded by straight lines through the points $A$ and $A'$ from one side and through $B$ and $B'$ from the other side (comp. Fig. 71) is uniaxially compressed by the stress $\sigma_2 = -2k$. Outside this rectangle the material is stress-free. Thus the lower bound is $p = -2k$. This value is represented by the segment $AC$ in Fig. 72.

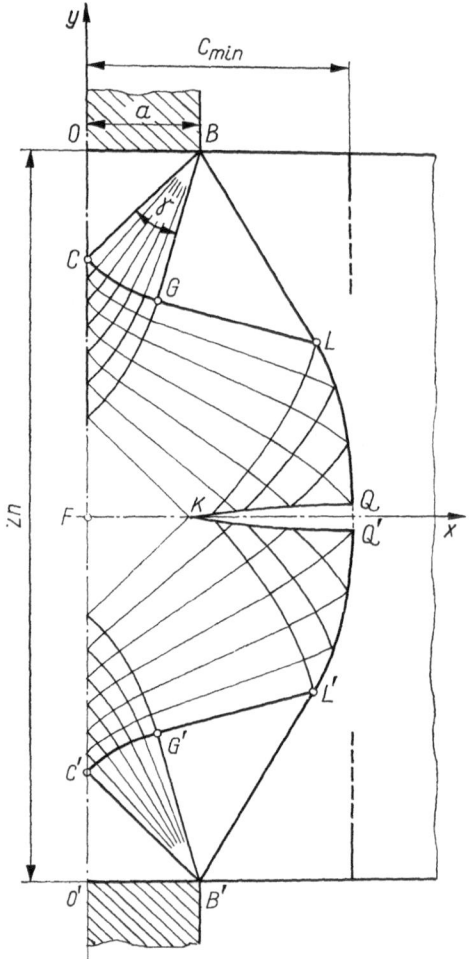

Fig. 75. Statically admissible stress field for $h/a = 3.24$

It is seen that the difference between the lower and upper bounds is small, except in the vicinity of the ratio $h/a = 2$, where it rises to a maximum value of 21 percent. Taking the mean value between the two bounds (dashed line in Fig. 72) we reduce the error to the half.

In constructing extensions of the slip-line fields for $h/a \geqslant 5.298$, or statically admissible stress fields for $h/a < 5.298$ we have assumed that the material outside these fields is stress free. Thus the line $BLQ$ representing a hypothetical free boundary gives us an important information on how

wide the indented block should be so that the described deformation mode
could be realized. The stress field must lie entirely inside the actual contour
of the block. If the block is rectangular, its minimum half-width $c_{min}$ is
equal to the distance of the point $Q$ from the vertical axis $OF$. This is
schematically shown in Figs. 74 and 75. Figure 76 shows the value of $c_{min}$

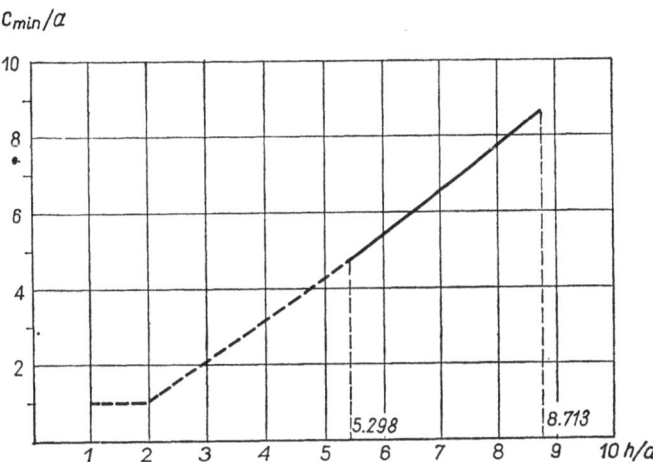

Fig. 76. Minimum width $c_{min}$ of a block indented by two punches

for various ratios $h/a$. For $h/a < 5.298$ the relation is represented by a
dashed line, because $c_{min}$ in this range has been obtained from the stress
fields which are only statically admissible. Thus the exact value of $c_{min}$
will be probably slightly higher in that part of the diagram.

Let us now consider the deformation of the indented block. All figures
are made for the particular case $h_0/a = 7.1$. Figure 77 represents the initial
situation when indentation depth is zero. We presently describe the deformation of the square grid shown in the figure as it occurs during the
more advanced stage of indentation. The continuous lines in Fig. 77a
represent the slip-line net. Plastic deformation takes place within the region
*BDFG*.

The punch moves downwards with velocity $v_0$. The two rigid parts
of the block outside the slip-line *BDF* and the symmetrical line on the
other side of the axis of symmetry move with the horizontal velocities equal
to $v_0 a/h$. The triangular region *BCG* moves as a rigid whole with the
punch with velocity $v_0$. The line *BDF* represents the line of velocity dis-

119

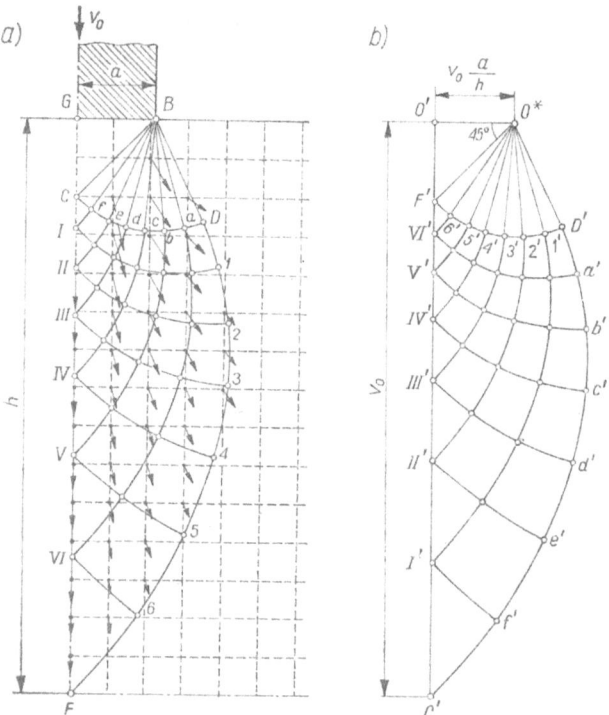

Fig. 77. Geometrical construction of deformation of indented block; a) slip-line net and displacement vectors of nodal points of a square grid, b) velocity hodograph

continuity. In the velocity plane (Fig. 77b) the velocity of the punch is represented by the vector $\overrightarrow{O'C'}$, and the velocity of the right-hand rigid part of the block, by the vector $\overrightarrow{O'O^*}$. The speed of the material to the right of the discontinuity line $BDF$ is represented by a single point $O^*$. Thus, according to the properties of the velocity hodograph discussed in Section 4.9, the left-hand side of the line of discontinuity is mapped onto the arc of the circle with the central point $O^*$. The radius of this circle is determined by the compatibility conditions at the point $F$. The two rigid parts moving horizontally in opposite directions with speed $v_0 a/h$ form a gap which must be filled by the material flowing from the deforming region. Hence the velocity of the plastically deformed material at the point $F$ must also be equal to $v_0 a/h$ and it must be directed downwards. Thus the image $F'$ of the point $F$ determines the radius $O^*F'$ of the circle

$F'D'$. The points $1, 2, ..., 6$ on the line of discontinuity are represented by the points $1', 2', ..., 6'$ in the velocity plane. The vectors $\overrightarrow{O'D'}$, $\overrightarrow{O'1'}$, ..., $\overrightarrow{O'F'}$ correspond to the velocities of these points. The velocity jumps at various points of the line of discontinuity are represented by the vectors $\overrightarrow{O*D'}$, $\overrightarrow{O*1'}$, ..., $\overrightarrow{O*F'}$. The remaining part of the hodograph is constructed graphically. The vectors $\overrightarrow{O'a'}$, $\overrightarrow{O'b'}$, ..., $\overrightarrow{O'f'}$ represent the velocities of the points $a, b, ..., f$. The velocities of the points $I, II, ..., VI$ located on the vertical axis of symmetry $GF$ are given by the vectors $\overrightarrow{O'I'}$, $\overrightarrow{O'II'}$, and so on. The net of the velocity hodograph is in our case identical with the slip-line net.

Having found the hodograph, we may easily obtain the velocity of an arbitrary point of the deforming region by finding the image of that point in the velocity plane. In Fig. 77a the arrows represent the velocities of the nodal points of a square grid. The originally computed slip-line net and the net of the hodograph were twice as dense as those shown in Fig. 77. For the sake of clarity a number of lines have not been shown in the nets.

Let the entire path $h_0$ of the punch be divided into a number of small increments $\Delta h$. If these increments are sufficiently small, one can assume that during each time increment $\Delta t = \Delta h / v_0$ the velocities of the material particles do not change. We shall assume that these velocities are equal to the mean values of the velocities at the beginning and at the end of each time increment $\Delta t$. The path of a specific point is now found by multiplying the mean velocity by the time increment $\Delta t$.

In order to obtain the velocities for each of the consecutive positions of the punch, we must construct the successive nets of slip-lines and hodographs. The net for each step will be similar to the initial situation shown in Fig. 77. As the indentation process advances the fan angle at the point $B$ decreases. For each step the net of the velocity hodograph is identical with the slip-line net.

The theoretical deformation of the square grid, as it occurs in two consecutive steps, is shown in Fig. 78. The deformation can be analysed in the same way also for more advanced indentation processes.

The experimental set-up is shown in Fig. 79. An aluminium block composed of two parts is placed between two immovable steel blocks $B$. Block $A$ is indented by two steel punches $C$. A square grid is marked on the contact surface of one of the aluminium blocks $A$ with the other block

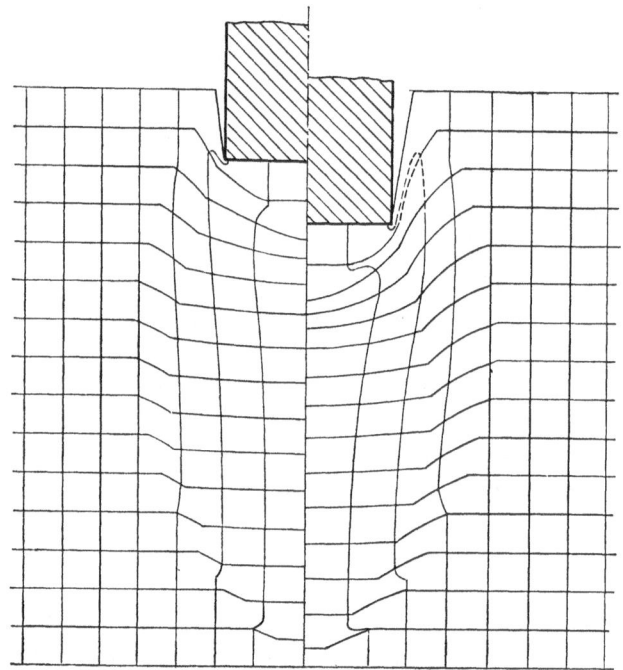

Fig. 78. Two stages of deformation of indented block [128]

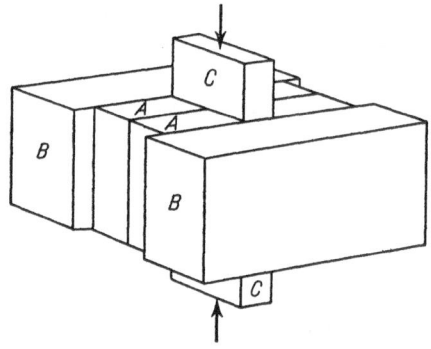

Fig. 79. Experimental set-up for indentation test

*A.* Under this set-up the plane strain conditions do not hold for the entire material, because in the vicinity of the horizontal axis of symmetry there appear tensile stresses causing a local lateral contraction which cannot be prevented by the steel blocks *B.* However, if the indentation depth is sufficiently small, one can assume that the deformation will be almost as in plane flow conditions, mainly because the deformation takes place predominantly in the vicinity of the punches, where the conditions of plane flow are strictly observed.

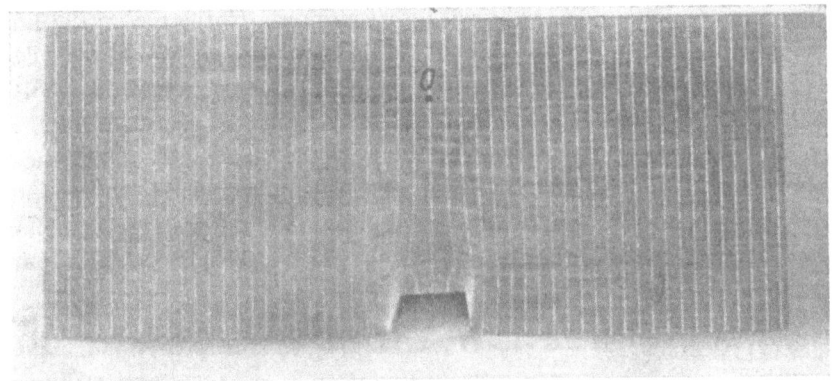

Fig. 80. Deformed grid in the indented aluminium block [128]

Figure 80 shows the deformed grid in a block of aluminium [128]. The position of the central point of the indented block is denoted by $O$. The rigid region under the punch is clearly visible. The general picture of the deformation mode is close the one predicted by the theoretical solution, except that some sharp bends of the grid lines appear in the latter which are not observed in the experimentally deformed grid. In real metals instead of the lines of velocity discontinuity there appear narrow strips of finite width across which the velocity vector changes gradually its direction.

The theoretical analysis of deformation can easily be made more complete by the determining the shape of the contour of the material above the punch. The condition of constant volume implies the equality $a \, dy = y \, dx$, where $y = y(x)$ is the equation of the contour. Let $h$ denote the distance of the punch from the centre of the block. After integration we obtain

$$y = he^{(x-a)/a}.$$

The integration constant has been found from the condition $y = h$ for $x = a$.

The theoretical deformation mode very well corresponds to the one observed under conditions of cutting and free forging of metals at high temperatures when the strain-hardening is very slight. An interesting experimental technique has been proposed by W. Johnson *et al.* [65]. Steel blocks heated up to a temperature of 680°C were compressed between two narrow punches on a drop-hammer. The energy dissipation is larger along the strips of strong deformations, corresponding to the theoretical lines of velocity discontinuity, than in the adjacent regions. Simple calculations indicate that within these strips the temperature will increase by 50–100°C. Thus they become clearly visible as bright high-temperature lines.

If the punches have different widths, the slip-line field has the form shown in Fig. 81. This particular case was investigated by W. Johnson [64]. For the unknown parameters defining the geometry of the slip-line net we shall take the angles $\gamma_1$ and $\gamma_2$. The unknown value of $\chi$ at the point $K$ will be denoted by $\chi_0$.

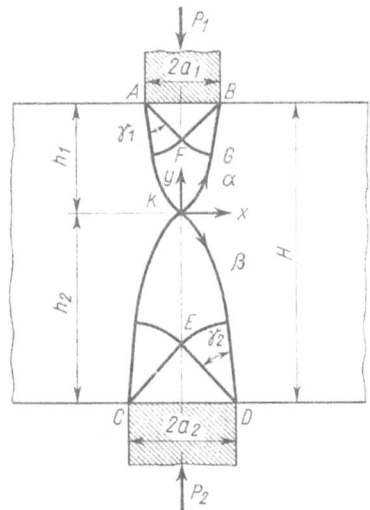

Fig. 81. Indentation of a block by two punches of different width

The pressure $p_1$ uniformly distributed along the contact line $AB$ is given by

$$p_1 = -2k\chi_0 + k(1 + 4\gamma_1), \tag{a}$$

and similarly, for the pressure along $CD$ we have

$$p_2 = -2k\chi_0 + k(1 + 4\gamma_2).\tag{b}$$

These expressions have been obtained directly from the equations (4.16) which must be satisfied along the slip-lines. Passing from the point $K$, where the angle enclosed between the $\beta$-slip-line and the $x$-axis is $(-\frac{1}{4}\pi)$, to the point $G$, where $\vartheta = -\frac{1}{4}\pi + \gamma_1$, we have $\chi_G = \chi_0 - \gamma_1$. Now, moving from $G$ to the point $F$, where $\vartheta = -\frac{1}{4}\pi$, we obtain $\chi_F = \chi_0 - 2\gamma_1$. This value of $\chi_F$ applies everywhere within the triangle $ABC$ and on the contact line $AB$ as well. Finally we get $p_1 = -2k\chi_F + k$, and substituting the above value of $\chi_F$, we arrive at (a).

The condition of equilibrium $p_1 a_1 = p_2 a_2$ implies the relation

$$\chi_0 = \frac{2(\gamma_2 n - \gamma_1)}{n - 1} + 0.5,$$

where $n = a_2/a_1$. The last relation is obtained by equating to zero the resultant horizontal force acting on the vertical cross-section through $K$. One obtains solutions for particular cases by a process of trial.

### 5.4   Compression of a plastic block between two flat rough plates

Let a plastic block be compressed between two flat rigid plates (Fig. 82). It is supposed that there occurs no deformation in the direction perpendicular to the plane of the figure. Thus the plane strain conditions are in force. A solution to this problem was first given by L. Prandtl [99]. A number

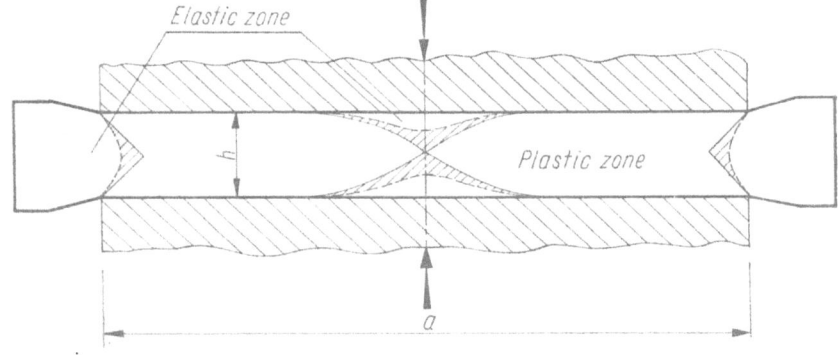

Fig. 82. Compression of a plastic block between two rigid plates

of numerical results was proposed by V. V. Sokolovsky [110] who showed
that the ratio of the block's width $a$ to its thickness $h$ is of significant
influence on the slip-line pattern and the stress distribution. The friction
conditions on the block-plate interface were also analysed. The complete
solution involving the velocity field was given in [54] and it may also be
found in [50].

In this section we shall consider the problem for the case when the
surface of the plates is sufficiently rough so that the frictional stress on
the contact line is equal to the yield stress in pure shear. We shall show
later that in such a case the friction coefficient $\mu$ must be at least 0.39.
For $\mu = 0.39$ the frictional stress reaches the value $t = \mu p$, where $p$ is
the normal stress along a short segment $OB$ of the contact line (Fig. 83)

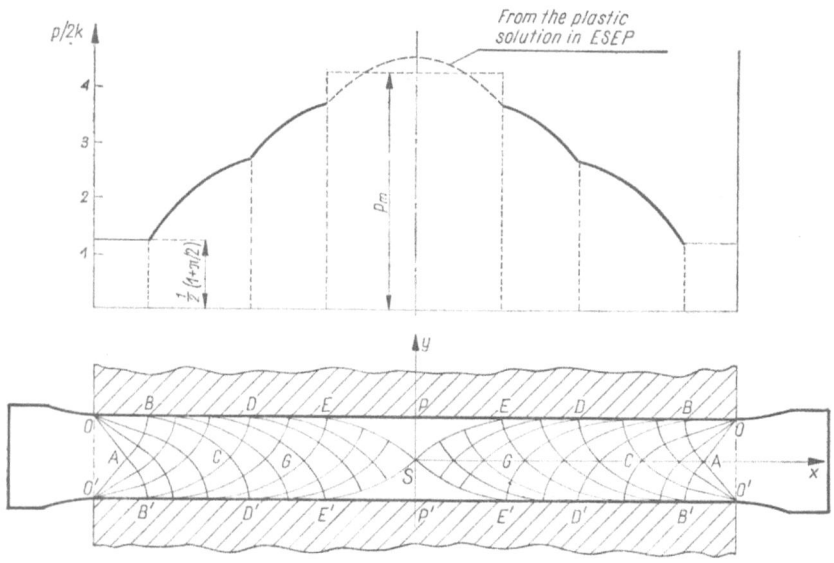

Fig. 83. Slip-line net and distribution of contact pressure for large friction ($\mu \geqslant 0.39$)
between the block and the plates

adjacent to the edge of the plate. This theoretical value of the friction
unit force cannot be reached on the remaining portion of the contact line,
since it would exceed there the yield locus in shear, and this is physically
inadmissible. The material will be assumed to be ideally plastic with no
strain-hardening.

## 5.4 Compression of a plastic block between two flat rough plates

Consider first an arbitrary instant of the process, characterized by the instantaneous value of the thickness $h$. We begin the construction of the slip-line field by assuming that the slip-lines $OA$ and $O'A$ are straight. We shall see that this assumption leads to the solution satisfying all static and kinematic conditions. On account of symmetry only the upper left-hand quarter of the block will be considered. The region to the left of $OA$ is rigid. The point $O$ is a singular one and therefore in the adjacent region $AOB$ we have a centred fan of slip-lines with the pole at $O$. The conditions of equilibrium of the rigid part to the left of $OA$ require that the horizontal stress is zero there. Thus we have $\sigma_x = 0$ on $OA$. Since the material is assumed to be in the plastic state, on $OA$ the other stress component must be equal to the yield locus in pure compression. Thus on $OA$ $\sigma_x = 0$, $\sigma_y = -2k$, and consequently $\chi = -0.5$. Since $\varphi = 0$ on $OA$, we get immediately $\chi + \varphi = -0.5$ along this line. The circular slip-lines in the fan $AOB$ belong to the first family of $\alpha$-lines. Thus along them the relation $\chi + \varphi = $ const must hold. On the contact line we have $\varphi = \frac{1}{4}\pi$. Hence the contact pressure along $OB$ has the constant value $p = k(1 + \frac{1}{2}\pi)$. The tangential stress component on $OB$ is obviously equal to $k$. We obtain the required smallest value of the friction coefficient $\mu$ by substituting the value of $p$ in the relation $p\mu = k$. Finally we get $\mu = (1 + \frac{1}{2}\pi)^{-1} \approx 0.39$.

Having found the stresses along the slip-line $AB$, we may determine the slip-line net and stress distribution within the curvilinear quadrangle $ABCB'$. The coordinates of the nodal points of this portion of the net are given in Table 1 in the foregoing chapter. The graphical method of constructing the net is shown in Fig. 56. In the curvilinear triangle $BCD$ the boundary conditions determine the mixed problem; all necessary magnitudes being already known along the slip-line $BC$ from the foregoing solution in the region $ABCB'$. Moreover, along the contact line the shear stress is equal to $\tau_{xy} = k$, and consequently $\varphi = \frac{1}{4}\pi$. These data suffice to find all the sought for magnitudes in the zone $BCD$. In Fig. 56 is also shown the graphical method of constructing this portion of the slip-line net. Proceeding in this way, one can find the solution until the moment when the slip-lines of the two families intersect each other at the central point of the block $S$. The shear stress $\tau_{xy}$ must vanish on the symmetry axis $PSP'$ and therefore the frictional stress $t$ at $P$ is zero. This leads to the conclusion that along the segment $EP$ the frictional stress gradually decreases from the value $k$ to zero. The region $ESEP$ is rigid and the stress distribution within it is undefined. In real material there occur elastic zones and transitory plastic zones where deformations are small. The latter have been schematically shown in Fig. 82 as dashed narrow strips. The extent of these transitory zones is not known. In Fig. 83 there is presented the distribution of the contact pressure, up to the point *E*. Along the central segment *EPE* the pressure distribution is not uniquely determined. However, its mean value $p_m$ can be found by considering the condition of equilibrium of the rigid part *ESEP*. This mean pressure multiplied by the length of the segment *EPE* must be in equilibrium with the stresses acting along *ESE*. An estimate of the pressure distribution on *EPE* can be obtained by extending the slip-line field into the rigid region *ESEP*. This can be done by solving the characteristic problem determined by the already known data along both slip-lines *ES*. Such a solution corresponds to the statically admissible extension of the stress field into the rigid region. E. P. Unksov [137] has shown experimentally that the distribution of contact pressure so obtained is in the central part close to the real one. In his experiments a block of lead was compressed between two plates made of a transparent photoelastic material. The pressure distribution was evaluated by means of standard methods of photoelasticity.

We have just found the statically admissible extension of the stress field into the rigid regions *ESEP*. Also the extension into the region to the left of *O'AO* can be constructed in a simple way. In the triangle *OAO'* the material is assumed to be uniaxially compressed by the stresses $\sigma_y = -2k$, $\sigma_x = 0$. The vertical line *O'O* represents the line of stress discontinuity. To the left of it the material is stress-free. Thus we have shown that the solution is statically admissible.

Consider now the velocity field for the right-hand half of the block, assuming that the upper plate moves downwards with velocity $v_0$, while the lower plate is shifted upwards with the same velocity. Both rigid regions *ESEP* move as rigid wholes with the plates. Thus the velocity of the upper rigid zone is $v_0$. *ES* is the line of velocity discontinuity, across which the normal velocity component must change continuously. Thus the normal velocity component is known along *ES*. The slip-line *ES* belongs to the first family of $\alpha$-lines with direction coefficient $dy/dx = \tan(\varphi + \frac{1}{4}\pi)$. Therefore the normal velocity component on *ES* has the value $v_\beta = v_0 \cos(\varphi + \frac{1}{4}\pi)$ obtained directly from (4.22) by substituting $v_x = 0$, $v_y = -v_0$ in the region *ESEP*. But the tangential velocity component is different on both sides of the discontinuity line *ES*. For the upper side we obtain it by decomposing the velocity of the rigid zone *ESEP* into the components tangent and normal to $ES$. The tangential component is given on $ES$ by $v_\alpha = -v_0 \sin(\varphi + \frac{1}{4}\pi)$. On the other side of $ES$ this component is determined by the first equation (4.23). Substituting the already known expression for the normal component $v_\beta = v_0 \cos(\varphi + \frac{1}{4}\pi)$, we obtain by integration that $v_\alpha = v_0[\sqrt{2} - \sin(\varphi + \frac{1}{4}\pi)]$ along $ES$. The constant of integration is determined by the symmetry condition $v_\alpha = v_\beta$ at the point $S$ ($\varphi = 0$).

Thus $ES$ represents the line of velocity discontinuity across which the velocity jump is $\sqrt{2}v_0$. This discontinuity is kinematically admissible, because it terminates on the plate-block interface. The velocity field is also kinematically admissible at the end-sections of the block. The normal velocity component is constant along $OB$, and consequently it is constant also along $OA$ since the radii of the fan $AOA$ are straight. This velocity distribution is compatible with our assumption that the region to the left of $OAO'$ moves outwards as a rigid whole. Such compatibility is possible only when the slip-line $AO$ is straight, as we have assumed.

To determine the velocity field inside the deforming region, we solve first the characteristic boundary value problem in the region $ESE'G$, determined by the known values of the normal velocity components along the characteristic lines $ES$ and $E'S$. The computed velocities on the characteristic $EG$ together with the known normal velocity component on the contact line $ED$ equal to the velocity $v_0$ of the plate yield a mixed boundary value problem. Solving this, the velocity solution can be found up to the line $BAB'$. This type of numerical procedure was used in [54]. We shall see later that the solving of practical problems is much simplified by constructing the velocity hodograph.

In a simple way one can find the equation of the contour $y = y(x)$ of the material squeezed out from between the plates. Assume that at the initial instant the block is wider than the plates (Fig. 84a). If the plates move by the distance $dy$, the constant volume condition requires the relation $\frac{1}{2}a\,dy = y\,dx$ to be satisfied. By integrating we get the equation of the contour between the points $O_1$ and $O_2$ (Fig. 84b):

$$y = \frac{h}{2}\exp\frac{2x-a}{a}.$$

The constant of integration is obtained from the condition $y = h/2$ for $x = a/2$.

The solution shown in Fig. 83 is valid for sufficiently large values of

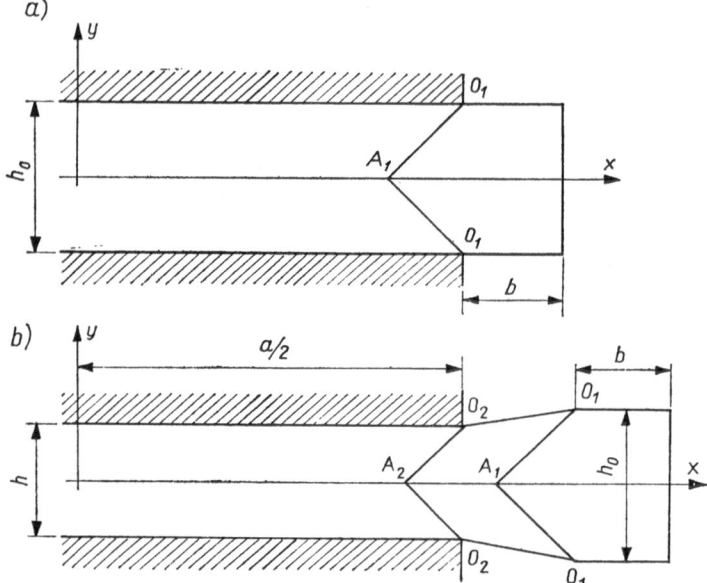

Fig. 84. Initial (a) and advanced (b) configuration of a compressed block. Here the block is wider than the plates

the ratio $a/h$. In the limit case the vertical symmetry axis passes through the point $C$. This corresponds to the ratio $a/h = 3.64$. For $1 < a/h < 3.64$ the rigid zone extends over the whole contact line.[1] The slip-line net is presented in Fig. 85a. The regions $OBSP$ are rigid and move together with the plates. The lines of velocity discontinuity coincide with the slip-lines $OBS$. The velocity jump propagates along them from the central point $S$ up to the end-points of the contact line $O$. The normal velocity component is constant along $OB$, and consequently also constant along $OA$. Thus the velocity field in the plastic zone is compatible with the rigid-body motion of the overhang.

The stress distribution in the rigid regions adjacent to the plates is not determined. Thus the exact distribution of the contact pressure is not known. However, V. V. Sokolovsky [110] has found an estimate of this distribution by assuming that the whole rigid region is also in the plastic state, and by solving the characteristic boundary value problem defined by the already known values of stresses along the slip-lines $OBS$. The

---

[1] Note that the case $a/h < 1$ has been analysed in Section 5.3.

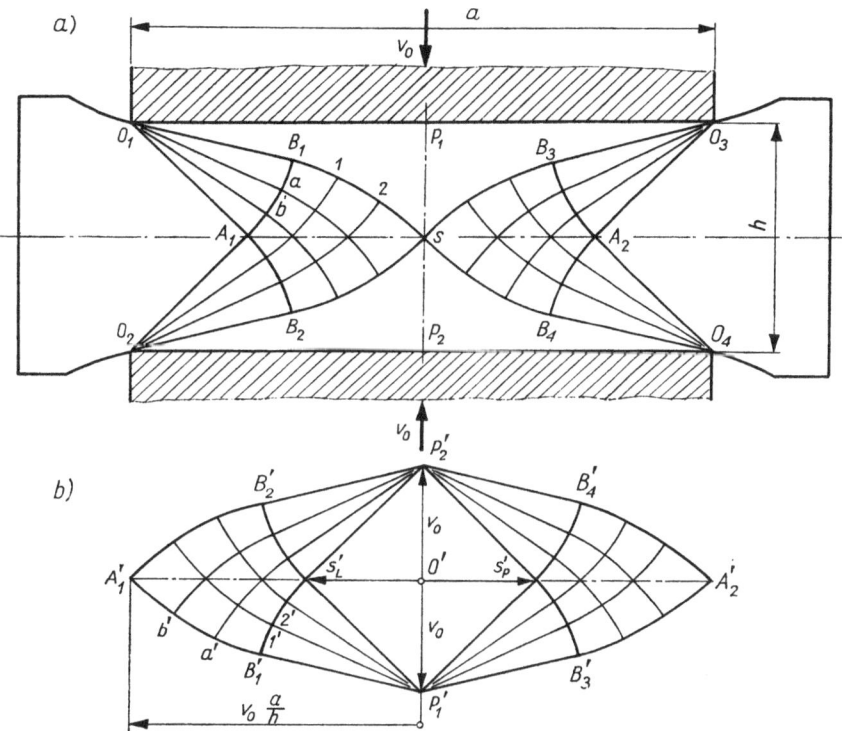

Fig. 85. Compression of a block. Prandtl's solution for $1 < a/h < 3.64$: a) slip-line net, b) velocity hodograph

largest value of the pressure is attained at the end-points $O$ of the contact line. However, the difference between it and the minimum value of the pressure at the point $P$ is small. In order to ensure that such an extension of the slip-line field into the rigid regions is statically admissible, the coefficient of friction on the contact line must be sufficiently large. In the extended stress field there appear tangential stress components on the interface which must be transmitted as the frictional forces. If this condition is satisfied, the solution is complete. If, however, the friction coefficient is too small, the solution is only kinematically admissible and gives only an upper bound on the compressive force. The velocity hodograph is presented in Fig. 85b. It consists of two identical parts on both sides of the vertical axis $P'_1 P'_2$. The net of lines in these two parts is identical with the slip-line net. The left-hand side of the hodograph corresponds to the left-hand

131

side of the slip-line field. Similarly, the right-hand half of the hodograph represents velocities for the right-hand half of the block. The upper half of the hodograph corresponds to the lower part of the block and similarly the lower part of the hodograph and upper part of the block correspond to each other. For example, the velocities on the discontinuity line $O_1 B_1 S$ are represented in the velocity plane by vectors from the pole $O'$ to respective points of the circular arc $B_1' S_L'$. Region $SB_1 A_1$ is mapped onto the region $S_L' B_1' A_1'$ in the velocity plane.[1]

The solution presented in Fig. 85 corresponds to an arbitrarily chosen moment during the process of compression. In the same way we can construct the slip-line net and the velocity hodograph for any other instantaneous value of the thickness $h$. We may analyse the entire process of deformation by splitting the total displacement of the plate into a number of small increments $\Delta h$ and by constructing the hodograph for each of the corresponding stages. During each of the stages considered the velocities are assumed to remain constant until the subsequent stage is reached. Proceeding according to the method described in Section 5.3, we found the deformation of a square grid for two subsequent stages of the compression of a block with the initial ratio $a/h_0 = 2.5$. The displacement of each plate in each stage was $\Delta h = 0.125 h_0$. Figure 86 shows the deformed grid after the second stage [129].

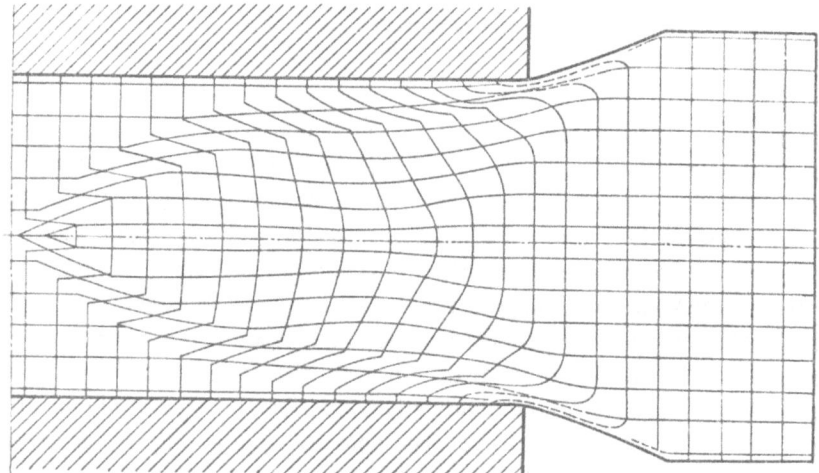

Fig. 86. Deformation of compressed block [129]

---

[1] Similarly, one can construct the velocity hodograph for large ratios $a/h$ (see Fig. 83).

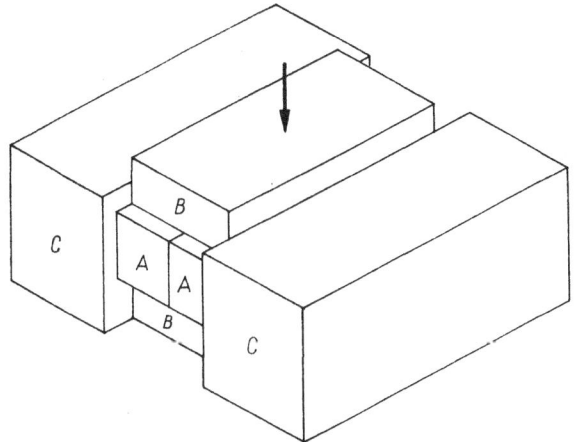

Fig. 87. Experimental set-up for compression tests

The experimental set-up is shown in Fig. 87. The compressed block of lead was composed of two parts *A*. On the surface of one of these parts a square grid was prepared. The block was compressed by two plates *B* with artifically roughened contact surfaces. The two steel blocks *C* prevented the material to contract laterally. Thus they ensured conditions of plane flow.

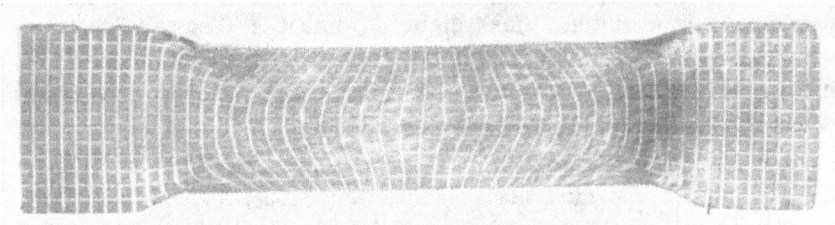

Fig. 88. Deformed grid in compressed lead block [129]

The deformed grid is shown in Fig. 88. The "rigid zones" at the ends of the block and in the regions adjacent to the plates are clearly visible, although in the latter significant plastic deformations are observed. The vertical lines of the grid have no sharp bends, contrary to what would be predicted by the theoretical solution since the strain hardening effect prevents the appearance of sharp discontinuities in real metals. However,

133

the general pattern of deformation is similar to that predicted by the theory.

Solutions presented in this section hold valid also in the case where the plates overlap the block. The free edge of the block coincides then with the vertical segment through the points $O_1$ and $O_2$ (Fig. 85). The region $O_1 A_1 O_2$ is in the plastic state and it is displaced outwards as a rigid whole. In Hill's book [50] the reader will find a discussion of the particular cases where the free edge of the block is convex or concave.

Solutions discussed here are in fairly good agreement with numerous practical compression operations. Nevertheless frequently the coefficient of friction is smaller than the limit value $\mu = 0.39$. To such cases the above solution does not apply. We shall discuss this problem in the following section.

### 5.5   Compression of a block between partially rough plates

Assume that the coefficient of friction on the contact line between the compressed material and the plate is sufficiently small, so that the friction force per unit length $t = \mu p$ is smaller than the limit value $t = k$. Then the solution to the problem of compression is different from the one discussed above (see V. V. Sokolovsky [110]). The net of slip-lines for the particular case $\mu = 0.15$ is presented in Fig. 89. Also in this case we begin the solution by assuming the straight slip-line $OA$. The slip-line fan at $O$ is limited by the angle $\gamma$, which is determined by the friction condition on the contact line $\tau_{xy} = -\mu\sigma_y$. Substituting (4.10), we get the condition which has to be satisfied on the contact line

$$2\mu\chi = \mu\cos 2\varphi - \sin 2\varphi. \tag{a}$$

Along the slip-lines $\alpha$ of the curvilinear family in the fan $AOB$ we have $\chi + \varphi = -0.5$, as before. On the slip-line $OB$ we have $\varphi = \frac{1}{4}\pi - \gamma$. The angle $\varphi$ has the same value also in the triangle $OBD$, where the slip-line net consists of two families of straight lines. By substituting the values $\varphi = \frac{1}{4}\pi - \gamma$ and $\chi = -0.5 - \frac{1}{4}\pi + \gamma$ into (a) we obtain the equation from which the angle $\gamma$ can be found. For $\mu = 0.15$ we get $\gamma \approx 35°$. The contact pressure and the friction unit force are constant along $OD$. The states of stress in the regions $ABCB'$ and $BCED$ can be determined by solving successively the characteristic boundary value problems, defined by the already

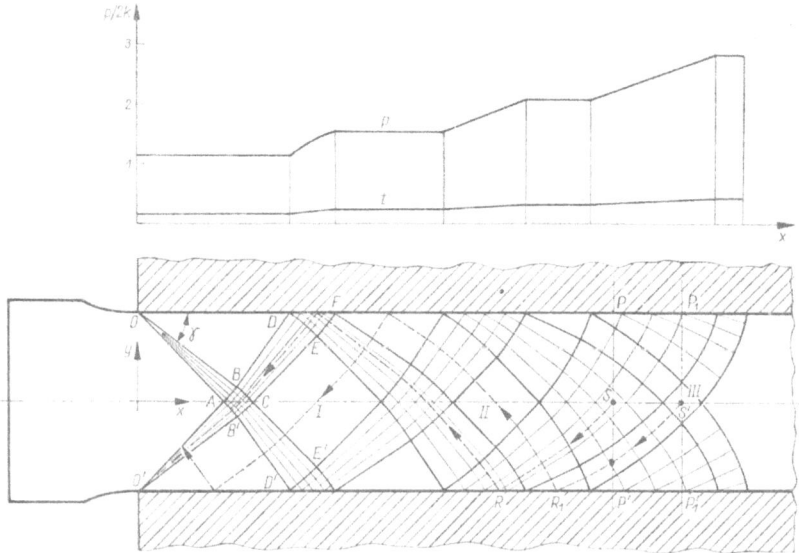

Fig. 89. Slip-line net for compression of a block between partially rough plates. Sokolovsky's solution. Friction coefficient on the contact surface $\mu = 0.15$

known data on $ABD$ and $AB'$. In the triangle $DEF$ we have a mixed boundary value problem, determined by the data on the slip-line $DE$ and by the relation (a) on $DF$. In this way the solution can be extended to the remaining part of the compressed block. The peculiar feature of this solution is the appearance of zones with homogeneous states of stress. These zones have the form of triangles adjacent to the contact lines and the squares $I, II, III, \ldots$ The further we are from the point $O$, the smaller are the zones of homogeneous stress state, and finally at a certain distance from $O$ they vanish. This happens when the friction force $t = \mu p$ reaches the value $k$. The contact line becomes the envelope of slip-lines and the net of slip-lines takes a form similar to that in Fig. 83. The distribution of the contact pressure and of the friction forces is shown in Fig. 89.

Consider now the velocity solution. On the basis of the slip-line net in Fig. 89 we may construct velocity solutions for compressed blocks of various ratios of length to thickness. Assume first that the central point of the block $S$ (compare Fig. 83) is not situated inside any of the squares $I, II, III, \ldots$ The central zone $SRP'$ bounded by the slip-line $SR$, the axis of symmetry $PP'$ and the contact line $RP'$ is rigid. The slip-line

135

*SR* is the line of velocity discontinuity, similarly as in the cases discussed in the previous section. However, now the discontinuity does not meet the plate tangentially as in Fig. 83, but it intersects the contact line at a certain angle $\delta$. The velocity jump is reflected from the plate and it propagates along another slip-line, being multiplied by the factor $\tan \delta$ at each reflection (comp. Section 4.8, Chapter 4). Thus the discontinuity progressively diminishes but it does not vanish. In our case, after two reflections the discontinuity terminates at the edge of the plate. Such a solution is obviously kinematically admissible. On the other hand, if the central point of the block lies inside one of the squares *I, II, III, ...*, say in *III* (point *S'*), then the velocity discontinuity initiated at *S'* terminates on the slip-line *AO'*. This is incompatible with the rigid-body movement of the region to the left of *AO'*. Thus the solution is not correct. Note, however, that if the plates are longer than the compressed block the solution is kinematically admissible, since the velocity discontinuity terminates on the free boundary of the block.[1]

V. V. Sokolovsky [110] considered another variant of the compression of a block, when the frictional stresses have a constant value $t \leqslant k$ along the entire contact line.

If there is no friction on the contact line, there can be distinguished two cases. When the plates are longer than the compressed block, we have uniaxial compression everywhere given by the stresses $\sigma_y = -2k$, $\sigma_x = 0$. But if the block is longer than the plates, this simple stress field is kinematically admissible only for integral values of the length-thickness ratio. For other values of this ratio the lines of velocity discontinuity would terminate on the boundaries between the plastic and the rigid zones. In correct solutions the exit slip-lines are curvilinear.

A. P. Green [39] has given slip-line nets for the ratios $\sqrt{2} \geqslant a/h \geqslant 1$ and $2 \geqslant a/h \geqslant \sqrt{2}$. For $a/h = \sqrt{2}$ the mean contact pressure has the largest value $p = 1.038 \cdot 2k$. For $a/h = 1$ and $a/h = 2$ we have obviously $p = 2k$. For other ratios the lower and upper bounds were found. For $a/h > 2$ the upper bounds do not exceed the lower bounds by more than 2 percent.

---

[1] For a more extended discussion see Hill's book [50] and the paper by J. Salençon [103].

### 5.6 Wedge indentation

Let us consider now a peculiar group of problems which can be solved by making use of the so-called unit diagram.[1] This diagram will be of use in the present and the following two sections. The problems are characterized by the geometric similarity of deformations throughout the entire process of the plastic flow. Let us note, however, that such problems may also be solved by means of the general method of slip-lines, by dividing the deformation process into a number of consecutive steps, thus using the standard procedure described in the foregoing sections.

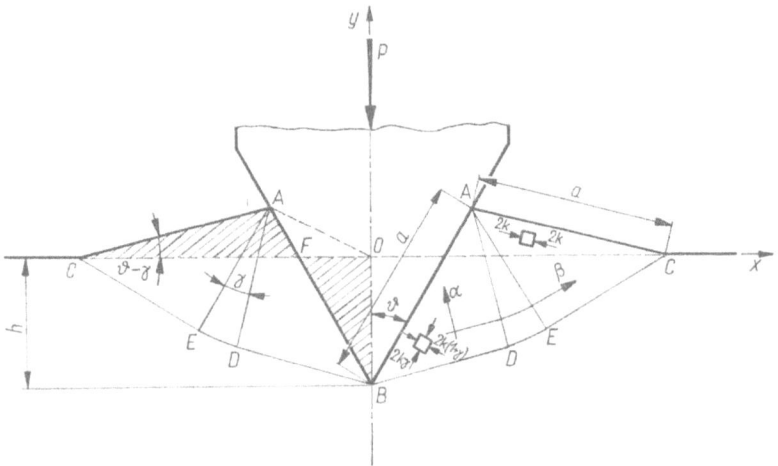

Fig. 90. Indentation of a half-space by a rigid wedge [53]. Slip-line field

Let a plastic half-space be penetrated by a rigid wedge of angle $2\vartheta$ (Fig. 90). The friction on the contact surface will be disregarded. The solution to this problem has been given by R. Hill, E. H. Lee and S. J. Tupper [53]. Any instantaneous stage of the process is characterized by the penetration depth $h$. The configuration of deformation remains geometrically similar during the whole process, the dimensions being at any instant proportional to $h$. The material outflows locally on both sides of the wedge, and the volume of the material displaced above the original surface must be equal to the volume occupied by the wedge. Assume that the free

---

[1] This method has been proposed by R. Hill, E. H. Lee and S. J. Tupper [53]. See also Hill's book [50].

surface of the displaced material is flat. We shall see later on that this assumption leads to the solution satisfying all conditions of the problem. The slip-line net is similar to the net in Fig. 68 and it consists of two rectangular triangles $ABD$ and $AEC$ with homogeneous states of stress, and of the centred fan $DAE$ with angle $\gamma$. To determine the configuration of the net we must know the length $a$ of the edges $AB$ and $AC$ and the value of the angle $\gamma$. The edge $AC$ is inclined to the horizontal at the angle $\vartheta-\gamma$; hence by projecting the segments $AB$ and $AC$ on the axis of the wedge we get

$$a = \frac{h}{\cos\vartheta - \sin(\vartheta-\gamma)}. \tag{i}$$

The angle $\gamma$ is now calculated from the condition of incompressibility which requires that the two dashed regions have the same area. The condition of equality of both areas takes the form

$$a^2[\cos(\vartheta-\gamma)+\sin\vartheta]\sin(\vartheta-\gamma) = ah\sin\vartheta.$$

Eliminating $a$ by means of the relation (i), we obtain finally the equation for the angle $\gamma$

$$\cos(2\vartheta-\gamma) = \tan(\tfrac{1}{4}\pi - \tfrac{1}{2}\gamma). \tag{ii}$$

Let us now examine the velocity field. Along the contact line $AB$ the normal velocity component must be compatible with the movement of the wedge. Thus it must be equal to $v_0\sin\vartheta$, where $v_0$ stands for the velocity of the wedge. Along the slip-line $BDEC$ the normal velocity component must be zero, because below it the material remains at rest. Thus from Geiringer's equations (4.23) we obtain immediately that everywhere in the region $ABDEC$ the velocity component $v_\alpha$ must be zero. The other component $v_\beta$ is determined by the condition on $AB$. It has the constant value $v_\beta = \sqrt{2}v_0\sin\vartheta$ inside the triangle $ABD$. Bearing in mind that everywhere $v_\alpha = 0$, we find from (4.23) that along the $\beta$-lines $v_\beta$ is constant, having the value $v_\beta = \sqrt{2}v_0\sin\vartheta$ and this equality holds also inside the fan $DAE$ and the triangle $ACE$. Thus all particles move along all $\beta$-lines with the same velocity. The triangle $ABD$ is shifted as a rigid whole along $BD$. Similarly, the triangle $AEC$ moves as a rigid block along $EC$. The slip-line $BDEC$ represents the line of velocity discontinuity. The free edge $AC$ is displaced in such a way that it is always parallel to its previous direction. Thus the geometric similarity is preserved, and the assumption that the free edge

*AC* is straight proved to be the correct one leading to the kinematically admissible velocity solution.

The slip-line field can be extended into the rigid region below the slip-line *BDEC* in a manner similar to that shown in Fig. 70. A centred fan of slip-lines will appear now at the point *B*. The angle of the fan is limited by the straight slip-line from *B* forming the angle of 45° with the symmetry axis. Two such slip-lines on both sides of this axis form the edges of a square region with the homogeneous stress state.

The contact pressure *p* is uniformly distributed along *AB*. Its size can be found by an elementary procedure, similarly as in the case of indentation by a flat punch (Fig. 68). In the region *ACE* we have $\sigma_1 = 0$ and $\sigma_2 = -2k$; hence $\chi = -0.5$. The angle $\varphi$ between the direction of the larger principal stress and the *x*-axis, decreases by the angle $\gamma$ as we pass from the triangle *ACE* to the triangle *ABD*. Thus we have $\chi = -0.5 - \gamma$ in *ABD* since along the $\beta$-slip-lines the relation $\chi - \varphi = \text{const}$ must hold. Therefore the principal stress $\sigma_2$ normal to the contact line *AB*, is equal to $\sigma_2 = -2k(1+\gamma)$. Thus finally we get

$$p = 2k(1+\gamma), \tag{5.2}$$

and for the indentation force,

$$P = 4ka(1+\gamma)\sin\vartheta.$$

Making use of the relations (i) and (ii) we may express this force as a function of the angle $\vartheta$ and of the penetration depth *h*.

Consider now deformations of the material. An ingenious method of solution has been proposed in [53]. The problem is analysed on the so-called unit diagram, which holds valid at any instant of the advancing penetration. Such an approach allows us to determine the paths of specific particles, and then to find the deformation of a square grid assumed in the material before penetration. The position of each particle can be defined by its radius-vector from the point *O*. Let the radius-vector **r** determine the position of an arbitrary point *M* at the instant when the penetration depth is *h* (Fig. 90). Instead of considering the problem in the physical plane it is more advantageous to introduce the concept of the unit diagram, where the point *M* will be represented by a point *M\** defined by the radius-vector

$$\mathbf{r}^* = \frac{1}{h}\mathbf{r}. \tag{5.3}$$

The configuration in the unit diagram is geometrically similar to the actual configuration in the physical plane. However, the unit diagram does not alter its dimensions and on this diagram the penetration depth is always equal to unity. The expansion of the deforming region has been replaced by the movement of particles. By assuming that the speed of the wedge is equal to unity we may take as a measure of time the depth of the impression $h$.

The velocity of a particle $M$ in the unit diagram is $\mathbf{v} = d\mathbf{r}^*/dh$, while the velocity of the particle $M$ in the physical plane is $\mathbf{v} = d\mathbf{r}/dh$. In the physical plane, particles located below the slip-line $BDEC$ remain at rest, and begin to move with the constant velocity $v_\beta = \sqrt{2}\sin\vartheta$ ($v_0 = 1$ is assumed) when they are reached by the still expanding deforming region. In the unit diagram, however, the corresponding points move below and above the line $B^*D^*E^*C^*$ (Fig. 91). By differentiating (5.3) with respect to $h$ we obtain

$$\mathbf{v}^* = \frac{1}{h}(\mathbf{r}^* - \mathbf{v}). \tag{5.4}$$

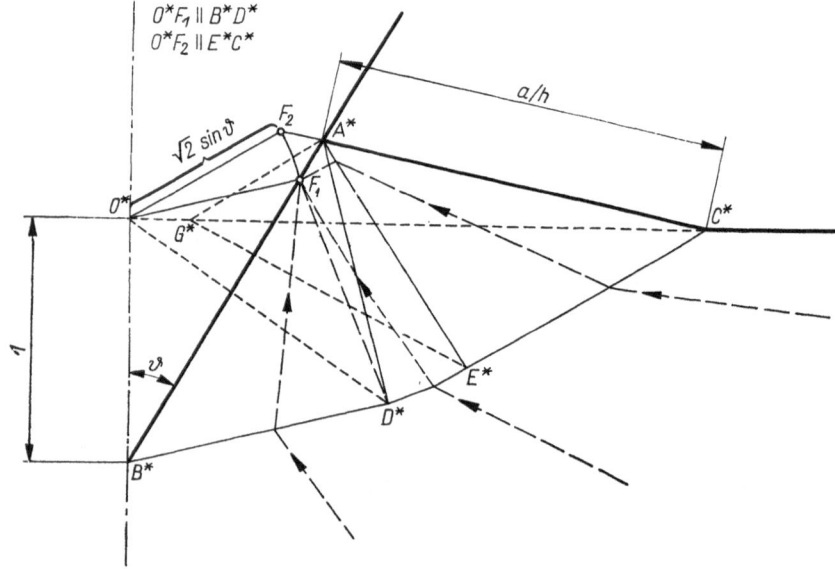

Fig. 91. Unit diagram and trajectories of elements for wedge-indentation (after [53])

Thus the points which are immovable in the physical plane ($v = 0$) have in the unit diagram the speed $v^* = -r^*/h$. The velocity vector $v^*$ is directed towards the pole $O^*$. If, however, a point is moving in the physical plane with the velocity $v$, then its velocity $v^*$ in the unit diagram is directed towards a pole determined by the radius-vector $v$. The magnitude of the velocity $v^*$ is equal to the distance from the point $M^*$ to that pole divided by the penetration depth $h$. In the triangle $ABD$ (Fig. 90) the velocities of all particles are of the same magnitude $\sqrt{2}\sin\vartheta$ and they are parallel to $BD$. The radius-vector $\sqrt{2}\sin\vartheta$ from $O^*$ parallel to $B^*D^*$ determines the pole $F_1$ for all points $M^*$ lying inside the triangle $A^*B^*D^*$. Similarly, the pole $F_2$ for the points of the triangle $A^*C^*E^*$ is situated at the end of the radius vector $\sqrt{2}\sin\vartheta$ parallel to $E^*C^*$. Velocities in the slip-line fan $ADE$ are of the same magnitude as in the two triangles. They change, however, their directions. Thus the poles for the region $A^*D^*E^*$ are situated on the circular arc $F_1F_2$.

Having found the poles, one can determine the images in the unit diagram of the paths of specific material particles. If a point $M$ is located below the slip-line $BDEC$, its image moves towards $O^*$ along a straight line. The direction of movement rapidly changes when the line $B^*D^*E^*C^*$ is reached. Any point crossing this line on the segment $B^*D^*$ begins to move along a straight line towards the pole $F_1$. The path of the points crossing the segment $D^*E^*$ is initially curvilinear. Instantaneous poles for this portion of the path are located on the arc $F_1F_2$. When reaching the line $A^*D^*$, the point moves along a straight path towards the pole $F_1$. The path of any point crossing the segment $E^*C^*$ consists of three parts. The first is straight and directed towards the pole $F_2$, the second, continued between the radii $A^*E^*$ and $A^*D^*$ is curvilinear, and the third is again straight and directed towards the pole $F_1$.

Now we can analyse the distortion of a square grid assumed in the material. It is seen that the portion of the material initially inside the triangle $OBD$ (Fig. 92) comes to lie after deformation in the triangle $FBD$. The deformation in $FBD$ is equivalent to pure shear parallel to $BD$. Similarly, the triangle $AEC$ had before deformation the form of the triangle $CGE$, and the material within it suffered pure shear parallel to $EC$. The material initially situated in the region $ODEG$ is finally in the region $AEDF$. Its deformation is not so simple as in the two triangles.

We shall carry out the analysis of distortion of a grid in the unit

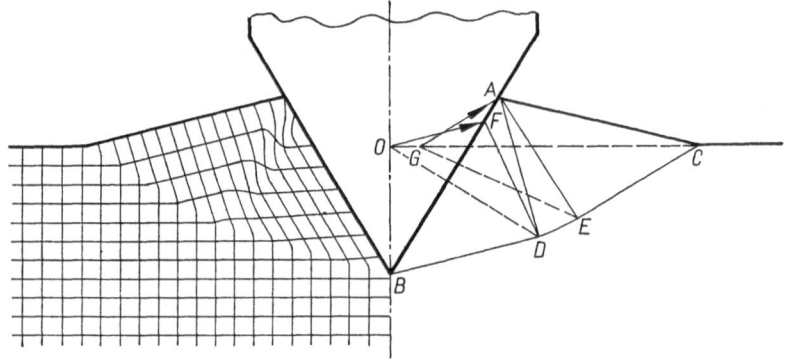

Fig. 92. Distortion of a square grid in wedge-indentation (after [53])

diagram. The problem will be solved when we shall find the final position of an arbitrary nodal point $M$ of the grid defined initially by the radius-vector $r_0$. Let $h_0$ be the penetration depth corresponding to the instant when the deforming zone reaches the point $M$. At this instant its image in the unit diagram is represented by the point $M^*$ of the radius-vector $r_0^* = r_0/h_0$. With the increasing penetration depth $h > h_0$ the displacement of the point $M^*$ along its path is equal to $s$. The velocity $v^*$ of that point can be expressed by the relation $v^* = ds/dh$. From (5.4) we obtain

$$\ln\frac{h}{h_0} = \int_0^s \frac{ds}{f(s)},\tag{iii}$$

where $f(s)$ stands for the distance from our point to the pole corresponding to its position. When the point is situated in any of the regions $F_1 B^* D^*$ or $A^* C^* E^*$, the poles are at $F_1$ and $F_2$, respectively, and the distance $f(s)$ is determined by the simple formula $f(s) = d-s$, where $d$ is the distance between $r_0^*$ and $F_1$ or $F_2$. Instead of (iii) we have now

$$\frac{h}{h_0} = \frac{d}{d-s}.$$

In the region $A^* E^* D^* F_1$, the integral must be calculated numerically. In Fig. 92 the distorted grid is presented for a wedge with angle $2\vartheta = 60°$. The strongly distorted zone corresponding to the region $AEDF$, and the two distorted triangles are clearly visible.

The same procedure can be applied in other problems of wedge

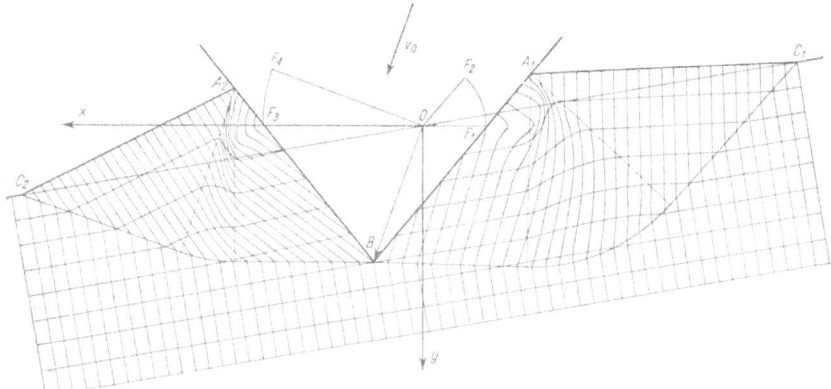

Fig. 93. Distortion of a square grid for a case of oblique penetration of a smooth wedge (after M. Hawrysz [46])

penetration. Figure 93 shows the distorted grid for the case of oblique penetration by a smooth wedge [46]. Solutions for a rough wedge were given by R. Hill [50]. For experimental results see [32, 44, 53].

## 5.7   Cutting of a strip with a knife-edged tool

A block of finite thickness $d$ resting on a rigid smooth foundation is cut with a knife-edged tool of total angle $2\vartheta$. The solution to this problem has been given by R. Hill [52].

At the beginning, the course of the cutting operation is identical with that analysed in the previous section (Fig. 90). When the depth of penetration $h$ is small, the thickness of the block is sufficiently great to carry the lateral force $pa\cos\vartheta$, resulting from the contact pressure on the tool surface. This initial stage lasts to the moment when the penetration depth will reach the critical value $h_1$, for which the lateral force is equal to the carrying capacity of the block in its narrowest section under the vertex of the wedge. This critical depth is for a smooth wedge determined by the relation

$$\frac{d}{h_1} = 1 + \frac{p\cos\vartheta}{2k} \cdot \frac{a}{h_1}. \tag{a}$$

The ratio $a/h_1$ is defined for a given angle $\vartheta$ by formulae (i) and (ii) in the foregoing section.

143

Further penetration of the wedge is accompanied by the plastic flow of the entire cross-section of the block below the cutting tool. From this moment on there begins the second stage of the cutting process. The slip-line net for the transitory instant is shown in Fig. 94a. The deforming region is limited to the triangle $OST$ in which we have the state of uniaxial tension created by the stresses $2k$ parallel to $ST$.

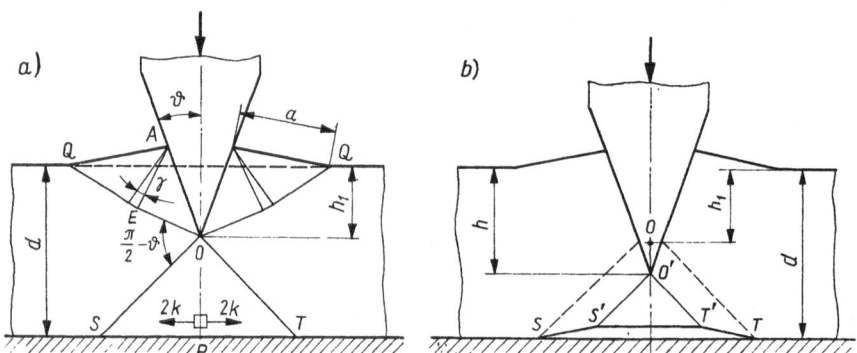

Fig. 94. Cutting of a strip with a knife-edged tool [52]; $\vartheta \leqslant 22°10'$

It can easily be shown that the triangular slip-line field $OST$ in Fig. 94a holds valid only for wedges with angle $\vartheta \leqslant 22°10'$. If $\vartheta > 22°10'$ in the region between the segments $EO$ and $SO$, the yield criterion is exceeded.[1] Hence for this range the slip-line field has the form shown in Fig. 95. In the square $OMNM$ the state of stress is determined by the components

$$\sigma_y = 2k(\tfrac{1}{2}\pi - \vartheta - \gamma),$$

$$\sigma_x = 2k(\tfrac{1}{2}\pi - 1 - \vartheta - \gamma).$$

The stress component $\sigma_y$ is compressive for $\vartheta > 50°40'$. The triangle $STS$ is loaded by uniaxial tension by the stresses $\sigma_y = 2k$. For $\vartheta = 30°$ the triangle $STS$ vanishes, and the point $T$ lies on the contact line.

Our considerations will be limited here to the practically more significant case of the sharp tool, thus to the case where $\vartheta < 22°10'$ (Fig. 94b). During the second stage of deformation when the penetration depth is greater than the critical value given by the relation (a), plastic deformation takes place only in the lower part of the block. If the penetration

---

[1] For the proof of this limitation see the original paper [52].

144

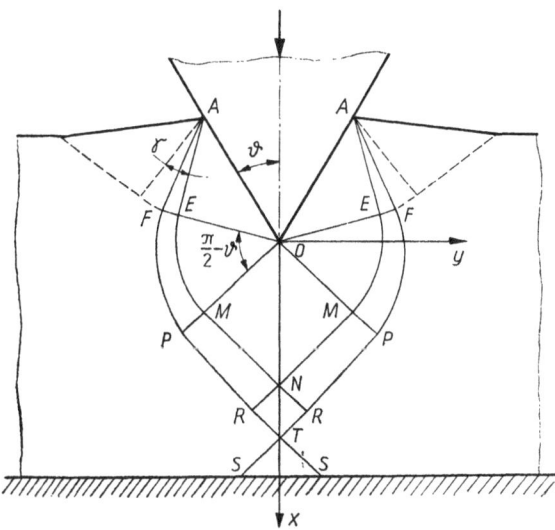

Fig. 95. Slip-line field for the cutting problem. $\vartheta > 22°10'$

depth is $h > h_1$, the two halves of the block move horizontally outwards by the distance $(h-h_1)\tan\vartheta$.

Between the lower edge of the block and the foundation there is formed a gap of width $(h-h_1)\tan\vartheta$. The deforming region is now limited to the triangle $O'S'T'$. As stated in the original paper [52], the existence of the gap can be observed experimentally. The state of deformation in the zones $OO'S'S$ and $OO'T'T$ is equivalent to pure shear. Each material element is deformed when passing across the displacing lines $O'S'$ and $O'T'$. The process is concluded at the instant when the points $S'$ and $T'$ meet each other at the vertex of the tool. This happens when $h$ reaches the magnitude

$$h_2 = \frac{h_1\tan\vartheta+d}{1+\tan\vartheta}.$$

The force applied to the wedge during the second stage ($h_1 \leqslant h \leqslant h_2$) has the value

$$P = 4k(h_2-h)(1+\tan\vartheta)\tan\vartheta.$$

Thus the largest value of the force is attained for $h = h_1$.

Finally, let us mention the allied operation where a plastic strip resting on a smooth flat foundation is indented by a narrow flat punch. This problem has been analysed by R. Hill [50], who assumed that the strip remains in contact with the foundation at all points. His solution is identical with that for the upper half of the block indented by two opposite punches (see Fig. 71). But experiments demonstrate that for certain ratios of punch width to strip thickness the ends of the strip rotate towards the punch. P. Dewhurst [22] presented slip-line solutions which show that a rigid-perfectly plastic strip can exhibit this rotation when indented by a punch.

### 5.8   Compression of a plastic wedge by a flat plate

The unit diagram method may be used also in other cases, as for instance in the case of compression of a plastic wedge by a flat plate. This problem is of practical significance since it corresponds to certain forging operations (see, for instance, Fig. 96).

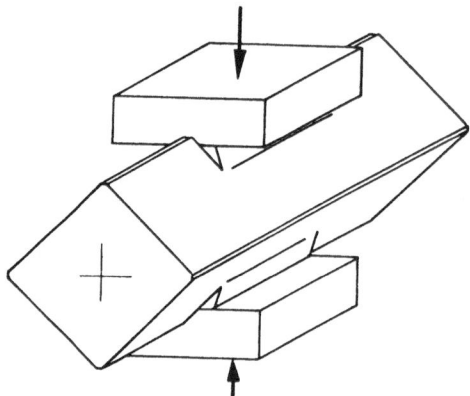

Fig. 96. Forging of a slab

Let a plastic wedge of total angle $2\vartheta$ be symmetrically flattened by a rigid plate [48, 50]. The plate is assumed to be perfectly smooth. Figure 97 shows an admissible slip-line field which can be constructed for any $\vartheta > 14° 2'$. The displacement $h$ of the plate can be taken as the time measurement. The free boundary $AD$ of the deformed material is assumed to be straight. Thus the region of plastic flow preserves geometric similarity at any instant. The slip-line field consists of the triangles $ABA'$ and $ADC$ and a centred fan $CAB$. In the triangles, the state of stress is homogeneous. The requirement that the point $D$ must lie on the surface can be written in the form

$$a + 2a\sin\gamma = (h + 2a\cos\gamma)\tan\vartheta. \tag{i}$$

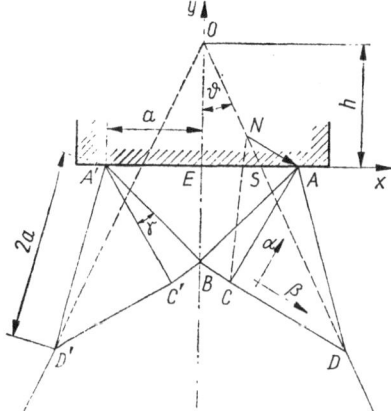

Fig. 97. Compression of a plastic wedge. Hill's solution [48]

The other equation, which together with (i) determines the length $a$ and the angle $\gamma$ for given $\vartheta$ and $h$, can be obtained from the condition of constant volume of the deformed material,

$$h^2\tan\vartheta = (a-h\tan\vartheta)2a\cos\gamma.\tag{ii}$$

Finally, the relation between the angles $\gamma$ and $\vartheta$ takes the form

$$\tan\vartheta = \frac{(1+2\sin\gamma)^2}{4\cos\gamma(1+\sin\gamma)},\tag{iii}$$

and similarly, for the length $a$, we have

$$a = h\frac{1+2\sin\gamma}{2\cos\gamma}.\tag{iv}$$

The region $ACD$ is uniaxially compressed by the stress $-2k$ since the other principal stress must be zero in order to satisfy the boundary conditions on $AD$. The uniformly distributed pressure on the contact line $AA'$ can be found in an elementary way analogous to that described in the foregoing section. This pressure is given by the formula

$$p = 2k(1+\gamma).$$

For any stage of the process, the extension of the actual slip-line field into the rigid region below $BCD$ can be constructed similarly as in Sections 5.2 and 5.3. The extended slip-line net lies entirely within the contour of the rigid part of the wedge. Thus the bearing capacity of this part is sufficient.

147

The velocity solution is similar to Prandtl's solution for the indentation by a punch (Fig. 67). To preserve the geometric similarity, the triangle *ACD* must move as a rigid whole along *CD*. The triangle *ABA'* is displaced downwards as a rigid block together with the plate. Within the entire zone *ABCD* we have $v_\alpha = 0$, $v_\beta = v_0/\sqrt{2}$, where $v_0$ is the speed of the plate. The slip-lines *A'BCD* and *ABC'D'* are the lines of velocity discontinuity.

To analyse the deformation of the wedge, the unit diagram will be used (cf. Section 5.7). We find the poles of the trajectories for the specific regions by plotting the vectors of the respective velocities from the original vertex *O\** of the wedge (Fig. 98). The plate is assumed to move with unit

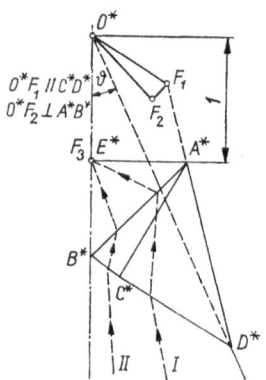

Fig. 98. Unit diagram for compression of a plastic wedge

velocity. For the region *A\*C\*D\**, where the velocity is equal to $1/\sqrt{2}$ and parallel to *C\*D\**, the pole is located at the point $F_1$ (*O\*F_1‖C\*D\** and $O^*F_1 = 1/\sqrt{2}$). For the fan *B\*A\*C\**, the poles are situated on the circular arc $F_1 F_2$. The point $F_2$ is determined by the vector from *O\** perpendicular to *A\*B\**. For the triangle *A\*B\*E\**, where the unit velocity is directed vertically, the pole $F_3$ coincides with *E\**. In Fig. 98 there are shown two trajectories of particles entering the deforming region across various segments of the line *B\*C\*D\**. An analysis of the trajectories shows that the triangle *ACD* arises from the triangle *CDN* (Fig. 97) by a simple shearing deformation. The region *ACBE* suffers more complex deformation. Knowing the trajectories in the unit diagram, we may find the deformation of a square grid assumed in the material.

148

## 5.8 Compression of a plastic wedge by a flat plate

Figure 99 shows the theoretical deformation of the grid for a wedge of total angle $2\vartheta = 100°$. The experimentally obtained deformation is presented in Fig. 100. The wedge was made of aluminium and the surface of contact with the steel plate was not lubricated. A fairly good agreement between the two deformed grids is visible, although the actual deforming region is larger than that predicted by the theory. This is attributed to the strain-hardening effect which is disregarded by the theoretical analysis.

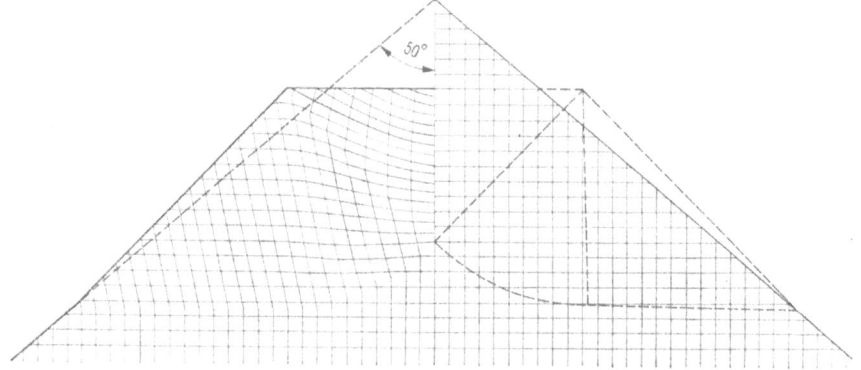

Fig. 99. Distortion of a square grid in compression of a plastic wedge [129]

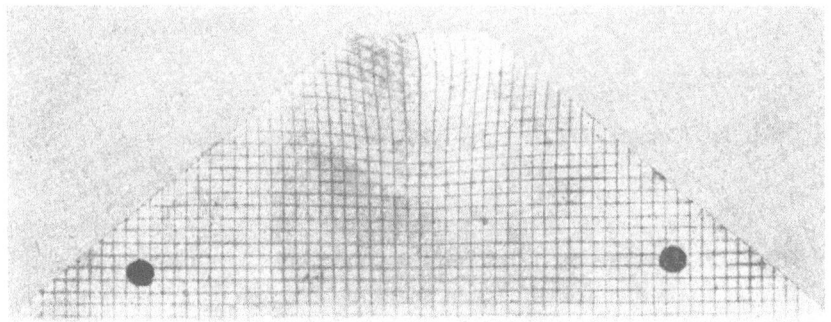

Fig. 100. Distortion of an aluminium wedge compressed by a flat plate [129]

The solution presented above is valid for $\vartheta > 14° 2'$. For the limit value $\vartheta = 14° 2'$, the angle $\gamma$ of the fan in the slip-line net in Fig. 97 reduces to zero. The solution has then a very simple form since the points $C$ and $C'$ coincide with the point $B$. The state of stress reduces to uniaxial

149

compression in the whole deforming region. The contact pressure is equal to $2k$.

As a practical example consider a slab of square cross-section, compressed diagonally between two rigid plates (Fig. 101; compare also Fig. 96). At the beginning, when the displacement of the plates is small, only local plastic deformation takes place according to the just described solution (Fig. 101a). At more advanced stages of compression when

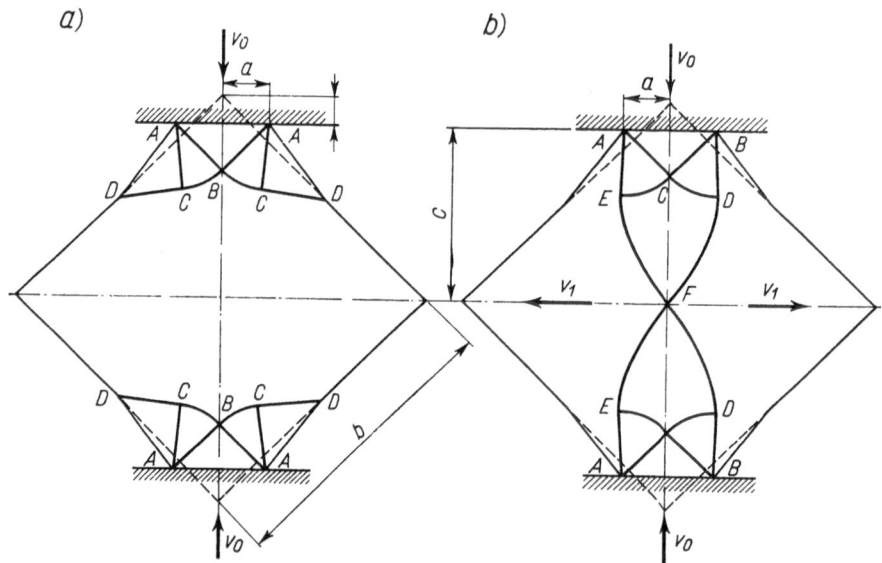

Fig. 101. Compression of a slab between two rigid plates

$c/a < 3.82$, the slab begins to deform according to the mechanism shown in Fig. 101b, because the force required to cause such a deformation mode is smaller than that for the local yielding. This mechanism is analogous to that discussed in Section 5.3 in the case of indentation by two opposite punches (compare Fig. 71). The deforming region is limited by the slip-lines $BDF$ and $AEF$. The material outside these lines moves outwards as a rigid whole. Both velocity solutions shown in Fig. 101 are kinematically admissible, only because the extensions of the stress fields into the rigid regions are unknown.

Hill [48, 50] proposed also another solution to the problem of flattening a plastic wedge by a smooth plate. The slip-line field is shown in Fig. 102. Instead of (iii) we have now

$$\tan\vartheta = \frac{(1+\sin\gamma)^2}{\cos\gamma(2+\sin\gamma)},$$ (v)

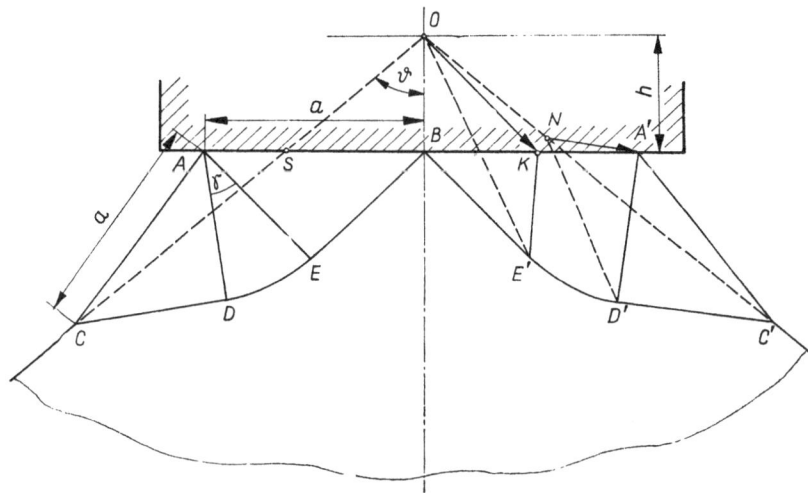

Fig. 102. Another Hill's solution to the problem of compression of a plastic wedge (after Hill [50])

and the length $a$ is given by

$$a = h\frac{1+\sin\gamma}{\cos\gamma}.$$ (vi)

The velocity field is similar to that for the indentation of a flat punch (comp. Fig. 68). The triangles $BE'K$ and $A'D'C'$ are formed by distortion of the triangles $OBE'$ and $ND'C'$. It is readily seen that this solution, which is valid for any $\vartheta > 26° \, 36'$, is possible only in the case where there is no friction on the contact surface.

The latter solution, although as correct as the previous one from the point of view of the theory of perfectly plastic solids, has little significance for real metals. According to this type of solution, the material below the line $EBE'$ remains underformed. This, however, was never confirmed by experiments.

151

Finally, let us compare the forces required for flattening which are obtained from the two solutions. From the first type of solution (Fig. 97) we get

$$P = 2kh \frac{1+2\sin\gamma}{2\cos\gamma}(1+\gamma),$$

and from the second type (Fig. 102),

$$P = 2kh \frac{1+\sin\gamma}{\cos\gamma}(1+\gamma).$$

Note that these formulae cannot be compared directly, because the angle $\gamma$ is for a given $\vartheta$ different in the two solutions. The comparison has been done numerically. The result is shown in Fig. 103. The solution

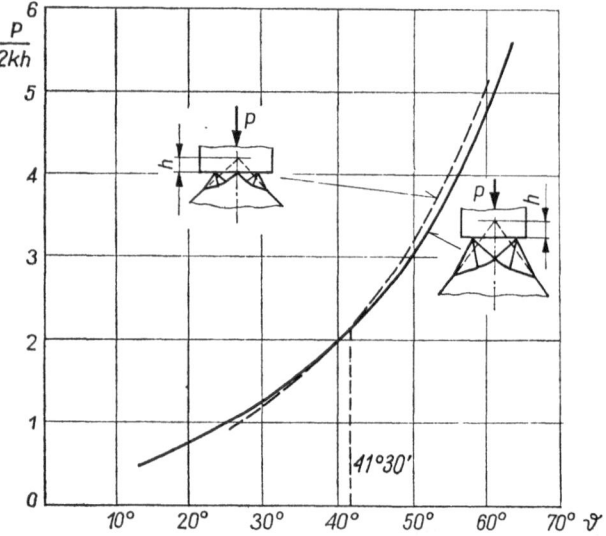

Fig. 103. Force versus semi-angle in the compression of a plastic wedge

of the first type predicts for the practically most important range $\vartheta > 41°\ 30'$ a smaller force than that predicted by the second type. Thus also from this point of view, the solution of the first type seems to be of more practical significance for real conditions.

# 6

# *Two-dimensional steady-state operations*

## 6.1 General remarks

In this chapter we shall consider the very important group of problems in which stresses and deformations do not change at some specific point. Examples of steady-state operations are, for instance, the strip drawing, rolling, and extrusion. The state of deformation is characterized by the appearance of a relatively small region of plastic yielding adjacent to the working surfaces of the tool. The material outside the deforming region remains rigid and moves as a rigid whole, entering from one side into the region of plastic flow, and leaving it from the other side with changed thickness. All particles entering the deforming region at a given point undergo the same deformation history and leave this region also at a common point. Thus the problem reduces to determining the stresses and velocities at any point of the deforming region. These depend obviously only on the position of the point, and not on time. All problems are statically undetermined since the restrictions imposed on the velocity field on the lines of contact with the tool impose also certain conditions on the stress field. In most cases, this difficulty can be overcome by the trial and error procedure, by assuming a certain slip-line field and then checking whether it is kinematically admissible. We shall see, however, that in some cases this method fails. Solving such problems, we shall consider both the stress and velocity field simultaneously.

## 6.2 Sheet drawing through a smooth wedge-shaped die

Theoretical solutions to the problem of sheet drawing under condition o plane flow are of little practical significance, because this method is rarely used in practice. However, they can give a useful estimate of forces required for wire drawing. The applicability of plane flow solutions to the analysis of axially symmetric problems will be discussed in Chapter 8.

Let a sheet be drawn through a rigid die with perfectly smooth surface. The stress and velocity distributions depend strongly on the reduction of the thickness which is defined as the ratio $(H-h)/h$, where $H$ is the original thickness and $h$ stands for the thickness of the drawn sheet. In this section we shall consider the case of large reduction of the thickness (Fig. 104).[1]

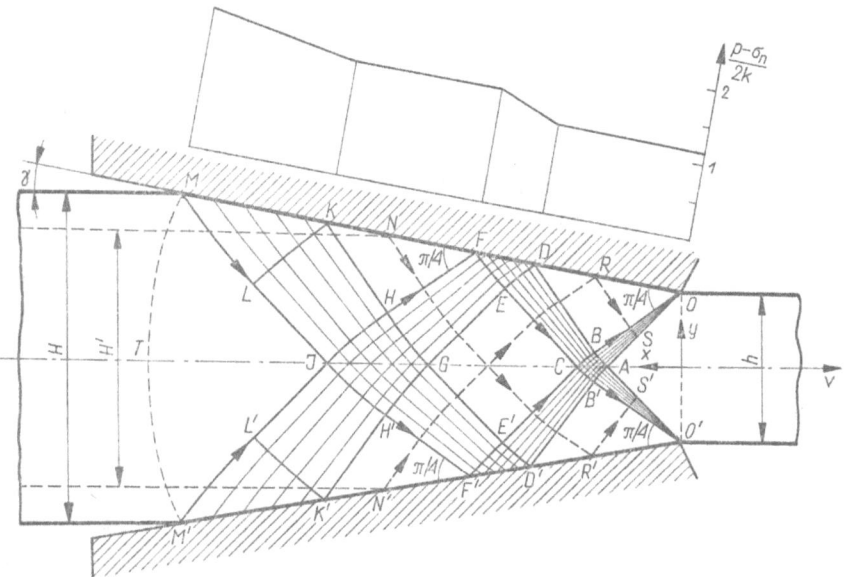

Fig. 104. Sokolovsky's solution for the drawing through a smooth wedge-shaped die

Similarly as in the case of the compression of a block between rigid plates, discussed in the foregoing chapter, the slip-line net depends on the conditions of friction on the sheet-die interface. We begin from the case where there is no friction between the die and the sheet (perfectly smooth die). At the exit, where the thickness is $h$, the sheet is pulled by a force whose value is to be found. We begin constructing the solution by assuming two straight starting slip-lines $OA$ through the edges $O$. We shall show that this assumption leads to the solution which satisfies, with the exception of

---

[1] The slip-line field for this problem is due to V. V. Sokolovsky [110]. The velocity field and its limitations have been analysed by R. Hill [50].

certain reduction ratios, all static and kinematic requirements. Assuming straight slip-lines $OA$, we assume, in other words, that everywhere to the right of them we have the homogeneous state of stress, with principal axes parallel to the coordinate axes $x$ and $y$. The stress component $\sigma_x$ (and also the other component $\sigma_y$) is not known and will be determined later from the condition of equilibrium of the whole sheet. The unknown value of $\sigma_x$ to the right of $OA$ will be denoted by $p$. The other stress component $\sigma_y$ is defined by the yield condition. Thus we have $\sigma_y = p - 2k$. Let the constant $\sigma_0$ in the relations (4.10) be $\sigma_0 = p$. From (4.10) we obtain that $\chi = -0.5$ in the region situated to the right of $OAO'$ since the larger principal stress $\sigma_x$ is parallel to the $x$-axis there. The angle of the slip-line fan $AOB$ is equal to the angle of the die $\gamma$. In the triangle $OBD$ we have $\varphi = \gamma$, because the slip-lines must intersect the smooth surface of the die at the angle of $\pi/4$. The value of $\chi$ is in this region defined by the relation $\chi + \varphi = $ const, which must hold along the slip-lines of the first family (one of them is $AB$). Hence we obtain $\chi = -0.5 - \gamma$ in $OBD$. Therefore the normal stress component $\sigma_n$ is given on $OD$ by

$$\sigma_n = p - 2k(1+\gamma). \tag{6.1}$$

In the curvilinear quadrangle $ABCB'$ we have a characteristic boundary value problem since the values of $\varphi$ and $\chi$ are already known along $AB$ and $AB'$. Thus the solution in this region can be found. Similarly, the state of stress in the region $CBDE$ is determined by a characteristic problem, for along $BD$ we have $\varphi = \gamma$ and $\chi = -0.5 - \gamma$ and along $CB$ the values of $\varphi$ and $\chi$ are already known from the solution in $ABCB'$. In $CBDE$ the slip-lines of the first family are straight. Passing now to the triangle $DEF$, we already know the functions $\varphi$ and $\chi$ on $DE$. Along the contact line $DF$, whose equation is $y = x\tan\gamma + 0.5h$, we have $\varphi = \gamma$. These data define the mixed boundary value problem, enabling us to find the solution in the region $DEF$. In the region $CEGE'$ the stresses are distributed uniformly, and the values of the auxiliary functions are $\varphi = 0$ and $\chi = -0.5 - 2\gamma$. From (4.10) we get the stress components $\sigma_x = p - 4k\gamma$, $\sigma_y = p - 2k(1+2\gamma)$. Extending this procedure, we find the stresses in the remaining regions of the slip-line net. In the triangle $FKH$ the homogeneous state of stress is defined by the values $\varphi = \gamma$ and $\chi = -0.5 - 3\gamma$. The normal stress component on the contact line $FK$ is

$$\sigma_n = p - 2k(1 + 3\gamma).$$

155

At the point $M$ we have $\varphi = \gamma$ and $\chi = -0.5 - 5\gamma$, and consequently

$$\sigma_n = p - 2k(1 + 5\gamma).$$

The material to the left of $MLJL'M'$ is rigid.

The value of the tractions $p$ applied at the exit can be computed from the condition of equilibrium of the sheet

$$ph + 2\int_0^L \sigma_n \sin\gamma \, ds = P_1, \tag{6.2}$$

where $L$ is the length of the contact line and $P_1$ stands for the so-called *back-pull force* applied at the entry to the die. Usually we have $P_1 = 0$. Evidently the tractions $p$ cannot be greater than the yield locus $2k$. This obvious condition imposes a limitation on the largest possible reduction of the thickness.

The slip-line net shown in Fig. 104 has been calculated numerically for the particular case $\gamma = 10°$. However, it can be obtained in a much simpler way graphically according to the procedure described in Chapter 4. The thickness reduction is in our case $(H - h)/h = 1.2$ and the pulling stress $p \approx 1.6k$. Figure 104 shows also the distribution of the contact pressure along the die.

Consider now the extensions of the stress field into the rigid regions. To the right of the slip-line $OA$ we may assume that in the triangle $AOO'$ we have $\sigma_x = p$ and $\sigma_y = p - 2k$. The vertical straight segment $OO'$ represents the line of stress discontinuity, to the right of which the strip is loaded by uniaxial tension $\sigma_x = p$. Thus the extension in this rigid region is statically admissible.

Estension into the rigid zone to the left of the slip-lines $MLJ$ and $JL'M'$ has not been as yet investigated. However, for a construction of such an extension the procedure used by B. A. Drujanov [30] in the analysis of strip-rolling may prove useful. According to this procedure, similar to Bishop's method described in Sections 5.2 and 5.3, the region adjacent to the slip-lines $MJ$ and $M'J$ is assumed to be in the plastic state. We find the stresses there by solving the characteristic boundary value problem defined by the data on $MJ$ and $M'J$. Next we begin to construct the line of stress discontinuity $MTM'$, starting from the point $M$ and assuming that to the left of this line only the stress component $\sigma_x$ is non-zero. Let $\theta$ be the angle at which the normal to the line of discontinuity intersects the $x$-axis. Then the normal and tangential stress components on the right-hand (plastic) side of the discontinuity line are

$$\sigma_n = 2k\chi + k\cos(\theta - \varphi) \text{ and } \tau_{nt} = k\sin 2(\theta - \varphi),$$

respectively. Note that the function $\chi$ and the angle $\varphi$, formed by the larger principal stress with the $x$-direction, are already known from the solution of the characteristic problem. Similarly on the left-hand side of the line of discontinuity we have

$$\sigma_n = \tfrac{1}{2}\sigma_x(1 + \cos 2\theta),$$

$$\tau_{nt} = \tfrac{1}{2}\sigma_x \sin 2\theta.$$

These stress components must be continuous across the line of stress discontinuity (see Section 4.8). Thus we can write

$$\begin{aligned}\sigma_x(1 + \cos 2\theta) &= 4k\chi + 2k\cos 2(\theta - \varphi),\\ \sigma_x \sin 2\theta &= 2k\sin 2(\theta - \varphi).\end{aligned} \tag{a}$$

By eliminating $\sigma_x$ we arrive at the equation for the angle $\theta$

$$\sin 2(\theta - \varphi) = 2\chi \sin 2\theta + \sin 2\theta \cos 2(\theta - \varphi). \tag{b}$$

This equation enables us to construct step by step the line of discontinuity $MTM'$, beginning from $M$. The horizontal stress $\sigma_x$ acting in the region to the left of it may be found from the second equation (a). The extension of the stress field is statically admissible, provided $\sigma_x$ does not exceed the yield locus and the strip is infinitely long. We begin the determination of the velocity field by solving the mixed boundary value problem in the triangle $KLM$. Along $ML$ the normal velocity component must be continuous. If $v_0$ denotes the velocity at the exit from the die, then the requirement of constant volume implies that the velocity of the sheet at the entry is $v_1 = v_0 h/H$. Along the segment $MK$ of the contact line the normal velocity component must be zero. Thus solving the mixed problem so defined, we find the velocities in $KLM$ and consequently also on the slip-line $KL$. Next, the velocities in the region $KLJH$ can be computed. The normal velocity component on the straight slip-line $LJ$ is constant. Thus, according to Geiringer's equations (4.23), both velocity components must be constant along each straight slip-line in this region. Hence the normal velocity component is constant also along $KH$. Since, moreover, the normal velocity is zero on $KF$, the triangle $FHK$ moves as a rigid whole along the contact line $KF$. Next we solve the characteristic problem inside the quadrangle $GHJH'$. Knowing the velocity normal to

157

*HG*, we can find the velocities in the region *EFHG*. By solving in this way the successive boundary value problems, the velocity field in the whole deforming region can be determined. In this field the square *CEGE'* moves horizontally as a rigid block, and the triangle *OBD* is shifted also as a rigid block along *OD*. This implies that the velocity is constant along the exit slip-line *OA*, as required by the rigid body motion of the strip outside the die. Thus our assumption that *OA* is straight proved correct. *MJF'CO* represents the line of velocity discontinuity, and so does its counterpart *M'JFCO'*. The velocity jumps terminate at the edges *O* and *O'* of the die. Thus the velocity field is kinematically admissible.

The same slip-line field may be used for other reductions in thickness. Each line terminating on the segment *KM* may be chosen as the limiting slip-line. The velocity discontinuity propagating along it will terminate at *O* and *O'*. If, however, the contact of the strip with the die surface began at any point *N* situated on the segment *FK* (see Fig. 104), the velocity discontinuity propagating along *NR'S'* will terminate on the boundary of the rigid part to the right of *OA*. Such a velocity solution is not kinematically admissible (for a discussion see [50] and [111]). One can expect that in such cases the exit slip-line is not straight. The schemes of slip-line nets with a curvilinear starting slip-line were proposed by A. P. Green [39].

The slip-line field in Fig. 104 may also be used when the slip-line bounding the deforming region terminates on the segment *FD*. The solution is similar to that shown in Fig. 108. Consider now the particular case shown in Fig. 105 for the limit reduction ratio

$$H/h = 1 + 2\sin\gamma. \tag{6.3}$$

The slip-line net consists of two triangles *OBD* and of the two centred fans *AOB*. The velocity discontinuity propagates along the lines *DBO'* and *D'B'O*. The pulling stress at the exit-section has the value

$$p = 2k(1+\gamma)\frac{H-h}{H}. \tag{6.4}$$

We shall show how to determine the deformation of the square grid inscribed in the strip caused by the passing through the die [50]. The velocity is constant in the triangle *OBD*. Its direction is parallel to *OD*. The magnitude of the velocity in *OBD* is found by comparing the normal components on *BD*; we get $v = \sqrt{2}v_1\cos(\frac{1}{4}\pi - \gamma)$ in *OBD*, where $v_1$ is the velocity of the strip at the entry into the die. On the segment *AB* the normal velocity component is $v_n = v_1 \cos(\frac{1}{4}\pi - \omega)$, where $\omega$ is the angle defining the position of the specific point on $AB$. This velocity component is constant along each of the straight slip-lines of the $\beta$-family in the fan $AOB$. From

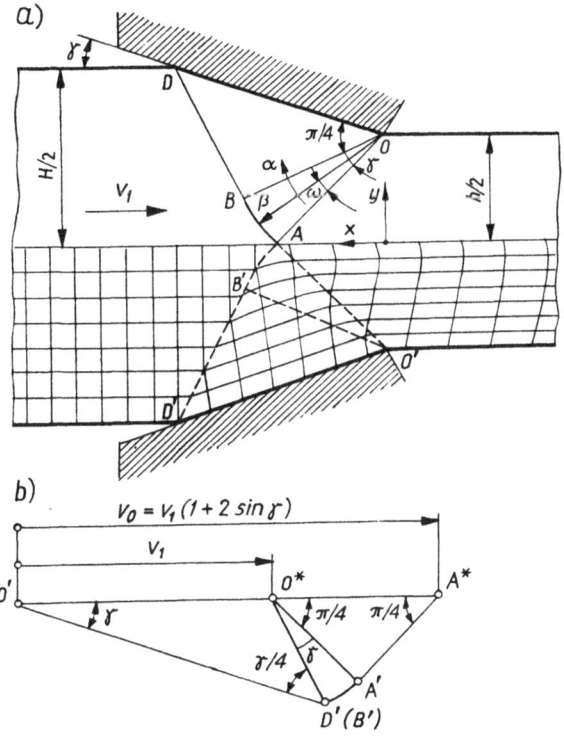

Fig. 105. Slip-line field and distortion of a square grid in drawing through a smooth die

(4.23) we conclude that along the $\alpha$-lines of the fan the following relation must hold:

$$dv_\alpha - v_1 \cos(\tfrac{1}{4}\pi - \omega)\,d\omega = 0.$$

By integrating we finally get the velocities in $AOB$

$$v_\alpha = -v_1[\sin(\tfrac{1}{4}\pi - \omega) + \sqrt{2}\sin\gamma],$$
$$v_\beta = -v_1\cos(\tfrac{1}{4}\pi - \omega),$$

$$(6.5)$$

159

where the integration constant is found from the condition that normal velocities must change continuously across *OB*.

The relations (6.5) can be directly obtained from the hodograph (Fig. 105b). The velocity $v_1$ at the entry is represented by the vector $\overrightarrow{O'O^*}$. Thus the left-hand side of the line of discontinuity *DBA* is represented by the single point $O^*$. The constant velocity in the triangle *DBO* is equal to the vector $\overrightarrow{O'D'}$ parallel to the die profile. The segment $O^*D'$ is parallel to *DB*. The velocities of the particles to the right of *BA* are represented by respective points on the circular arc $D'A'$ with centre $O^*$. The velocity jump across the line of discontinuity *AO* corresponds to the vector $\overrightarrow{A'A^*}$. The velocity of the rigid part of the strip at the exit $v_0 = v_1 H/h = v_1 \times$ $\times (1+2\sin\gamma)$ is represented by the vector $\overrightarrow{O'A^*}$.

Knowing the velocities, we can find the trajectories of specific particles. Outside the deforming region the trajectories are obviously parallel to the symmetry axis. In the triangle *OBD* they have the form of straight lines parallel to *OD*. In the slip-line fan *AOB* we find them by dividing the total angle $\gamma$ into sufficiently large number of angle increments and determining, for each of the so obtained radii, the direction of velocity from (6.5) or from the hodograph. If the latter method is used, we divide the fan $A'O^*D'$ into the same number of angle increments as the fan *AOB* in the physical plane. The points on the arc $D'A'$ obtained in this way determine the velocities on the respective radii of the fan *AOB*. Knowing the directions of these velocities, we can plot graphically the trajectories of the particles. On the lines of velocity discontinuity *ABD* and *AO* there appear sharp bends of the trajectories. The trajectories represent also one family of lines of the deformed square grid. The deformation of vertical lines of the grid can be found in the following way. Let $s_0$ be the coordinate of a nodal point of the grid, measured along the trajectory at the instant $t_0$. Then the new position of that point at the instant $t_1$ can be obtained from the equation

$$s_1 = s_0 + \int_{t_0}^{t_1} v\,\mathrm{d}t,$$

where $v$ is the velocity along the trajectory. This velocity is constant, except between the slip-lines *OA* and *OB*, and only in this region the integral must be computed numerically. In Fig. 105 there is presented the deformed grid for $\gamma = 20°$.

Slip-line solutions discussed in this section are completely satisfying all static and kinematic conditions provided the strip on the entry-side is infinitely long, since only in such a case the extension of the stress field into the rigid region can be constructed. But if the length of the strip is finite, the solution must be regarded as merely kinematically admissible; thus it furnishes the upper bound on the unknown exact value of the pulling force. A lower bound on this force can be obtained for instance from the elementary statically admissible stress field shown in Fig. 106.

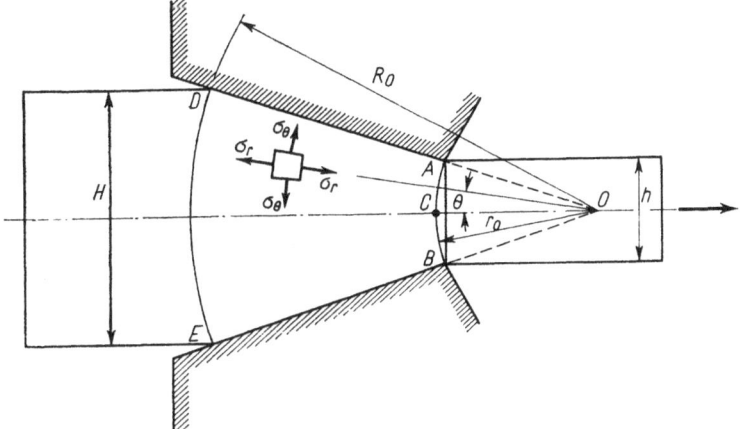

Fig. 106. Statically admissible stress field for lower estimate of the drawing force

In the region *ACBED* bounded by two circular arcs with the centre at *O* the state of stress is defined in the polar coordinate system $r, \theta$ by the components

$$\sigma_r = 2k\ln\frac{R_0}{r} \quad \text{and} \quad \sigma_\theta = -2k\left(1-\ln\frac{R_0}{r}\right).$$

Within the region *ACB* we have biaxial tension by the stresses equal to $2k\ln(R_0/r_0)$. The arc *ACB* and the vertical straight segment are the lines of stress discontinuity. The strip to the right of *AB* is uniaxially loaded by uniaxial tensions equal to $p = 2k\ln(H/h)$. For relatively large reductions in the thickness the lower bound so obtained is only slightly smaller than the upper bound. For instance, in the particular case shown in Fig. 105 the difference between the two bounds is less than 5.5 percent.

161

### 6.3   Drawing through a rough die

The friction on the sheet-die interface has a significant influence upon the slip-line net, stress distribution and velocity field. The simplest case arises when the friction forces reach their maximum possible value equal to the yield locus in shear $k$. An example of the slip-line net for such a case is shown in Fig. 107. $DCB'AO$ and its counterpart $D'CBAO'$ are the lines

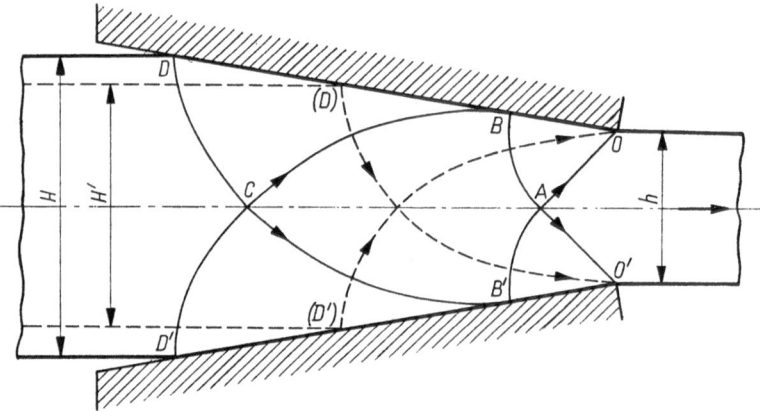

Fig. 107. Slip-line field for the case of drawing through a rough die

of velocity discontinuity. It is readily seen that for the arbitrary reduction in thickness the velocity jumps terminate at the points $O$ and $O'$. Thus the solution is correct for any position of the point $D$, where the sheet enters into the die. The dashed lines represent the discontinuity lines for one of the possible position of $D$.

Solutions assuming the Coulomb friction $t = \mu\sigma_n$ on the contact line are much more difficult to obtain, since the normal stress $\sigma_n$ depends on the unknown pulling tractions $p$ which depend on the condition of equilibrium of the whole sheet.

V. V. Sokolovsky [110] has given the solution under the assumption that the friction unit force $t < k$ is constant over the entire contact surface. The slip-line field for the particular case $\gamma = 12°$ and $t = 0.2k$ is shown in Fig. 108. Here the angle $BOD$ is equal to $\frac{1}{4}\pi - \delta$, whereas in the case of a perfectly smooth die its value is $\frac{1}{4}\pi$. The angle $\delta$ is determined by the relation $\sin 2\delta = t/h$. Thus the angle $\varphi$ contained between the direction of the greater principal stress and the $x$-axis is given in the triangle

*OBD* by $\varphi = \gamma + \delta$. $\varphi$ preserves this value along the remaining portion of the contact line *DF*. The angle of the centred fan *AOB* is $\gamma + \delta$. Assuming that $\sigma_0$ [see (4.10)] is equal to the tensile stress at the exit ($\sigma_0 = p$), we obtain $\chi = -0.5$ in the region *OAO'*. In the triangle *OBD* we get $\chi = -0.5 - \gamma - \delta$ since $\varphi = 0$ in *OAO'* and $\varphi = \gamma + \delta$ in *OBD*. The normal stress component on the contact line is determined by the relation

$$\sigma_n = p + 2k\chi - k\cos 2\delta.$$

Thus on the segment *OD* it retains the constant value

$$\sigma_n = p - 2k(0.5 + \gamma + \delta) - k\cos 2\delta. \tag{6.6}$$

The absolute value of the contact pressure is now larger than in the case of a smooth die [comp. (6.1)].

The remaining part of the slip-line net can be determined in the same way as in the previous section. Note, however, that we have now $\varphi = \gamma + \delta$ along *FD*, while for the smooth contact there was $\varphi = \gamma$.

We can find the mean drawing stress $p$ at the exit from the condition of equilibrium of the whole sheet

$$ph + 2\int_0^L \sigma_n \sin\gamma \, ds - 2tL\cos\gamma = P_1, \tag{6.7}$$

where $L$ is the length of the contact line, and $P_1$ stands for the back-pull force. The reader can find in [9] some numerical results concerning the influence of back tension.

The hodograph is shown in Fig. 108b. Let $v_1$ be the speed of the sheet entering into the die, and let $v_2$ be the speed at the exit. On account of symmetry only the lower half of the hodograph corresponding to the upper half of the strip will be considered.

The left-hand side of the line of discontinuity *FEHG* is represented by a single point $O^*$. The velocities on its right-hand side are represented by the circular arc $F'E'G^*$ with the centre at $O^*$. The constant velocity on *EH* is determined by the vector $\overline{O'E'}$. The segment $O'F'$ is parallel to the die surface *FDO*, and the segment $O^*F'$ is parallel to the tangent to *FE* at *F*. The vector $\overline{O'D'}$ represents the velocity of the triangle *DBO*, which moves as a rigid block. By plotting in the triangle $F'E'D'$ the mesh of lines orthogonal to the corresponding slip-lines covering the triangle *FED* in the physical plane, we find the mapping of the velocities in *FED*. Similarly, we obtain the region $E'G'K'D'$, determining the velocities in the quadrangle *BHGK*. The segment $G'K'$ represents the mapping of the region

163

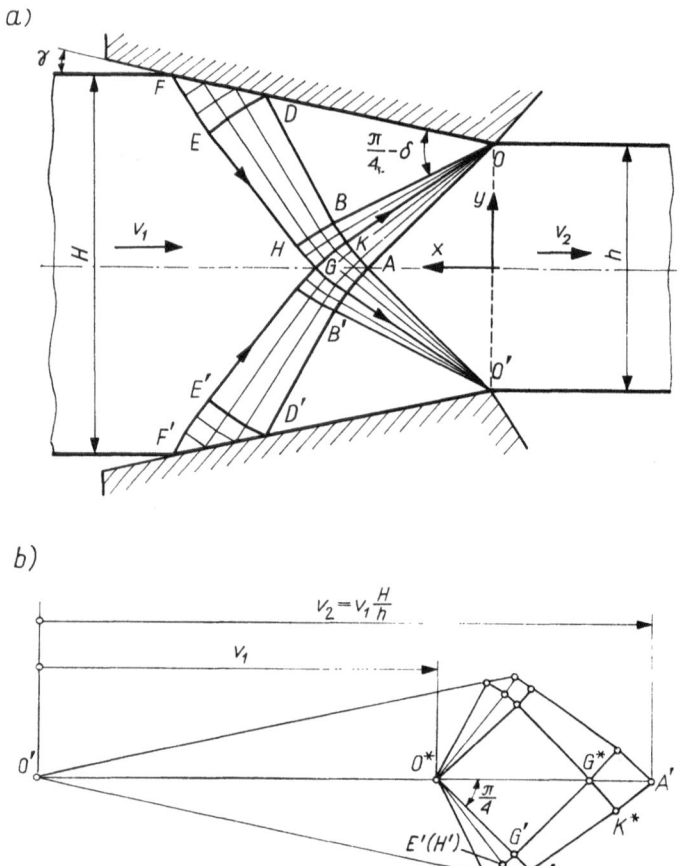

Fig. 108. Drawing through a partially rough die with constant frictional stresses;
a) Sokolovsky's slip-line field, b) velocity hodograph

at the left-hand side of the line of velocity discontinuity $GK$. The velocities
on its right-hand side have to be represented by a line equidistant from $G'K'$.
For reasons of symmetry the velocity of the point $G$ must be directed
horizontally. The vector of the velocity jump $\overrightarrow{G'G^*}$ has to be parallel to
the tangent to $GK$ at $G$. Similarly, the velocity jump $\overrightarrow{K'K^*}$ is parallel to
the tangent to $GK$ at $K$. This jump propagates along the radius $KO$ of the
slip-line fan and terminates at $O$. The region $G^*K^*A'$ formed by a net

of lines orthogonal to the slip-lines in the triangle *GKA*, determines the velocity field in this region.

Similarly as in the case of a perfectly smooth die, there are certain reduction ratios for which there cannot be constructed a kinematically admissible velocity field associated with the type of slip-line net discussed here. The range of these non-admissible reductions gets smaller as the unit friction force *t* on the contact line increases. One can suppose that for these reduction ratios slip-line nets with a curvilinear starting slip-line *AO* will lead to kinematically admissible solutions.

Following the procedure described in this section one can construct the slip-line field for any arbitrary law of friction *t(s)* on the segment *DF* of the contact line. Solving the mixed boundary value problem in *FED* we have now $\varphi = \gamma + 0.5 \arcsin(t/k)$ on the segment *DF*.

When the friction force *t* is proportional to the normal stress, the problem can be solved by means of the method of successive approximations. As the first approximation the solution which neglects friction may by taken. Knowing the distribution of $\sigma_n$ along the die-sheet interface, we can calculate the first approximation of the distribution of the frictional forces $t' = |\mu\sigma_n'|$. This distribution permits us to compute the second approximation by solving the problem with a known variation of *t* along the contact line. Having calculated the second approximation we may find the corrected distribution of *t*, and then start to compute the third approximation and so on. This procedure proved to be quickly converging also when applied to other problems.

## 6.4   Sheet drawing with small reduction in thickness

The slip-line fields discussed in the foregoing sections do not apply when the reduction ratio is smaller than that given by (6.3) for the case of frictionless drawing. The solution to the problem of drawing with small reductions in thickness is due to R. Hill and S. J. Tupper [55]. Suppose that the die is perfectly smooth. Since, as in the previous cases, the number of data is not sufficient to formulate the boundary value problems for the stress field, we begin the solution by assuming in advance certain starting slip-lines. In the next section we shall discuss a general method of solution in which we begin the procedure by constructing the velocity hodograph. In this method no assumptions concerning the geometry of slip-lines are necessary. However, in most practical cases one simplifies the solution by assuming that certain geometrical features of the slip-line net hold, and then checking whether all static and kinematic requirements are fulfilled. This applies particularly to the cases where the die profile is straight.

Returning to the drawing through a smooth wedge-shaped die, we assume in advance that the distribution of contact pressure on *AB* is uniform (Fig. 109). The unknown value of this pressure will be denoted by *p*.

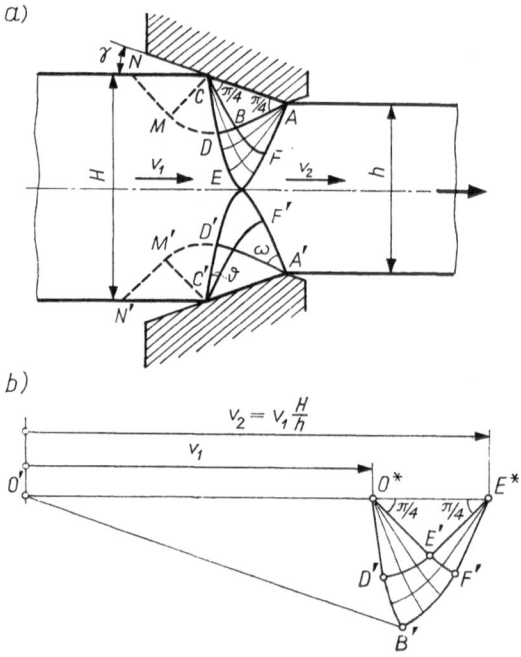

Fig. 109. Sheet drawing with small reduction in thickness; a) slip-line field (R. Hill and S. J. Tupper [55]), b) velocity hodograph

Regarding *p* as the constant parameter, we have the elementary Cauchy boundary value problem in the triangle *ABC*. The slip-line net in this triangle consists of two families of straight lines intersecting *AC* at 45°. The angles $\vartheta$ and $\omega$ of the two centred fans *BCD* and *BAF* will be determined later. The slip-lines *BD* and *BF* define the slip-line net in the region *BDEF* (compare Fig. 50). The slip-lines *DE* and *EF* must meet the centre line at 45°. This symmetry requirement determines the angles $\vartheta$ and $\omega$.

166

It can easily be shown that the deforming region, bounded by the outer slip-lines *AFE*, *CDE* and their counterparts below the symmetry axis, satisfies all kinematic conditions. Let the speed of the sheet at the exit be $v_2$. The preservation of volume requires the speed at the entry to be $v_1 = v_2 h/H$. Hence the normal velocity components on the slip-lines *CDE* and *AFE* are known. In Fig. 109b there is presented the velocity hodograph for the upper part of the strip. The velocities to the right of the discontinuity line *DE* are determined by the arc *D'E'*. Similarly, the other discontinuity line *FE* is mapped onto the arc *F'E'* in the velocity plane. The triangle *ABC* moves as a rigid whole with the speed represented by the vector $\overrightarrow{O'B'}$ parallel to the wall *AC*. The region *BDEF* is mapped onto *B'D'E'F'*.

When no back-pull is applied to the left-hand part of the strip, the resultant force of stresses acting on *CDE* and *C'D'E* must be equated to zero. This condition determines the contact pressure *p* on *AC*.

For a given thickness ratio the range of admissible values of the die angle $\gamma$ is limited, because when $\gamma$ is too large, the material deforms plastically according to the scheme shown by dashed lines in Fig. 109. The deforming region is bounded by the discontinuity line *ABDMN* and its counterpart in the lower part. Plastic flow in this region is similar to that for the indentation problem.

When the friction forces on the sheet-die interface are accounted for, the slip-line field is similar to that in Fig. 109. The angle *CAB* has the value $\frac{1}{4}\pi - \delta$, where $\delta$ is defined in the same way as in the foregoing section.

## 6.5 Dynamic effects in sheet drawing

In the foregoing sections we have considered sheet drawing as a quasi-static process in which the inertia forces have been neglected. Computational difficulties have forced us to neglect also the strain-hardening of the material and its sensitivity to the rate of deformation. As we have already observed, a dynamic problem of plane flow reduces to the solution of the system of five equations (4.1), (4.7) and (4.8) in five unknowns $\sigma_x$, $\sigma_y$, $\tau_{xy}$ and $v_x$, $v_y$. As we have stated in Chapter 4, an explicit solution of such a system of equations is at the present still unknown. However, the problem of drawing can be treated in an approximate way as a dynamic one, as presented below.

167

Assume a cylindrical coordinate system $r, \theta$ with the origin at $O$ (Fig. 106). Let the velocity $v_0$ at the exit from the die be a given function of the time $t$. Thus we have $v_0 = v_0(t)$. If there is no friction on the die-sheet interface (perfectly smooth die), we may assume that $\tau_{r\theta} = 0$ everywhere, and that all the significant magnitudes depend on the radius $r$ and the time $t$ only. At every point the radial stress $\sigma_r(r, t)$, the circumferential stress $\sigma_\theta(r, t)$ and the radial velocity $v(r, t)$ are the three unknowns to be found. We have a system of three equations which consists of the equation of motion [compare (2.20)]

$$\frac{\partial \sigma_r}{\partial r} - \frac{\partial \sigma_\theta - \partial \sigma_r}{r} - \varrho \left( \frac{\partial v}{\partial t} + v \frac{\partial v}{\partial r} \right) = 0, \tag{i}$$

the incompressibility condition

$$\frac{\partial v}{\partial r} + \frac{v}{r} = 0, \tag{ii}$$

and the yield condition [compare (4.6)]

$$\sigma_r - \sigma_\theta = 2k. \tag{iii}$$

By integrating (ii) with the boundary condition $v = -v_0(t)$ for $r = r_0$ we obtain the velocity distribution

$$v = -\frac{r_0}{r} v_0(t). \tag{iv}$$

By substituting (iii) and (iv) into the equation of motion (i) we get an ordinary differential equation with one sought for function $\sigma_r$. This equation can be integrated in the elementary way. However, certain difficulties arise connected with the determination of the integration constant. Suppose that the sheet to the left of the die has mass $M$ per unit width. If now the drawing velocity undergoes change defined by the acceleration of drawing $dv_0/dt$, then the acceleration of the mass $M$ must be equal to $(r_0/R_0)(dv_0/dt)$, as required by (iv). Such an acceleration of the sheet is possible if along the cross-section $DE$ at the entry ($r = R_0$) there is applied a stress of the magnitude

$$\sigma_r = \frac{M}{H} \frac{r_0}{R_0} \frac{dv_0}{dt}.$$

If the sheet is long enough, the required magnitude of this stress may be quite large. Thus rapid changes in the drawing velocity may lead to a significant increase of the stresses in the deforming region and even to a fracture of the drawn sheet.

Thus for solving specific problems we must know the mass $M$ of the sheet to enter the die. When the end of the sheet approaches the entry into the die, we have the boundary condition $\sigma_r = 0$ for $r = R_0$, and integrating the equation of motion we arrive finally at the expression

$$\sigma_r = 2k \ln \frac{R_0}{r} + \frac{\varrho v_0^2}{2}\left(\frac{r_0^2}{r^2} - \frac{r_0^2}{R_0^2}\right) + \frac{dv_0}{dt}r_0 \varrho \ln \frac{R_0}{r}. \qquad (v)$$

The circumferential stress $\sigma_\theta$ can be obtained directly from (iii).

The first term in (v) represents the quasi-static solution. The second term represents the influence of the acceleration of material particles occurring inside the deforming region, which is non-zero even for the steady-state ($v_0 = $ const) process. It is readily seen that this term has a significant magnitude only for very fast drawing speeds, which are rarely used in practice. Let us compute the so-called *limit reduction ratio*, obtained from the condition that the stress $\sigma_r$ reaches for $r = r_0$ the value $2k$. For a steady-state process this leads to the equation

$$\ln\left(\frac{R_0}{r_0}\right)^* - 1 - \frac{\varrho v_0^2}{4k}\left\{1 - \left[\left(\frac{r_0}{R_0}\right)^*\right]^2\right\} = 0.$$

If the dynamic term is neglected, we obtain the limit reduction ratio $[(H-h)/H]^* \approx 0.632$. This value changes by 5 percent assuming the value 0.600 for very fast drawing speeds $v_0 = 1800$ m/min for the copper, and $v_0 = 2500$ m/min for a mild steel. Thus it is evident that the quasi-static approach assumed in the foregoing sections is fully justified.

The third term in (v) shows how the stresses change due to rapid changes in the drawing velocity. Let us examine the value of the stress $\sigma_r$ at the exit. Neglecting the second term in (v) we obtain

$$\sigma_r = 2k\left(1 + \frac{\varrho}{2k}\frac{h}{2\sin\gamma}\frac{dv_0}{dt}\right)\ln\frac{R_0}{r_0} \qquad \text{for} \qquad r = r_0.$$

It is seen that with increasing thickness of the sheet the influence of the variation of the drawing velocity gains in significance. Figure 110 shows for the reduction ratio $(H-h)/h = 0.5$ how the stress at the exit depends on the thickness and acceleration. For a wide range of acceleration its influence is negligible. In such cases only the inertia force of the sheet outside the die should be taken into account.

169

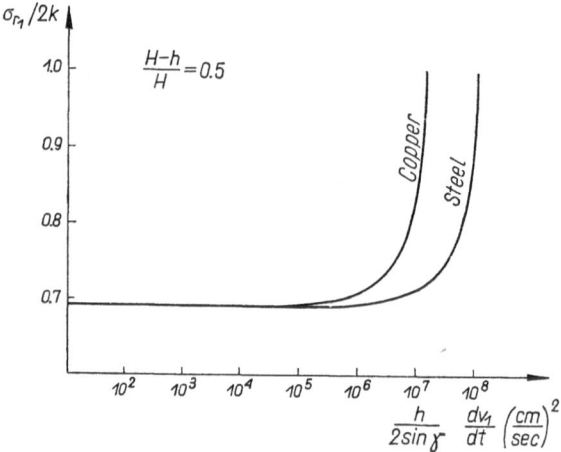

Fig. 110. The relationship between drawing stress, thickness $h$, and acceleration, in the final stage of drawing

Let us examine now how the stresses are affected by the viscous properties of the material. Our considerations will be limited to the steady-state drawing with relatively slow velocity ($v_0 < 50$ m/min). Thus in the equation of motion (i) all terms containing $v$ will be neglected. Assume the simplest linear relation between the yield point in simple tension and the rate of deformation

$$\sigma = \sigma_0 + D\epsilon.$$

One can generalize this relation to the cases of complex states of stress by assuming, for instance, a fixed relation between the equivalent stresses $\sigma_i$ and the equivalent rate of strain $\epsilon_i$, depending on the given material. Thus we have $\sigma_i = \sigma_i(\epsilon_i)$. The strain rate components are

$$\epsilon_r = \frac{\partial v}{\partial r} = \frac{r_0}{r^2} v_0,$$

$$\epsilon_\theta = \frac{v}{r} = -\frac{r_0}{r^2} v_0,$$

$$\epsilon_z = 0.$$

The strain rate intensity can be computed from the general relation

$$\epsilon_i = \sqrt{\tfrac{1}{6}[(\epsilon_r - \epsilon_\theta)^2 + (\epsilon_\theta - \epsilon_h)^2 + (\epsilon_h - \epsilon_z)^2]}$$

170

analogous to (2.12). Thus, finally, we obtain

$$\epsilon_i = \frac{r_0}{\sqrt{3}} \frac{v_0}{r^2},$$

and instead of (iii) we can write

$$\sigma_r - \sigma_\theta = 2k\left(1 + \frac{2D}{3\sqrt{3k}} \frac{r_0}{r^2} v_0\right). \tag{vi}$$

Integrating now the equation of equilibrium, we obtain the expression for the radial stresses

$$\sigma_r = 2k \ln \frac{R_0}{r} + \frac{2}{3\sqrt{3}} D \frac{v_0}{r_0}\left(\frac{r_0^2}{r^2} - \frac{r_0^2}{R_0^2}\right). \tag{vii}$$

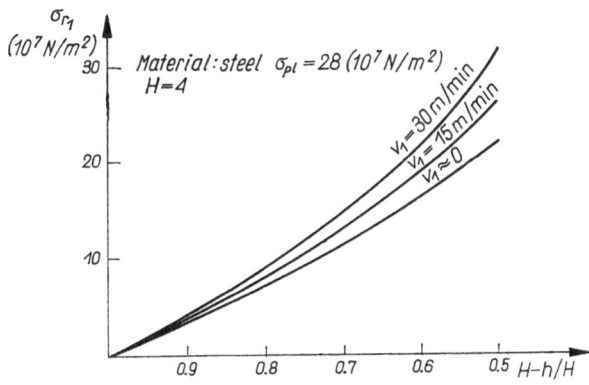

Fig. 111. The relationship between drawing stress and drawing velocity

Figure 111 shows this relation computed for a mild steel, for which the constant $D$ is approximately equal to $D = 24$ kpsec/cm$^2$.

In a similar manner the strain-hardening effect can be accounted for.

An analogous approximate approach can be used in the analysis of the mechanics of wire drawing. Instead of (i) we have then the equation of motion (2.21), and instead of (ii) we have to write

$$\frac{\partial v}{\partial r} + \frac{2v}{r} = 0.$$

The yield condition (iii) remains unchanged.

### 6.6   Drawing through a curvilinear die

All problems analysed in the previous sections were statically indeterminate, because the data were insufficient to properly formulate the boundary value problem for the stress field. This difficulty was overcome by assuming *a priori* that the contact pressure is distributed uniformly on the entire wall of the die, or on a part of it. This assumption, justified by the geometrical simplicity of the profile, proved correct, satisfying all the kinematic requirements.

If, however, the profile of the die is curvilinear such an assumption would be evidently incorrect. On the other hand, the known stress boundary conditions alone do not suffice to formulate uniquely the boundary value problems needed to define the slip-line field. We shall discuss in this section a particular example of drawing through a curvilinear die. The methods which we shall use can also be applied to analyse other problems, where the slip-line field cannot be found in advance.

Let a sheet be drawn with a small reduction in thickness through a die of an arbitrary curvilinear profile, and let the contact on the die surface be perfectly smooth. A solution to this problem has been given by V. V. Sokolovsky [115, 116]. The analytical method used by him seems to be too complex for practical purposes. But Sokolovsky's solution can also be obtained by a graphical method, beginning from the construction of the velocity hodograph, and then constructing the slip-line net. First we assume in advance the general pattern of this net (Fig. 112a), by analogy with the particular case of drawing through a wedge-shaped die (Fig. 109a). Unfortunately, the exact configuration cannot be directly defined since the distribution of the pressure on the contact line $AC$ is not known. It is readily seen, however, that the initial data concerning the velocity field suffice to construct the velocity hodograph without the need to consider the state of stress. The speeds $v_1$ and $v_2 = v_1 H/h$ of the rigid parts of the strip outside the die are known. Along the slip-lines $CDE$ and $AFE$ the velocities must suffer a jump. Thus both lines will be mapped in the velocity plane onto circular arcs with the centres at the end-points of the vectors $v_1$ and $v_2$, respectively (Fig. 112b). The radii $O^*E'$ and $E^*E'$ meet the horizontal axis of the hodograph at an angle of 45°, as required by compatibility conditions at the point $E$, where the slip-lines must intersect the axis of the strip also at 45°. The vectors of velocity jumps at $E$, represented by the vectors $\overrightarrow{O^*E'}$ and $\overrightarrow{E^*E'}$, must, therefore, meet

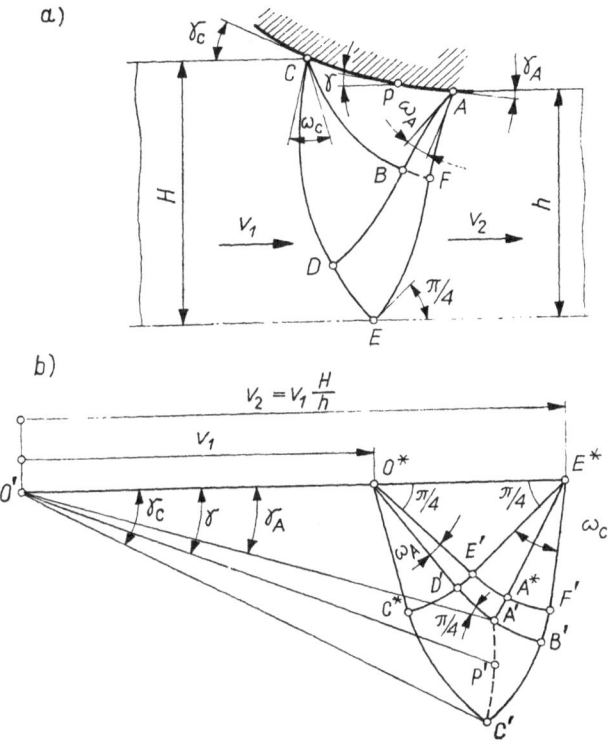

Fig. 112. Drawing through a curvilinear die; a) slip-line field (Sokolovsky [115, 116]), b) velocity hodograph

$O'E^*$ at $45°$. At this stage of solution the angles of fans at $O^*$ and $E^*$ are not defined. However, by assuming arbitrary values for these angles, we can construct the hodograph graphically, starting from the two arcs $E'C^*$ and $E'F'$. The construction can be made in the same manner as in the case of a slip-line net (compare Fig. 56), since Hencky's first theorem (see Section 4.3) applies also to the net of the velocity hodograph. To establish the hitherto undetermined geometrical parameters of the hodograph, we shall use the last remaining velocity condition which states that the flow velocity of the particles adjacent to the wall $CA$ of the die must be parallel to its profile. This condition makes it possible to find the mapping of the contact line $AC$ onto the velocity plane. In order to find the image of the point $A$, we draw through the origin $O'$ of the hodograph a straight line parallel to the tangent line to the die profile at $A$;

173

this is the line making the angle $\gamma_A$ with the axis $O'E^*$. Since frictionless contact is assumed, the slip-lines must meet the die profile at 45°. Hence, if the velocity is to be directed tangentially to the wall, its components in the $\alpha$ and $\beta$ directions must be equal to each other. Thus we have $v_\alpha = v_\beta$ on $AC$. Therefore the end-point $A'$ of the velocity vector $\overrightarrow{O'A'}$ must be situated at a point, where the straight line through the origin $O'$ inclined at the angle $\gamma_A$ intersects the lines of the hodograph net at 45°. The point $A'$ is in this way uniquely determined, and it can be found with sufficient accuracy provided the net of the hodograph is dense enough. In the same manner, the image of the point $C$ can be obtained by plotting through $O'$ a line inclined at $\gamma_C$ to $O'E^*$. The point $C'$, where $O'C'$ intersects the lines of the mesh at 45°, defines the flow velocity at $C$. Thus $C'F'E^*$ and $C'C^*O^*$ represent the bounding lines of the net of the hodograph. In the same way the image of any arbitrary point $P$ situated on $AC$ can be found. Therefore the segment $A'P'C'$ represents the contact line $APC$.

The curvilinear triangle $A'C'B'$ corresponds to the triangle $ACB$ of the slip-line field. Knowing the net of lines in the velocity plane, we can now construct the slip-line net in this region in the physical plane, by choosing first a number of points on $AC$ and then finding their images on the arc $A'C'$ in the plane of the hodograph. Then we construct, step by step, the slip-lines in the region $ABC$, which must be orthogonal to the respective lines of the hodograph.

There is a singularity at $A$, where the velocity changes from the value defined in the velocity plane by $A'$ to the value given by $A^*$. The curvilinear triangle $ABF$ is mapped onto the region $A'A^*F'B'$. Since the net of the hodograph has the same geometrical properties as the slip-line net, the angle between the tangents to the hodograph lines at $A'$ and $A^*$ is of the same magnitude as the angle at the vertex $A$ of the fan $E'O'D'$. Thus its magnitude is $\omega_A$. Now the net of slip-lines in the region $ABF$ can easily be constructed graphically by plotting step by step the slip-lines orthogonal to the respective lines of the hodograph.

The region $B'C'C^*D'$ corresponds to the slip-line fan $BCD$ in the physical plane. It is readily seen, therefore, that certain points of the region $A'B'C'$ represent the velocities of two different material particles of the strip. We shall see later that this takes place also in the region $A'A^*F'B'$. The angle $\omega_C$ of the fan $BCD$ at $C$ is of the same magnitude as the angle $E'E^*F'$ in the velocity plane. Now, knowing the angle $\omega_C$, we can easily construct the mesh of slip-lines forming the region $BCD$.

The curvilinear quadrangle $BFED$ is mapped onto the region $B'F'E'D'$. Its part $A'A*F'B'$ represents also the velocity field inside the triangle $ABF$. On the basis of the just found slip-lines $BD$ and $BF$, one can complete now the solution by constructing the mesh for the region $BDEF$. The point $E$ should be situated on the axis of the strip. This condition may be used to check whether our solution is correct.

We can find the stresses, the pressure on the die-strip interface and the drawing force by assuming, for instance, that at the point $E$, $\chi_E = \chi_0$ and $\sigma_0 = 0$ [see (4.10)]. Having already found the net of slip-lines, we compute the auxiliary function $\chi$ at the nodal points of the net. The unknown value $\chi_0$ is treated at this stage of computation as a constant parameter. Then the stresses can be determined. The value of $\chi_0$ can be found by equating to zero the resultant force of stresses acting on $CDE$, provided there is no back-pull.

Note that the drawing through a die with a straight wall (Section 6.4) constitutes a particular case of the general solution. The image of the die profile in the velocity plane reduces to a single point, radically simplifying the solution. The assumption that the slip-line $AB$ is straight, which has been previously introduced *a priori*, is now obtained as a consequence of the rectilinearity of the die profile.

The graphical method demonstrated here is very simple. However, it lacks the generality of the analytical method used in [115, 116]. The latter permits to establish the relation between the reduction ratio, and the limiting values of the angles $\gamma_A$ and $\gamma_C$ for which the slip-line field of the type shown in Fig. 112a is still valid. The angle $\omega_A$ becomes zero when

$$\frac{H-h}{H} = 2\sin\gamma_A,$$

and similarly for

$$\frac{H-h}{H} = \frac{2\sin\gamma_C}{1+2\sin\gamma_C}$$

the angle $\omega_C$ at the point $C$ vanishes. For given values of $\gamma_A$ and $\gamma_C$ we get from these relations the range of reduction ratios for which this type of solution can be constructed.

A working example has been solved for a particular case where the die profile is formed by a circular arc of radius $R = 0.475h$, and the

175

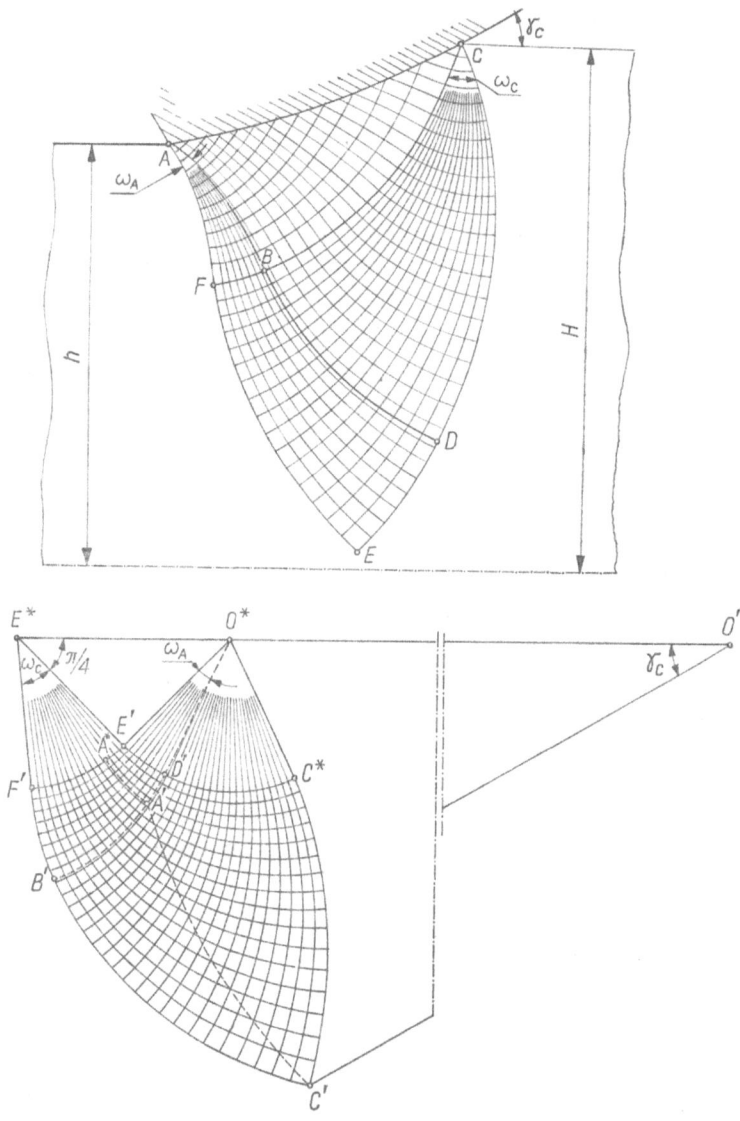

Fig. 113. Drawing through a curvilinear die; a) slip-line net, b) velocity hodograph

reduction ratio is $(1 - h/H) = 0.2$ (Fig. 113). The slope of the die profile at the points $A$ and $B$ is defined by the angles $\gamma_A = 10°$ and $\gamma_C = 30°$, respectively. Both nets have been constructed graphically with angular

176

intervals between the lines equal to 2° 30'. A numerical solution to this problem has been given in [115].

It is readily seen that the point $E$ is not situated on the axis of the strip (Fig. 113a). B. A. Drujanov [29] has shown that such an inconsistency is due to the guessing in advance of the type of slip-line net, which evidently should be somewhat different for the assumed profile of the die. Nevertheless for practical purposes our solution is sufficiently accurate.

The case where the angle $\gamma_A$ (comp. Fig. 112a) at the exit is zero has been examined by T. C. Firbank and P. R. Lancaster [36] and B. A. Drujanov [31]. In the first work the problem was solved by a procedure of successive approximations, while in the latter an analytical treatment of the problem was presented.

### 6.7  Extrusion operations

Let us examine the extrusion of a billet of plastic material held in a container and forced through a die. In the vicinity of the die hole there appears a region characterized by large plastic deformations. The description of the mechanics of such processes is due to R. Hill [47]. An extensive and detailed treatment of this subject is given in the book by W. Johnson and H. Kudo [66].

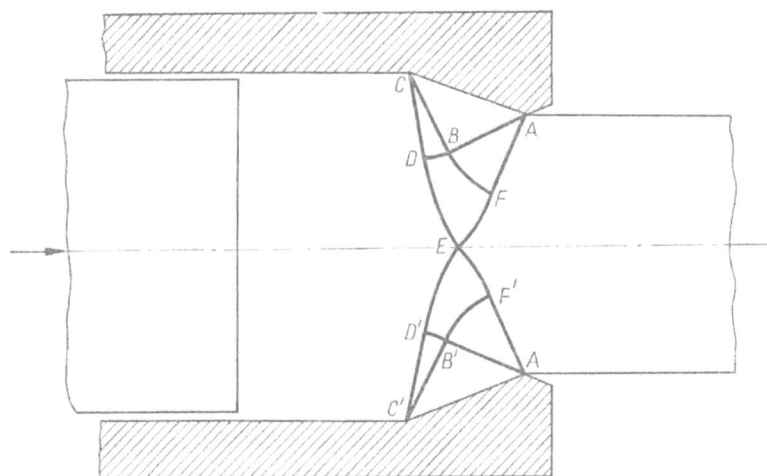

Fig. 114. Slip-line field for extrusion through a smooth wedge-shaped die (R. Hill [47])

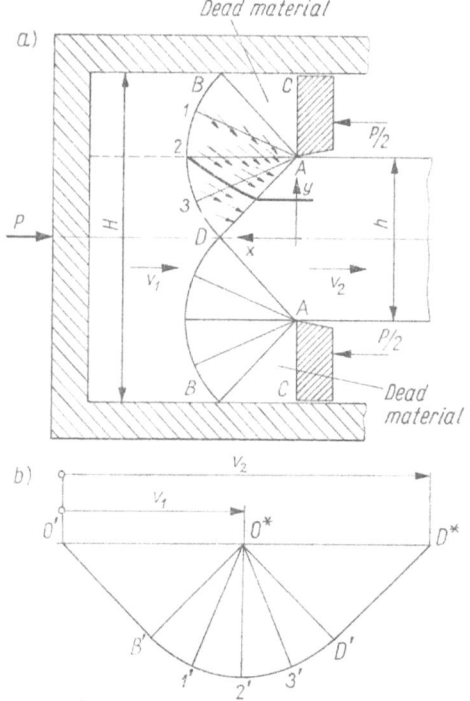

Fig. 115. Inverted extrusion; a) slip-line field, b) velocity hodograph

For a wedge-shaped die, the slip-line net (Fig. 114) has a form identical
to that for the sheet drawing (comp. Fig. 109) provided that in both cases
the friction on the contact surface is neglected. The stress distribution
will of course be different because the boundary conditions are different.
In the exit section the resultant force must be zero. The pushing force
from the left-hand side must equate the stresses acting on the contact
surfaces. We shall not examine this problem in detail since the process
is similar to that of strip-drawing. The velocity field is identical in both
cases. Thus the hodograph shown in Fig. 109 holds valid also for the
extrusion process.

But the solution is completely different when the hole has not the
shape of a wedge and the reduction ratio is great. Let us examine first
a particular case of a 50 percent reduction in thickness. The slip-line field
presented in Fig. 115 has been given by R. Hill under the assumption that
there is no friction between the billet and the wall of the container. This

178

is particularly true for inverted extrusion, as shown in the figure, when the container is closed from one side and the material is extruded by a die with a hole. The slip-line net is composed of two centred fans $DAB$, covering the deforming region. The material contained in the triangles $ABC$ adjacent to the die remains rigid and moves together with the die. These regions of the so-called *dead metal* form a kind of wedge-shaped exit through which the metal is extruded. However, in the present case the edges of such an exit coincide with the slip-lines $AB$, and therefore the tangential stress there must be equal to the yield locus in pure shear $k$. There is a velocity jump propagating along these lines since the material situated to the right of them moves together with the die and the velocity of the particles on their left-hand side is parallel to $BA$.

We begin the determination of the stress field from the condition that the resultant force acting on the extruded material to the right of $ADA$ is zero. The stresses in the triangular region $ADA$ are assumed to be $\sigma_x = 0$ and $\sigma_y = -2k$. Hence from (4.10) we get $\chi = -0.5$ for $\sigma_0 = 0$. The $\sigma_y$ component is a principal one and therefore $\varphi = 0$ in $ADA$. Circular slip-lines in the upper fan belong to the first family of $\alpha$-lines along which the relation $\chi + \varphi = \text{const}$ must hold. Thus we obtain $\chi = -(\frac{1}{2} + \frac{1}{2}\pi)$ on the slip-line $AB$ where $\varphi = \frac{1}{2}\pi$. Now from (4.10) we calculate the stresses on $AB$:

$$\sigma_x = -(2+\pi)k,$$
$$\sigma_y = -\pi k.$$

The force required to maintain the process is

$$P = (2+\pi)kh.$$

The velocity field can be determined by solving a characteristic problem for the region $DAB$. Along the slip-line $DB$ the normal velocity component is equal to the normal component of the rigid dead zone shifted horizontally to the left (the die is assumed to be immovable, and the container moves to the right). Along the slip-line $AB$ the normal velocity component is zero. These data suffice to find the velocity distribution within the deforming region $DAB$. The lines $BDA$ are the lines of velocity discontinuity.

The velocity hodograph is shown in Fig. 115b. The velocities of particles situated on the straight slip-lines $AB, A1, A2, A3$ and $AD$ correspond to the vectors $\overrightarrow{O'B'}, \overrightarrow{O'1'}, \ldots, \overrightarrow{O'D'}$ in the velocity plane. Directions of motion are schematically shown in the figure by arrows. The trajectories of the particles may be found graphically by drawing curves tangent to the directions of motion. Now by calculating the distances covered in equal times by nodal points of a square grid marked on a section of the initial billet, we can find the distortion of the grid (Fig. 116a).

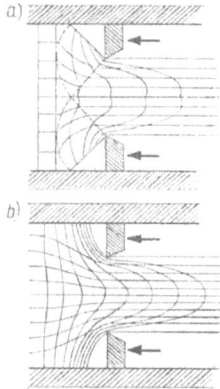

Fig. 116. Theoretical (a) and experimental (b) distortion of a square grid in inverted extrusion (after N. W. Purchase and S. J. Tupper [100])

The actual distortion of such a grid (Fig. 116b) obtained experimentally for extrusion of lead with the slightly smaller reduction of 46.7 percent [100] is similar to that calculated theoretically. However, the actual deforming region is larger than that predicted by the theory. This is rather obvious since, in a real metal displaying strain-hardening effect, the appearance of distinct boundaries between the rigid and plastic zones is impossible. The dead metal zones in the corners between the die and the wall of the container are smaller than those predicted by the theoretical solution. No sharp bends of the lines of the grid on the symmetry axis are experimentally observed.

For larger reductions $(H-h)/H > 0.5$, the net of slip-lines takes another form (Fig. 117a). It consists of centred fans $BAC$ and of parts of the type shown in Fig. 50. The direction of the radius $AC$ of the fan is defined by the condition that the slip-line $CD$ meets the wall of the container at 45° if the tangential stress is assumed to be zero there. We begin computations from the region $ABA$ where, as in the previous example, $\chi = -0.5$ and $\varphi = 0$. Then the values of $\chi$ may be determined

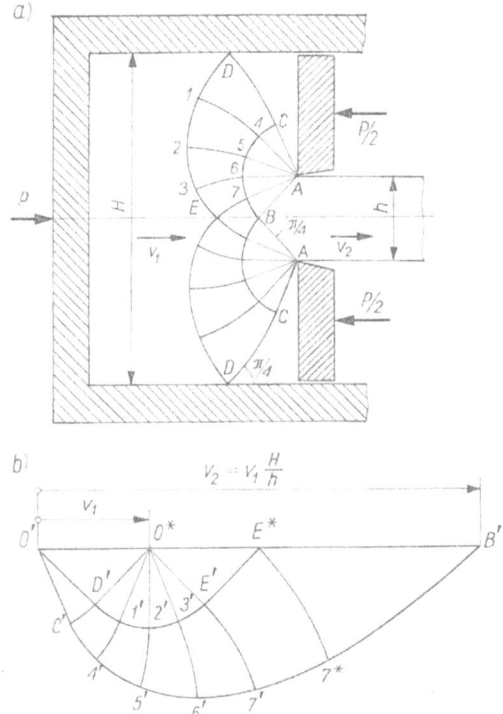

Fig. 117. Inverted extrusion with large reduction in thickness; a) slip-line field, b) velocity hodograph

along *BC*. Now the characteristic problems for the region *BE*7 and *CDE*7 can be solved. Positions of the nodal points of the net can be taken from Table 1 given in Chapter 4 (pp. 78-79). The net may be also determined by a graphical method. Thus the computations reduce to the determination of $\chi$ from the relations $\chi + \varphi = $ const and $\chi - \varphi = $ const, which must be fulfilled along the slip-lines.

Figure 117b presents the velocity hodograph for the upper half of the billet. Assume that the die with the hole is fixed and that the container moves with the speed $v_1$. The line of discontinuity *DE* is mapped onto the circular arc *D'E'* with centre at *O\**. Similarly, the discontinuity line *DCA* is represented by the arc *C'D'* with centre at *O'*, since the dead material to the right of *DCA* is immovable. Both sides of the line of discontinuity *E*7*A* are mapped onto the arcs *E'*7' and *E\**7\*. The velocity jumps at $E$ and 7 are represented by the segments $E'E^*$ and $7'7^*$, respectively.

For reduction ratios $(H-h)/H < 0.5$ the slip-line net may be assumed in the form shown in Fig. 118. Note that elements of this net are identical

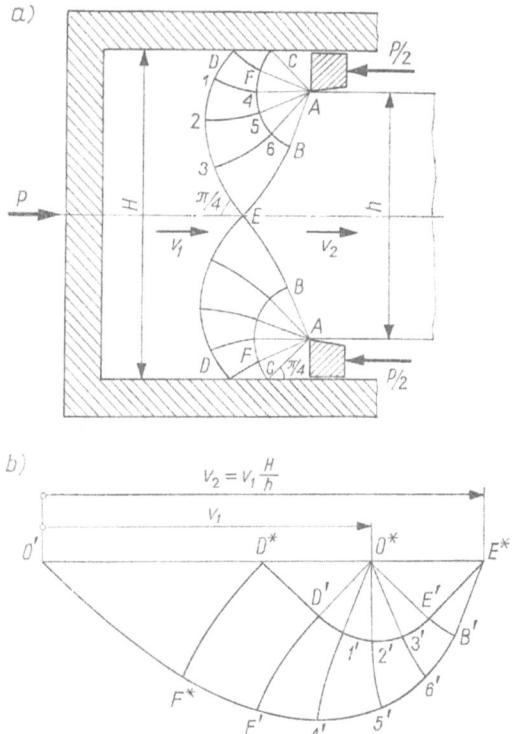

Fig. 118. Inverted extrusion with small reduction in thickness; a) slip-line field, b) velocity hodograph

with those of the net for large reductions (Fig. 117a). However, they are situated in a completely different manner relative to the axis of symmetry. The direction of the radii $AB$ is defined by the condition that the slip-lines $BE$ intersect the longitudinal axis at 45°. We begin the computations by assuming an arbitrary value for the auxiliary function $\chi$ at $E$ or at $C$. Then, having calculated the stresses with $\chi_0$ as a parameter, we find its value by requiring that the longitudinal resultant force acting on the extruded material is zero.

If there is no friction between the surface of the die and the extruded material, the other type of slip-line field is possible for large reduction ratios (Fig. 119a). The particular case of such a solution for the reduction $(H-h)/H = 2/3$ has been examined by E. H. Lee (see [97]), and generalized

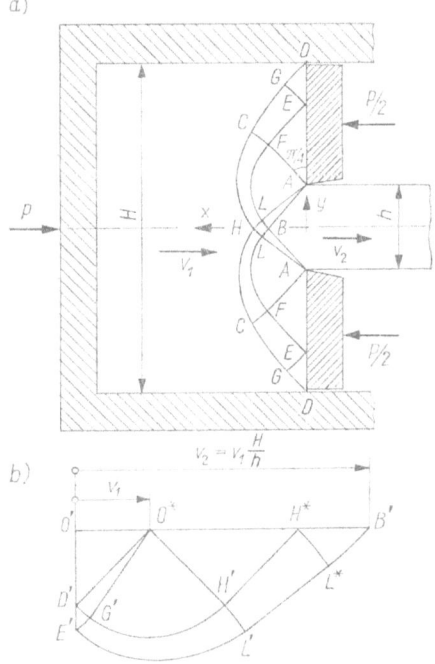

Fig. 119. Inverted extrusion through a smooth die; a) slip-line field, b) velocity hodograph

for larger reductions by W. Johnson [63]. The sequence of solutions of boundary value problems is similar to that for the foregoing case. But now a mixed boundary problem defined by frictionless contact on $AD$ ($\varphi = \frac{1}{2}\pi$) must be solved in the region $AFCD$. In contrast to the solution shown in Fig. 117 we have at present no dead material zones at the corners between the die and the wall of the container. The slip-lines $ALHCGD$ represent the lines of velocity discontinuity. In Fig. 119b there is presented the velocity hodograph for the upper half of the billet, constructed under the assumption that the die does not move. Thus the velocities of the points situated along the contact surface $AED$ are parallel to it.

183

W. A. Green [42] gave the slip-line field solution for smaller reduction ratios between 50 and 66.7 percent under the assumption that the die face and the container wall are smooth. The starting slip-lines are not straight as the lines *AB* in Fig. 119, but they are curved in order that the velocity discontinuities initiated at the corners between the die and the wall may reach the exit at its corner. These solutions were also examined in detail by I. F. Collins [18] who used the operational technique which proved very effective.

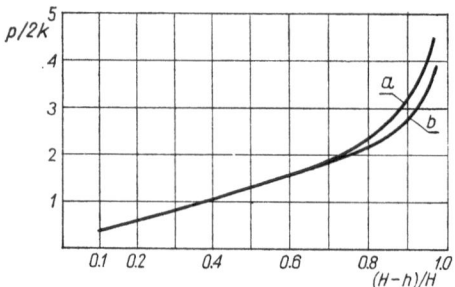

Fig. 120. The relationship between the mean pressure and the reduction ratio for the solution shown in Fig. 117 (curve *a*) and for the solution shown in Fig. 119 (curve *b*)

The two types of solution (Figs. 117 and 119) give different estimates of the mean pressure $p = P/H$ inside the container. The line *a* in Fig. 120 represents the pressure for the slip-line field shown in Fig. 117. The other slip-line field (Fig. 119) gives a smaller value for the pressure (curve *b* in Fig. 120). The first type of solution (Fig. 117), which holds valid also in the case where the face of the die is rough, is as yet only kinematically admissible, since we do not know the extension of the stress field into the zones of dead material at the corners. Thus it furnishes only an upper bound on the mean pressure. The second type of solution (Fig. 119) is complete, because the statically admissible extension of the stress field can be constructed.

This extension, due to J. M. Alexander [1], is based on the concept proposed by J. F. W. Bishop [10] and used by him for the analysis of other problems. Some applications of Bishop's procedure were presented in Sections 5.2 and 5.3 of the foregoing chapter.

To begin constructing the extension of the slip-line field bounded by

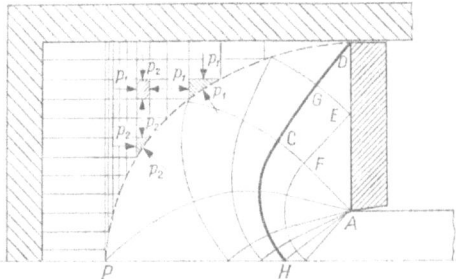

Fig. 121. Extension of the slip-line field into the rigid region (J. M. Alexander [1])

the line *DGCH* (Fig. 121), we start from the mixed boundary value problem, defined by the data on the slip-line *DGCH*, and by the value of the angle $\varphi = 0$ on the symmetry axis *HP*. Next, beginning from the corner *D*, we plot the trajectory of principal stresses, which is uniquely defined, for it must intersect the characteristics of the extension at the angle of 45°. Having found the trajectory, we divide it into a number of very small segments, along which the normal stresses may be assumed to preserve constant values. These segments are supported by columns and triangular elements. Let the triangular elements be loaded by hydrostatic compression of magnitude equal to the principal stress on the trajectory *DP*. In the columns we have biaxial compression which never exceeds the yield condition. When the number of the rectangular elements is increasing, our stress system tends to a continuous one satisfying all static conditions. Thus the solution is complete since the rigid material in the container is able to sustain the stresses on *DCH*. The extension into the rigid extruded zone can be done in the elementary way, similarly as in the case of strip drawing.

Such an extension can be constructed also for other reductions, in particular for the 50 percent reduction. Thus the slip-line solution shown in Fig. 115 is also complete.

The above discussed solutions hold valid also for direct extrusion in the cases where the wall of the container is perfectly smooth. But if the friction on the wall cannot be neglected, solutions for direct extrusion are different than those for the inverted process. For large reduction ratios the slip-line net is similar to that shown in Fig. 117. However the slip-lines *CD* meet now the wall at an angle $\vartheta < 45°$. The value of $\vartheta$ depends on the coefficient of friction $\mu$ if the frictional stress on the wall is $t = \mu\sigma_n$,

185

where $\sigma_n$ is the stress normal to the wall. We shall not discusss here this subject in detail, because we have already shown in Chapter 5 and in the foregoing sections of the present chapter how the frictional forces can be accounted for in the case of compression and in drawing operations. It should be emphasized, however, that the computation procedure is very laborious since the normal stress on the wall is obtained from the integral condition, requiring the resultant force at the exit to be zero. Therefore we do not know the angle $\vartheta$ in advance. We met a similar difficulty when solving the sheet-drawing process (see Section 6.3). The problem may be solved by the procedure of successive approximations assuming for instance the solution for frictionless contact, as the first approximation. Having found the contact pressure $\sigma_n$, we compute the first approximation of the frictional force and then the angle $\vartheta$. Next, the new slip-line net can be constructed, meeting the wall at the angle $\vartheta$, and so on, until the difference between two consecutive approximations becomes sufficiently small.

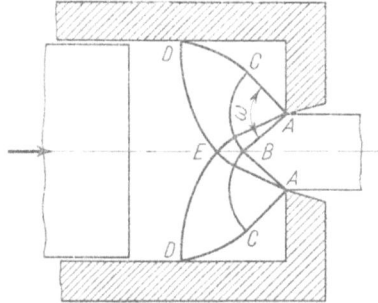

Fig. 122. Forward extrusion from a rough container

If the friction between the billet and the container's wall is so great that the frictional stress reaches the value $t = k$, the slip-lines $CD$ meet the wall tangentially. Then the slip-line net can be constructed directly (Fig. 122). It consists of the centred fans $ABC$ and of the part of the net shown in Fig. 50.

## 6.8   Piercing

Let a billet of a metal be held in a container and indented by a punch. When the foot of the punch is flat, the slip-line field (Fig. 123) is very similar to that for the extrusion process (comp. Fig. 117). Also the analysis of stress and velocity distributions is similar. The deforming region is

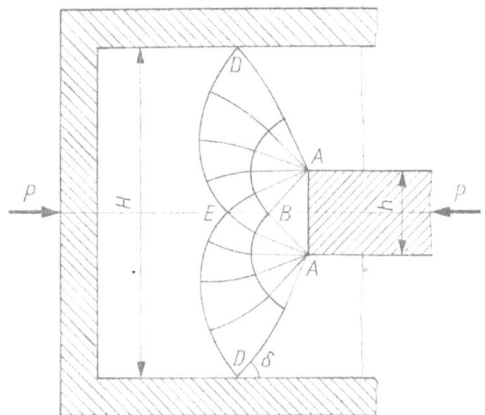

Fig. 123. Slip-line field for piercing (R. Hill [50])

bounded by the outer slip-lines which simultaneously represent the lines of velocity jump. The triangular region *ABA* does not deform and it remains attached to the punch. When the wall of the container is perfectly smooth (or lubricated) the slip-lines meet the wall at the point *D* at an angle of 45°. In the other case, when the contact is not perfectly smooth, they make with the wall an angle different than 45°.

In Fig. 124 there are presented slip-line nets for the ratio $h/H$ close to unity. The net shown below the longitudinal axis is constructed for the frictionless contact on the wall ($t = 0$). The upper net applies to the case when the frictional stresses on the wall reach the maximum possible value $t = k$. Numerous examples of slip-line solutions for various ratios $h/H$ and different friction conditions were given by W. Johnson and H. Kudo [100]. The reader will find there also solutions to the problems of piercing by a wedge-shaped punch.

All solutions mentioned above were concerned with piercing by punches with a contour formed by straight lines. As we have seen, the method of

187

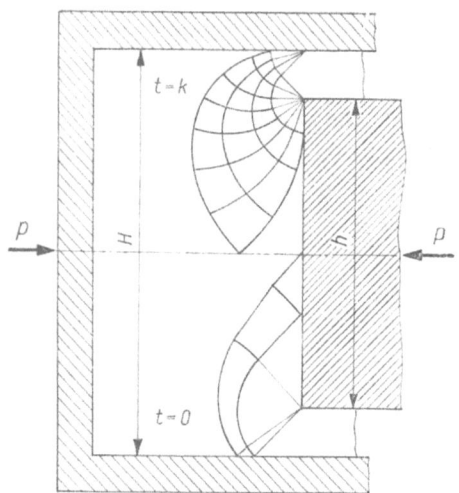

Fig. 124. Two solutions for piercing. Slip-line field below horizontal axis refers to the case of smooth punch and container. Above the axis is shown a slip-line field for rough punch and container

solution, in which a slip-line field has been guessed, and next examined whether it is compatible with the kinematical conditions, proved very effective in such cases. However, this method fails when the contour of the punch is curvilinear.

A solution for the cases when the punch is curvilinear was given by V. V. Sokolovsky [117]. The problem was solved by an analytical method used in the previous paper [115] for the problem of sheet-drawing. However, we shall examine here the problem by means of the graphical method similarly as we have done in the case of drawing.

Let the punch with curvilinear contour $AB$ be fixed, and let the container in which the billet is held move to the right with speed $v_1$. The velocity of the material in between the wall of the container and the punch is $v_2 = v_1 H/(H-h)$ because of incompressibility. Let us further assume that there is no friction on the walls of the container or of the punch.

We begin constructing the solution by guessing the scheme of the slip-line net, shown by heavy lines in Fig. 125a. The exact configuration of the net cannot be calculated directly, because the distribution of the contact stresses on the foot of the punch is not known. But the velocity hodograph (Fig. 125b) can easily be constructed.

188

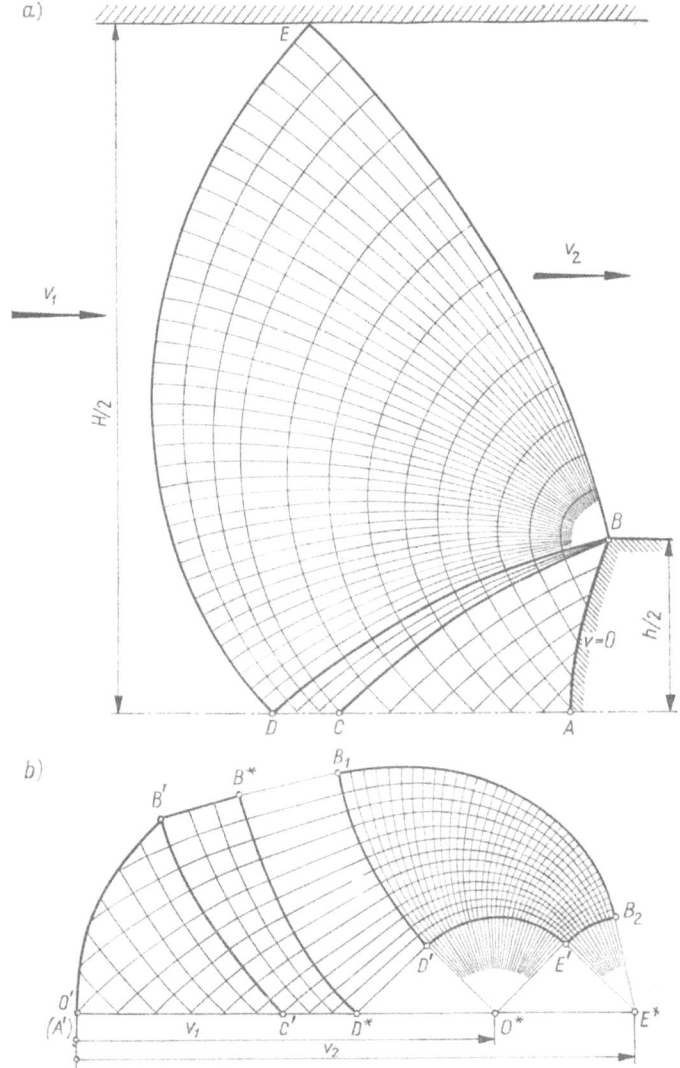

Fig. 125. Piercing by a curvilinear punch; a) slip-line field, b) velocity hodograph

The velocities $v_1$ and $v_2$ determine the positions of the points $O^*$ and $E^*$. Circular arcs $D'E'$ and $E'B_2$, their centres being located at $O^*$ and $E^*$, represent the velocities on the lines of discontinuity $DE$ and $EB$. The angle $E'O^*D'$ is obviously equal to $\frac{1}{2}\pi$, since the velocity jumps at

189

$E$ and $D$, represented in the velocity plane by the vectors $\overrightarrow{O^*E'}$ and $\overrightarrow{O^*D'}$, must meet the longitudinal axis of the hodograph at 45° (we have assumed perfectly smooth contact at $E$; hence $DE$ intersects the wall at 45°). The angle $E'E^*B_2$ of the centred fan is not known in advance. Assuming its value arbitrarily, we can construct graphically the net of lines within the region $D'E'B_2B_1$ mapping the triangular region $DEB$ into the velocity plane. There is a singularity at $B$, where velocity changes from that determined by the vector $\overrightarrow{O'B_2}$ to the other given by $\overrightarrow{O'B_1}$. Thus the point $B$ is mapped onto the arc $B_2B_1$. $DB$ is the line of velocity discontinuity. Its upper side is represented in the velocity plane by the line $B_1D'$, whereas the lower side is mapped onto $B^*D^*$. The segments $D'D^*$ and $B_1B^*$ represent velocity jumps at $D$ and $B$, respectively. $D'B_1$ and $B^*D^*$ are equidistant lines, the distance between them being equal to the velocity jump. The remaining part of the hodograph can be constructed graphically in the same way as the slip-line nets.

Having constructed the net of the hodograph, we have to find the image of the contour of the punch $AB$. Since the velocities of the particles adjacent to $AB$ must be directed tangentially to the punch surface, the velocity components $v_\alpha$ and $v_\beta$ directed along the slip-lines must be equal to each other (compare Section 6.6). For reasons of symmetry, the velocity at the point $A$ is zero. Hence this point is mapped onto the point $A'$ in the velocity plane coinciding with the origin $O'$ of the hodograph. We can begin now to determine the image of the contour $AB$ of the punch by plotting through the origin $O'$ a number of straight lines parallel to the tangents to $AB$ at chosen points. The image of each of these points in the velocity plane will correspond to that point of the hodograph at which the respective line intersects the lines of the net at 45°. In this manner the segment $O'B'$ can be found. Note that the point $B'$ uniquely determines all parameters of the hodograph, particularly the angle $E'E^*B_2$, hitherto left arbitrary.

The triangle $A'B'C'$ corresponds to the triangle $ABC$, and similarly the region $B'B^*D^*C'$ represents velocities within the triangular zone $BDC$. The change in velocity at $B$ within $BDC$ is represented by $B'B^*$.

Having found the velocity hodograph, we can construct the slip-line net according to the same procedure as that used in the case of strip-drawing through a curvilinear die (see Section 6.6). The slip-line net shown in Fig. 125a has been constructed by taking angular intervals equal to

2° 30′. It is worth noting that the accuracy obtained by the graphical method is very good.

The hodograph and the slip-line nets enable us to determine the flow velocity for any material particle. In Fig. 126 there are presented the velocity vectors in the deforming region. The streamlines and a distinct zone of slow motion near the punch are clearly visible.

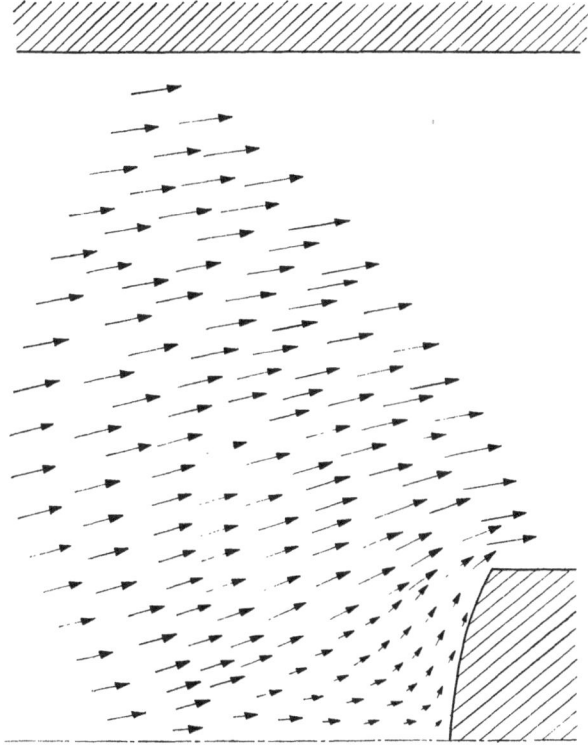

Fig. 126. Velocity field for piercing by a curvilinear punch

Note that the solution shown in Fig. 123 can be treated as a particular case of the solution discussed here. The straight contour of the punch is mapped onto a single point coinciding with the origin $O'$ of the hodograph. The net of the velocity hodograph is bounded by the line through the points $O'$ and $E^*$.

# 7

# *Some two-dimensional non-steady state operations*

## 7.1 Introduction

In this chapter we shall examine certain two-dimensional non-steady state forming operations. In these problems the stress and velocity fields are varying in time with respect to a fixed coordinate system. The analysis of deformation must be performed step by step by splitting the course of the process into a number of stages corresponding to certain chosen positions of the elements of the tool. For each of these stages, the slip-line net and the velocity hodograph are different. Next we shall assume that during the time increments between the consecutive positions of the tool the velocities do not change their magnitudes. In this way we may examine the path of deformation of an arbitrary particle of the material. This analysis can be supplemented by the determination of the forces acting on the tool at any stage of the advancing forming operation.

A certain class of such non-steady motion problems has been examined in Chapter 5. Each of these problems displayed some peculiar features simplifying the solution. In the case of wedge-indentation and flattening of a plastic wedge the slip-line net preserved geometrical similarity during the course of the process. The other example was the indentation of a plastic block by two opposite narrow punches, where the nets of slip-lines and of the hodograph were identical. However, in the case of indentation of a plastic half-space by a flat punch computational difficulties forced us to limit the analysis to the incipient flow.

We shall consider below the foundations of the theory of forging operations and some other complex processes of plastic forming. The slip-line solutions discussed in this chapter are aimed to present a more fundamental approach to the mechanics of these processes. In numerous cases they can replace semi-empirical solutions used in practical calculations and based on the simplified equations of equilibrium.

## 7.2 Press forging in dies

Consider first the simplest case when a rectangular block of initial thickness $H_0$ and width $b$ is placed between two parts of a die (Fig. 127). Let the initial gap between the two halves of the die be $h_0$, and let both parts of the die move towards each other with speed $v_0$, squeezing the metal

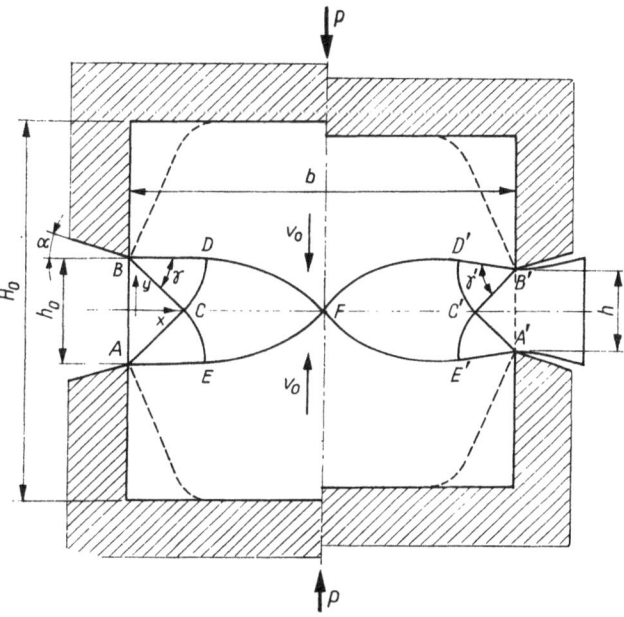

Fig. 127. Press forging in a die. On the left initial stage, on the right advanced stage of the process

into the gap. Assume further that the free surfaces of the dies are inclined to the horizontal at an angle $\alpha$. The angle $\alpha$ is suitably chosen, so that the squeezed flash does not touch the die. We shall examine later the case when the surfaces are horizontal and the flash is compressed as the dies move closer together.

The slip-line net for the initial instant of the plastic flow is schematically shown on the left-hand side of the figure. The deforming region is bounded by the outer lines *BDF* and *AEF*. In the triangle *ABC* we have $\sigma_x = 0$, $\sigma_y = -2k$, and consequently $\chi = -0.5$ and $\varphi = 0$. These data suffice to determine the stresses within the entire slip-line field, similarly

193

as in the case of compression of a block between two plates (see Section 5.4). The total force $P$ needed to deform the metal can be calculated by integrating the stresses $\sigma_y$ along the horizontal axis. On the right-hand side there is presented the slip-line net for a more advanced stage of the process when the distance between the dies decreases down to the value $h$, and part of the material is squeezed into the flash. Now we have $\chi = -0.5$, and $\varphi = 0$ also in the triangle $A'B'C'$. In the same way a slip-line net can be constructed for an arbitrary stage of the process, the only difference being in the value of the angle $\gamma$ of the centred fans at the corners. This angle is defined by the condition that the outer lines $D'F$ and $E'F$ of the deforming region must pass through the centre $F$.

The velocity field may be found by solving a characteristic boundary value problem for Geiringer's equations (4.23). Across the outer slip-lines the normal velocity component must be continuous. Since the velocity $v_0$ of the rigid zones is known, this condition suffices to determine the velocities in the whole deforming region. It is readily seen that the triangle $ABC$ moves horizontally as a rigid block, and similarly during the more advanced stages of the process the material particles which crossed $A'C'$ and $B'C'$ do not suffer further plastic deformations. In other words, the portion of the material to the right of $A'C'$ and $B'C'$ moves as a rigid whole with the speed $v_1 = v_0 b/h$, as required by the condition of volume preservation. The smaller the gap $h$, the faster moves the squeezed material.

Figure 128 shows the velocity hodograph and the velocities of nodal points of a square grid. The elements of the hodograph net are identical with those of the slip-line field. But their position is different. In order to determine the velocity of a specific point, one must find the image of this point in the velocity plane. For example, it is shown in the figure how the velocity of the point $R$ can be determined.

Let us now examine the distortion of a square grid drawn on a longitudinal section of the block [129]. We shall consider the instant when the distance between the dies is $h_s$, and each of the dies has moved by the distance $(h_0 - h_s)/2$ from its initial position. We divide the displacement of the die into a number of small steps $\Delta h$, and then for each instantaneous position of the die between the consecutive steps we construct the slip-line net and the velocity hodograph. Knowing the hodograph, we can determine the instantaneous velocities of the nodal points of the grid in the manner shown in Fig. 128. Now, assuming that the velocities remain constant during each step, we find the displacements of all nodal points of the

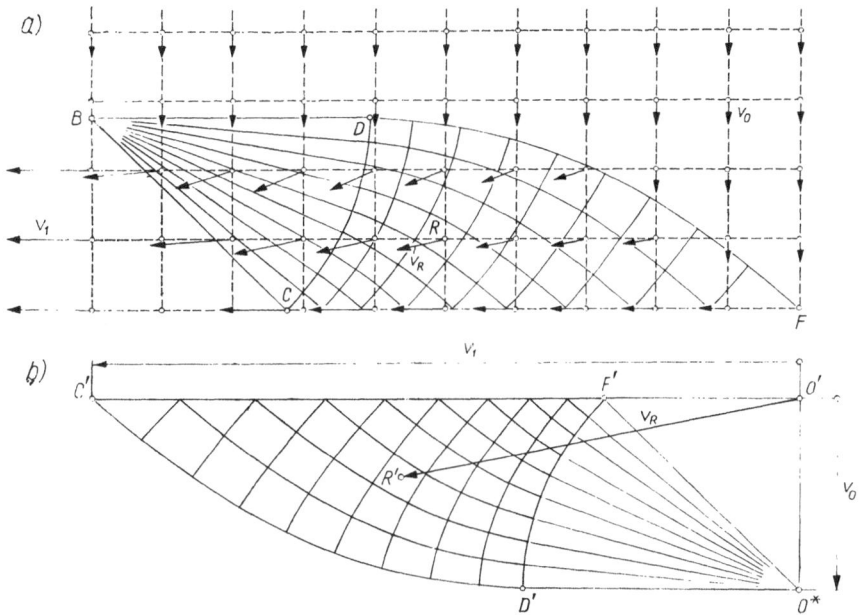

Fig. 128. Geometrical construction of deformation of forged block; a) slip-line net and displacement vectors of nodal points of a square grid, b) velocity hodograph

grid. Repeating this procedure for each of the steps, we obtain finally the distortion of the grid at a chosen instant. An example of the distorted grid for $h_s = 0.765h_0$ is presented in Fig. 129. The total displacement of the die was divided into two steps.

An experimentally deformed grid drawn on a block of lead is shown in Fig. 130. The block was compressed under the same conditions as those assumed in the theoretical analysis. The experimental set-up was similar to that shown in Fig. 87. The actual deforming region was greater than that predicted by the theory. However, the general mode of deformation is similar. As in the other cases, no sharp bends of the grid lines are observed in the deformed metal. These bends are connected with the appearance of the lines of velocity discontinuity in theoretical solutions. Such discontinuities are impossible in real metals displaying strain-hardening effects. The theoretical solution indicates that the extent of the deforming region does not depend on the shape of the interior of the die, provided the die is not too shallow. A possible configuration of the die is marked in

195

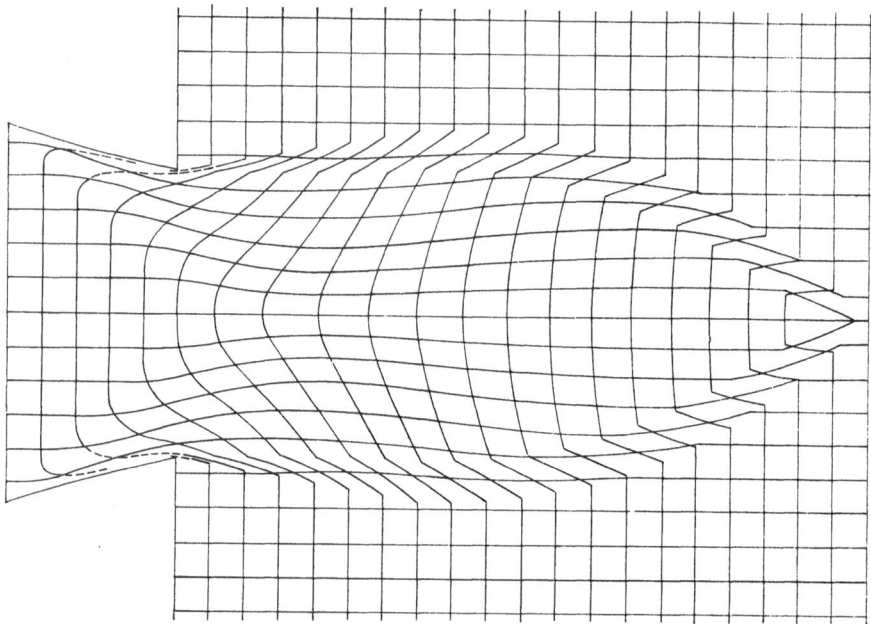

Fig. 129. Distortion of a square grid in a forged block

Fig. 127 by dashed lines. We must remember, however, that our solution is kinematically admissible only until a statically admissible extension of the stress field into the rigid regions is found. Such extensions have not been examined as yet. Nevertheless, it seems that they can be constructed in a way similar to the one in the case of extrusion (comp. Fig. 121). However, it is evident that the configuration of the die affects the difficulty of constructing the extensions. Thus the slip-line solutions of the type discussed above will prove complete only for some configurations of the die. For other configurations they will be only kinematically admissible.

Let us now examine the very important case of the forging in dies with parallel land surfaces ($\alpha = 0$). The squeezed metal is additionally compressed between the die lands.

If the surfaces of the die lands are perfectly smooth, there is no friction imposed on the compressed material by the die-land surfaces, and the slip-line net inside the die has the same configuration as in the foregoing case of inclined die surfaces. The squeezed material in the die lands is uniaxially compressed. In the previous case we had this state of stress only in the triangle $A'B'C'$. The straight segment $A'B'$ represented the line of stress discontinuity, to the right of which the material was stress-free. In the present case there is no discontinuity along such a segment. The analysis of the deformation process can be carried out in a manner similar to that used in the previous case. However, now the force required to cause the plastic flow is complemented by the resultant of normal stresses acting on the die land surfaces. This simplified case of press forging was investigated in [70] and [135].

Let us now examine, following the author's work [8], a case where the frictional stresses on the contact surfaces reach the maximum possible value $k$. The initial situation is identical with that shown on the left-hand side of Fig. 127. We shall arrive at the solution for the advanced stage of deformation by dividing the displacement of the dies into a number of sufficiently small steps, and constructing for each step the slip-line field and the velocity hodograph. To the first step there applies directly the solution shown in Fig. 128. Since the velocity of the material at the exit is constant, we arrive at the situation corresponding to the end of the first step (Fig. 131). In the stress field the singularities appear at the points $A$

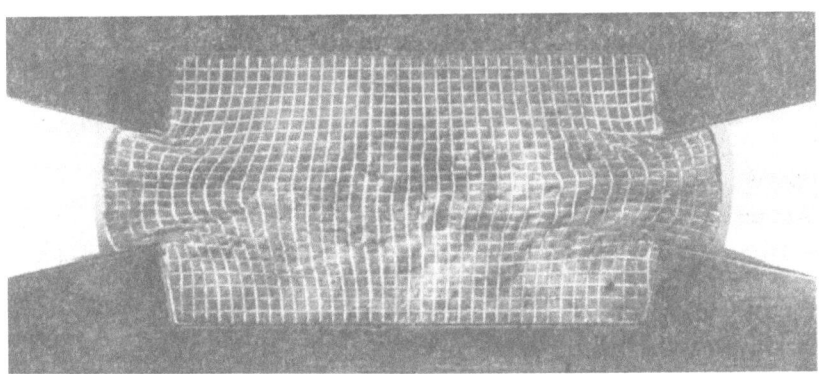

Fig. 130. Distortion of a square grid in a lead block

and $D$. Following now the standard procedure, we determine the stresses in the centred fan $ACS$ and in the triangle $ARC$, where $\sigma_y = -2k$, $\sigma_x = 0$. The radius $AS$ is parallel to the $x$-axis in order to satisfy the condition of friction $\tau_{xy} = k$ on $AD$. The position of the radius $DK$ limiting the other fan with origin at $D$, is determined by the condition that the slip-line $KGF$

must pass through the centre $F$. The velocity hodograph (Fig. 131b) is similar to that shown in Fig. 128b. But now the velocity discontinuity initiated at $F$ propagates along $FGD$ and then, after being reflected at $D$, propagates further along the slip-line $DHI$. Velocities at the singular point

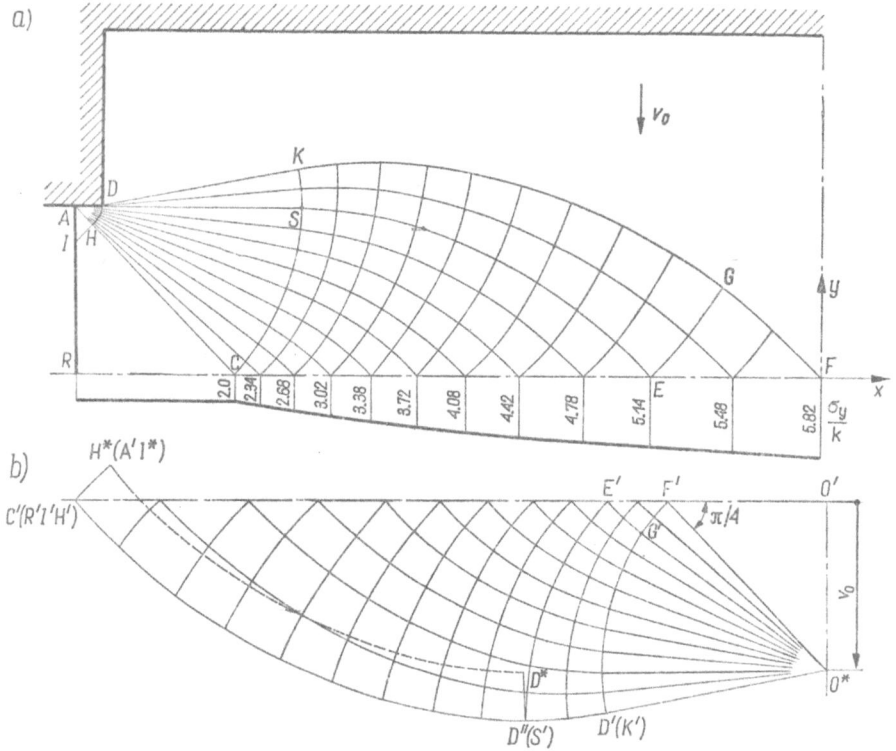

Fig. 131. Slip-line net (a) and velocity hodograph (b) for press forging in a die with parallel land surfaces

$D$ are represented by the vectors through the origin $O'$ and the respective points of the segment $D'D''$. The normal velocity component on $AD$ must be equal to $v_0$, and therefore there arises at $D$ a new velocity jump represented in the velocity plane by the segment $D''D^*$. This discontinuity is propagated along the slip-line $DHI$ which is mapped onto two equidistant lines $D''H'$ and $D^*H^*$ in the hodograph plane. The jump of velocity terminates at the point $I$ forming sharp bends on the stress free boundary, provided the velocities are assumed to be constant throughout each step. If the consecutive steps were chosen infinitely small, the deformation of the boundary would be regular without any singularity. Thus having found in the first step the deformed boundary with sharp corners, we replace it by a regular curve.

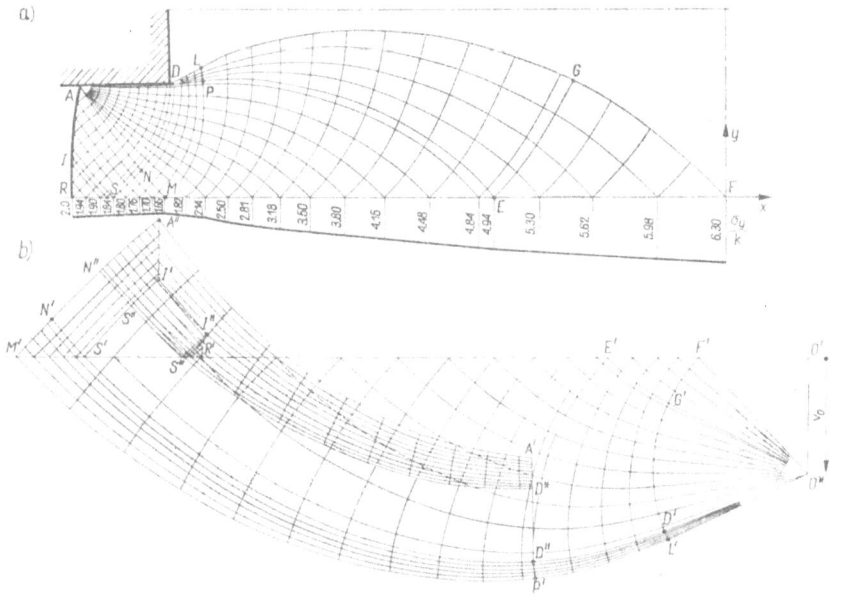

Fig. 132. Slip-line net (a) and velocity hodograph (b) for an advanced stage of press forging in a die with parallel land surfaces

Now the solution for the next step can be constructed. Fig. 132 presents the solution for the fifth step of deformation. We begin to determine the slip-line field by solving the Cauchy boundary value problem in the curvilinear triangle $ARM$ defined by the data on $AR$, where $\chi = -0.5$ and $\varphi$ is the angle contained between the normal to $AR$ and the $x$-axis. Having determined the auxiliary functions $\chi$ and $\varphi$ on the slip-line $AM$, we can solve numerically the characteristic boundary value problem with the

199

singularity at *A*. In the so obtained fan *MAP* all slip-lines are curvilinear. The outer line *AP* passing through *D* does not coincide with the die surface *AD*, but it bounds a very narrow region of the material adjacent to the die which moves downwards as a rigid body attached to the die. The remaining part of the slip-line net is similar to that constructed for the second step (Fig. 131).

The velocity hodograph (Fig. 132b) is also similar to that for the second step. The velocity discontinuity propagates along *FD*, then along *DS* and finally it terminates in the lower part below the horizontal axis at the point symmetrical to *I*. The velocities suffer also a jump across the slip-line *AD*, which is mapped onto the arc *A'D\** in the velocity plane. Velocities in the region *DAN* are represented by respective points of the part *D\*A'A''N''* of the hodograph. Note that the edge *AIR* of the material is represented in the velocity plane by two separate segments *A''I'* and *I''R'*. The jump of velocity at *I* is defined by the segment *I'I''*.

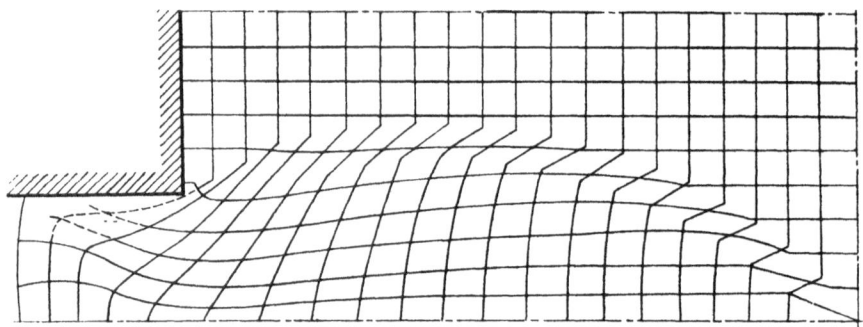

Fig. 133. Distortion of a square grid in a block forged in a die with parallel land surfaces

The final distortion of a grid after five steps is shown in Fig. 133. The total force required to squeeze the material can be calculated by integrating the normal stresses along the longitudinal axis. The distribution of these stresses has been shown in Figs 131 and 132. Figure 134 indicates how the total force increases with the displacement of the dies.

The analysis simplifies radically if we assume that the stress-free edge of the squeezed material remains straight in the course of the process. An example of such a simplified solution for the advanced stage of press

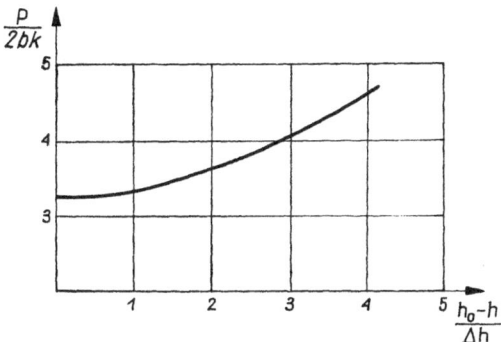

Fig. 134. The relation between total force and displacement of dies in press forging in a die with parallel land surfaces

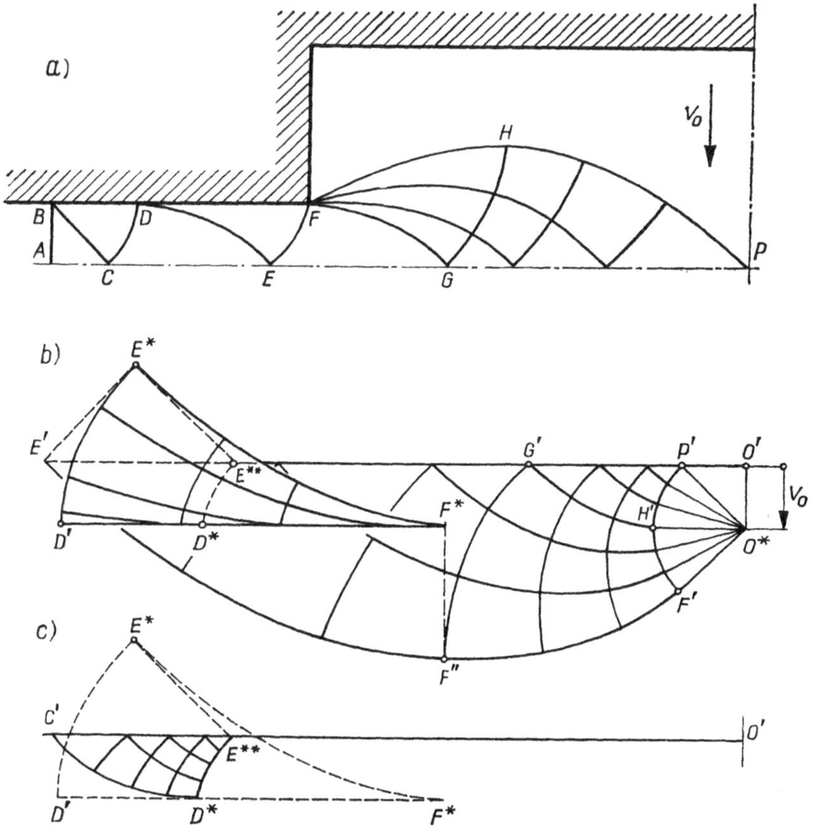

Fig. 135. A simplified solution for press forging in a die with parallel land surfaces; a) slip-line net, b) and c)—velocity hodograph

forging is presented in Fig. 135. The surface of the die is assumed to be perfectly rough. The figure is self-explanatory and does not require any additional comments. Similar solutions can be constructed for other friction conditions on the die-squeezed material interface. Other cases of approximate solutions were examined by L. A. Shofman [107], however he did not investigate the velocity field. Some simple particular cases were considered in [70] and [135].

When the die is shallow, the slip-line solution takes another form. An example of press forging in a shallow die with perfectly rough surface is shown in Fig. 136. The angle $\gamma$ of the centred fans with origins at $A$

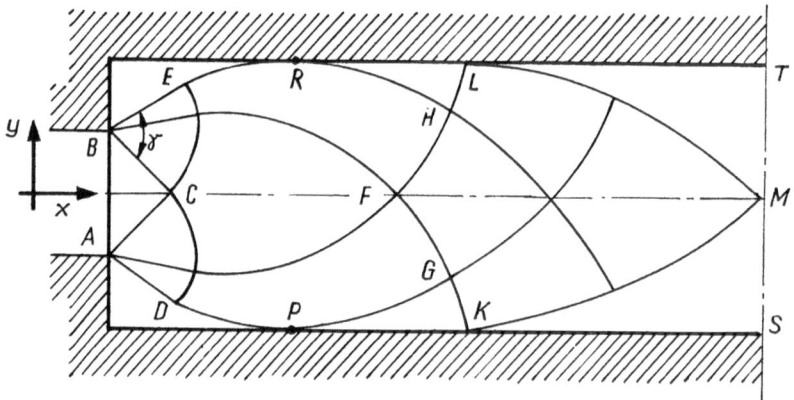

Fig. 136. Slip-line field for press forging in a shallow die

and $B$ is determined by the condition that the outer slip-lines $ADPG$ and $BERH$ meet the surface of the die tangentially at $P$ and $R$. The remaining part of the slip-line net may be constructed according to the procedure applied in the case of compression between rough plates in Chapter 5 (Sec. 5.4). In the present solution there appear zones of dead material in the corners of the die. Also the triangular regions $LMT$ and $KMS$ are rigid and move together with the die as if they were attached to it. In a similar manner one can construct the slip-line net for a more advanced stage of the process, when a portion of the material is squeezed into a flash.

Consider now a more complex case where in both halves of the die the holes are made on the vertical axis (see [107] and [64]). The initial stage of press forging in such a die is presented in Fig. 137. The initial

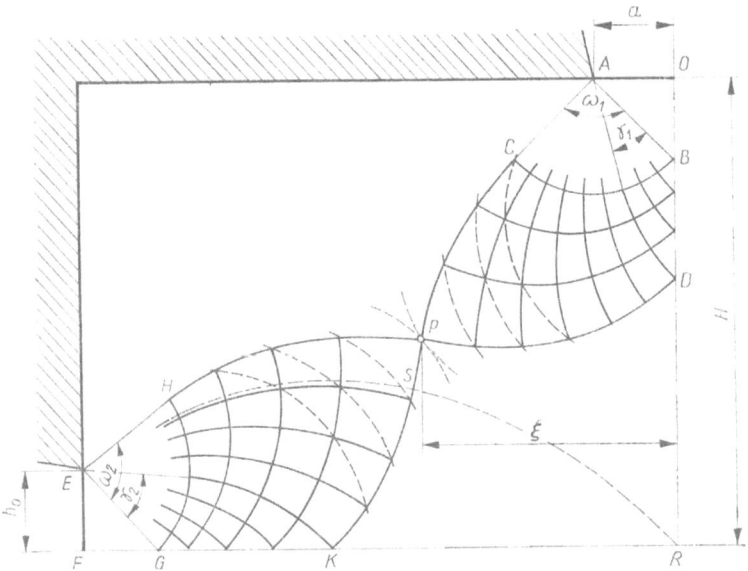

Fig. 137. Slip-line field for initial stage of press forging in a die with a hole

gap between the edges of the dies is $2h_0$. Since the problem displays two-fold symmetry, only one quadrant is shown in the figure. The slip-line net consists of two parts based on the centred fans $CAB$ and $GEH$. The solution of the problem consists in determining the position of the common point $P$ where the two parts come together. The value of the function $\chi$ calculated for each part of the net, must be obviously the same at $P$. From elementary considerations we get the relation

$$\omega_1 + \gamma_1 = \omega_2 + \gamma_2$$

which must be satisfied by the geometrical parameters of the two parts.

The most practical method of establishing the position of the point $P$ seems to be that proposed in [107]. We draw one of the slip-line nets on a transparent sheet and then we put it on the second net drawn on another sheet of the paper. The starting dimensions of the two nets are equal to the width $a$ of the hole and to the initial width $h_0$ of the gap, respectively. On both nets the lines of the constant values of $\chi$ (isobars) are plotted. It is easily seen that these lines, along which the mean stress is the same, constitute the net of diagonals of the slip-line net. The isobars are shown in Fig. 137 by dashed lines. Setting both nets in the positions corresponding to the edges of the die we find the point *P*, as the point of intersection of the isobars from different nets with same value of $\chi$.

Having determined the point *P*, we can calculate the stresses in the plastic zone and the total force. Since the statically admissible extension of the stress field into the rigid region is not known, this force can be treated as the upper bound only. Thus we must always examine whether the other kinematically admissible mechanism of flow along the horizontal axis gives a smaller bound on the force. Such a mechanism, shown in Fig. 137 by its outer slip-line *ESR*, is identical with that for a die without any hole (Fig. 127).

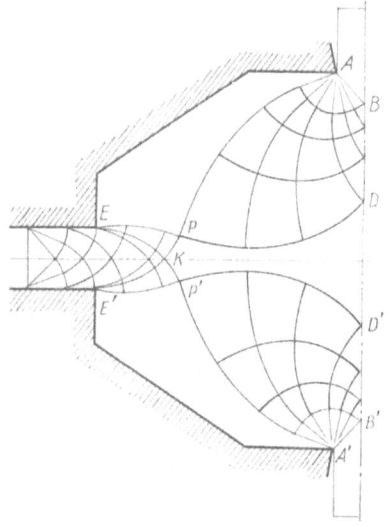

Fig. 138. Slip-line field for advanced stage of press forging in a die with a hole

A more advanced stage of press forging is shown in Fig. 138. The material in the gap between the two halves of the die is compressed by two parallel surfaces. The maximum frictional stress $t = k$ is assumed on the interface. This is an approximate solution assuming straight edges of the squeezed material. The position of the point *P* was found in the same manner as in the previous case.

## 7.3 Combined extrusion and piercing

The methods presented above allow us to analyse various combined processes of plastic forming. As an example, let us examine, following [107],
the process of combined extrusion and piercing.

Let a billet of the material be placed in a container with a hole in the
bottom of width $a$. The material is indented by a flat punch of width $h$
not exceeding the width $H$ of the container. If the billet is sufficiently long,
the left-hand portion of the material remains rigid in the case considered,
and the process reduces to simple piercing (see Section 6.8). In this
stage of the process there is no outflow of the material through the hole.
At the end of the process, when the punch approaches the bottom of the
container, a smaller force will be required to force the material to flow
according to the scheme presented in Fig. 139. The material simultaneously
flows through the hole in the container and through the gap between the
punch and container's wall. If a short billet of the material is placed in
the die, then there begins the incipient plastic flow according to the scheme
shown in Fig. 139.

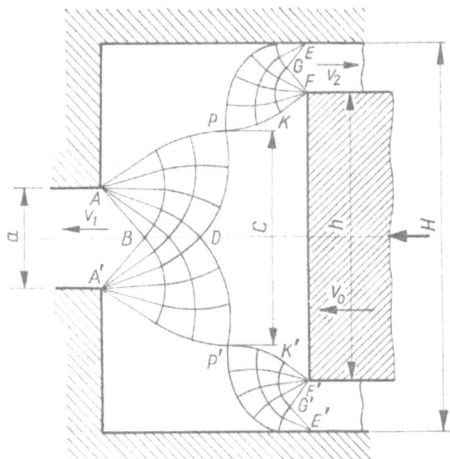

Fig. 139. Slip-line field for combined extrusion and piercing

The slip-line net is composed of elementary nets constructed on the
centred fans. It is assumed in the solution that the frictional stress on the
wall reaches the value $t = k$. Also in this case the position of the point $P$
was determined by using nets with isobars drawn on the transparent sheets.

Having found the distance $c$, one can calculate the velocities $v_1$ and $v_2$ of the squeezed parts of the material. The portion of the material situated within the plastic zone $ABDP$ moves towards the hole, while the material in the plastic region $EGFKP$ moves towards the gap between the punch and the wall. If the container is fixed and the punch moves with the speed $v_0$, we obtain from the condition of constant volume that

$$v_1 = v_0 \frac{c}{a} \quad \text{and} \quad v_2 = v_0 \frac{H-c}{H-h}.$$

The reader will find a similar solution assuming a perfectly smooth wall of the container in Johnson's and Kudo's book [66].

# Axially symmetric plastic flow

## 8.1 Introduction

In this chapter we shall consider the mechanics of plastic flow under conditions of axial symmetry. Most of the processes which have been examined in Chapters 6 and 7 under conditions of plane plastic flow, are used in industrial practice when producing the parts of an axially symmetric form. This applies especially to such processes as drawing, extrusion, piercing, press forging in dies, and other combined forming operations. However, the analysis of the equations of an axially symmetric plastic flow is much more difficult than that for the plane flow. Nevertheless recently some axially symmetric problems of plastic flow have been solved, although exact solutions to many other problems are still not available. In such cases one can obtain a fairly good estimate of the forces from the solution of the appropriate plane flow problem. We shall discuss this problem in more detail in Sections 8.7 and 8.8. Attention will be paid there to the possible error of such an approximation, and to the significant differences in velocity fields obtained for the same problem treated first as one of plane flow and then as being axially symmetric.

## 8.2 Basic relations

To discuss axially symmetric problems, we shall use a system of cylindrical coordinates $r, \theta, z$ such that the $z$-axis coincides with the axis of symmetry. Due to rotational symmetry, the stress components $\tau_{\theta z}, \tau_{\theta r}$ are zero; hence only the components $\sigma_r, \sigma_\theta, \sigma_z$ and $\tau_{rz}$, depending on the two variables $r$ and $z$, may exist. Thus the state of stress is determined by the components

$$\sigma_r = \sigma_r(r, z), \ \sigma_\theta = \sigma_\theta(r, z), \ \sigma_z = \sigma_z(r, z), \ \tau_{rz} = \tau_{rz}(r, z),$$
$$\text{and } \tau_{\theta z} = \tau_{\theta r} = 0.$$

207

Since a generic point can suffer displacement in the meridian plane only, the velocity $v_\theta$ in the circumferential direction must be zero. The other two velocity components depend on the $r$ and $z$ coordinates only,

$$v_r = v_r(r, z), \quad v_z = v_z(r, z) \quad \text{and} \quad v_\theta = 0.$$

Thus at a generic point four stress components and two components of the velocity are to be found. Let us assume, as it has been done in the case of plane plastic flow, that the material is rigid-perfectly plastic, which means that neither elastic strains nor strain-hardening are being taken into account. Moreover our considerations will be limited to the case of quasi-static flow. Thus the inertial forces in equations (2.20) will be neglected.

The equations of motion (2.20) reduce to the equations of equilibrium

$$\frac{\partial \sigma_r}{\partial r} + \frac{\partial \tau_{rz}}{\partial z} + \frac{\sigma_r - \sigma_\theta}{r} = 0,$$

$$\frac{\partial \tau_{rz}}{\partial r} + \frac{\partial \sigma_z}{\partial z} + \frac{\tau_{rz}}{r} = 0.$$

$$(8.1)$$

The Huber–Mises yield condition (3.5) takes now the form

$$(\sigma_r - \sigma_\theta)^2 + (\sigma_\theta - \sigma_z)^2 + (\sigma_z - \sigma_r)^2 + 6\tau_{rz}^2 = 6k^2. \tag{8.2}$$

According to the flow law associated with (8.2) we obtain [comp. (3.15a)]

$$\frac{\epsilon_r}{2\sigma_r - \sigma_\theta - \sigma_z} = \frac{\epsilon_\theta}{2\sigma_\theta - \sigma_z - \sigma_r} = \frac{\epsilon_z}{2\sigma_z - \sigma_r - \sigma_\theta} = \frac{\epsilon_{rz}}{3\tau_{rz}},$$

and substituting the kinematical relations [comp. (2.16)]

$$\epsilon_r = \frac{\partial v_r}{\partial r}, \quad \epsilon_\theta = \frac{v}{r},$$

$$\epsilon_z = \frac{\partial v_z}{\partial z}, \quad \epsilon_{rz} = \frac{1}{2}\left(\frac{\partial v_r}{\partial z} + \frac{\partial v_z}{\partial r}\right),$$

$$(8.3)$$

we finally arrive at the equations

$$\frac{\partial v_r/\partial r}{2\sigma_r - \sigma_\theta - \sigma_z} = \frac{v_r/r}{2\sigma_\theta - \sigma_r - \sigma_z} = \frac{\partial v_z/\partial z}{2\sigma_z - \sigma_r - \sigma_\theta}$$

$$= \frac{(\partial v_r/\partial z) + (\partial v_z/\partial r)}{6\tau_{rz}}. \tag{8.4}$$

The six equations (8.1), (8.2), and (8.4) contain six unknown functions which must be found. Unfortunately, no effective method of solving this system of equations exists. Hill [50] demonstrated that this system is not a hyperbolic one, hence the theory of characteristics cannot be used here.

However, the problem may be effectively solved provided the Tresca yield condition and the associated flow law are employed. An extensive study of this subject was given by Shield [105]. In this chapter we shall confine ourselves to the description of the axi-symmetrical state of a plastic body based on Shield's results. Though it must be remembered here that the Tresca yield condition and particularly the plastic potential flow law associated with it do not describe the actual behaviour of metals very accurately. Nevertheless, this approach enabled us to solve a number of problems which were of practical importance.

The Tresca yield condition and the associated flow law can be most conveniently represented in the principal stress space $(\sigma_1, \sigma_2, \sigma_3)$. If a material particle is stressed to the yield point, its stress state is represented by a point on the surface of an infinite prism with hexagonal cross-section shown in Fig. 24. The projections of the Tresca yield surface on the $\sigma_1 \sigma_2$ and $\sigma_2 \sigma_3$-planes are shown in Fig. 140. Now let us take a plane $\alpha$ with the equation

$$\sigma_1 + \sigma_2 + \sigma_3 = 3\sigma_m, \tag{8.5}$$

where $\sigma_m$ stands for the mean normal stress at a considered point. It is readily seen that this plane $\alpha$ is equally inclined to all three axes $\sigma_1, \sigma_2, \sigma_3$, and, moreover it is perpendicular to the axis and to the faces of the Tresca prism. The intersections of the plane (8.5) with the $\sigma_1 \sigma_2$ and $\sigma_2 \sigma_3$-planes are denoted by $v_\alpha$ and $h_\alpha$, respectively. The intersection of the Tresca prism with the plane $\alpha$ yields a regular hexagon *ABCDEF* whose rabattement in the $\sigma_2 \sigma_3$-plane is denoted by $A^0 B^0 C^0 D^0 E^0 F^0$.

The flow law associated with the Tresca yield condition can be obtained from the general relations (3.13) identifying the yield condition with the plastic potential. This leads to the conclusion that the strain rate vectors in the stress space have to be normal to the yield surface. In our case they lie in the plane $\alpha$ if the stress state is characterized by a point on one of the sides of the hexagon *ABCDEF*. The strain-rate vector is always normal to the respective side of the hexagon. When the stress state happens to be represented by one of the corners of the hexagon, the direction of the strain rate vector is not uniquely determined. For instance, at a point *B*

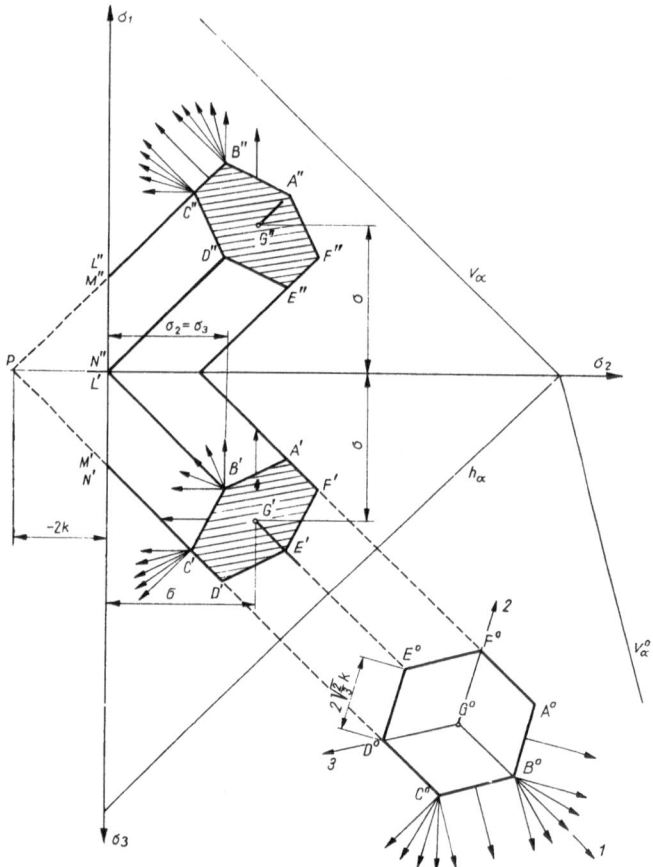

Fig. 140. Tresca yield surface in the space of principal stresses $\sigma_1$, $\sigma_2$, $\sigma_3$

this vector can be oriented arbitrarily, the only restriction being that it must lie between the normals to the sides $AB$ and $AC$. Figure 140 shows a few possible directions of the strain rate vector.

Consider the stress states represented by points situated on one of the edges of the hexagon $ABCDEF$. If the state of stress corresponds to the sides $BC$ or $EF$, Fig. 140 immediately shows that $\epsilon_3 = \epsilon_\theta = 0$, [see (8.13)]. Then the equation $\epsilon_\theta = v_r/r$, and the incompressibility condition (8.11) imply $v_r = 0$ and $v_z = v_z(r)$. Substituting both these velocities in the equations (8.3), we find that the only non-vanishing component of the

210

strain rate tensor is $\epsilon_{rz} = \dfrac{1}{2}\dfrac{\partial v_z}{\partial r}$. Thus the problem is kinematically determinate since the velocities have been obtained without any analysis of stresses.

Similarly kinematically determinate are the stress states corresponding to the edges *AB*, *AF*, *ED*, and *CD*. For example, for the edge *AB* we have $\epsilon_2 = 0$. Introducing (8.3) into (8.13) one obtains

$$\left(\frac{\partial v_r}{\partial r} + \frac{\partial v_z}{\partial z}\right)^2 = \left(\frac{\partial v_r}{\partial r} - \frac{\partial v_z}{\partial z}\right)^2 + \left(\frac{\partial v_r}{\partial z} + \frac{\partial v_z}{\partial r}\right)^2,$$

which together with the incompressibility condition (8.11) constitute the system of two equations in two unknowns $v_r$ and $v_z$. It has been shown ([19, 78]) that this system is hyperbolic and it has two families of characteristics coinciding with the trajectories of principal stresses $\sigma_1$ and $\sigma_2$.

Consider now the case which is most important for practical applications when the stress state in the whole region in question is represented by the points lying on one of the edges of the Tresca prism, for example on the edge *BL*. Then the equality

$$\sigma_2 = \sigma_3 \tag{8.6}$$

holds and clearly $\sigma_1 > \sigma_2$.

When the stress state corresponds to the points of the edge *CM*, we have

$$\sigma_1 = \sigma_3, \tag{8.7}$$

and as previously, $\sigma_1 > \sigma_2$.

Both equalities (8.6) and (8.7) represent the Haar–Kármán postulate [45]. In these cases the Tresca yield condition reduces to the form

$$\sigma_1 - \sigma_2 = 2k, \tag{8.8}$$

or to

$$(\sigma_z - \sigma_r)^2 + 4\tau_{rz}^2 = 4k^2. \tag{8.9}$$

The additional condition (8.6) or (8.7) makes the solution of the problem considerably easier. Replacing in it the principal stresses by the components $\sigma_z$, $\sigma_r$, $\sigma_\theta$, and $\tau_{rz}$, we obtain a set of four equations in four unknowns consisting of the yield condition (8.9) and the equilibrium equations (8.1). Thus, as concerns the stress boundary conditions, the problem may be looked upon as statically determinate, similarly as it was in the case of plane plastic flow.

Since the direction of the strain rate vector is not uniquely determined on the edge of the Tresca prism, one cannot write a uniquely solvable set of relations between the flow velocities and the stresses. The only relation we have at our disposal is the condition of isotropy, i.e. the requirement that the principal axes of the strain rate tensor and those of the stress tensor coincide. The circumferential direction is by definition a principal one for the stresses, as well as for the strain rates. Therefore the stress $\sigma_\theta = \sigma_3$ and the strain rate $\epsilon_\theta$ prove also to be principal components. Now the condition of isotropy is immediately obtained by considering Mohr's circles for stresses and strain rates

$$\frac{\epsilon_{rz}}{\epsilon_r - \epsilon_z} = \frac{\tau_{rz}}{\sigma_r - \sigma_z}. \tag{8.10}$$

Lastly, on account of the incompressibility of the material, we have the condition $\epsilon_r + \epsilon_z + \epsilon_\theta = 0$, which may be also written as

$$\frac{\partial v_r}{\partial r} + \frac{\partial v_z}{\partial z} + \frac{v_r}{r} = 0. \tag{8.11}$$

Equations (8.10) and (8.11) enable us to find the velocities $v_r$ and $v_z$ once the stress state has been determined. As can readily be seen from Fig. 140, the fact that the stress point lies on the edge of Tresca's prism implies the following conditions on the strain rates:

$$\epsilon_1 \geqslant 0, \quad \epsilon_2 \leqslant 0, \quad \epsilon_3 \leqslant 0 \quad \text{on the edge } BL \tag{8.12a}$$

and

$$\epsilon_1 \geqslant 0, \quad \epsilon_2 \leqslant 0, \quad \epsilon_3 \geqslant 0 \quad \text{on the edge } CM. \tag{8.12b}$$

If the conditions (8.12) are not satisfied, the strain rates obtained by solving the kinematical problem will not be associated with the stress field at any given point and the rate of energy dissipation might prove negative, which is inadmissible.

When the velocities are found, one should check whether they satisfy the conditions (8.12). The principal strain rates $\epsilon_1, \epsilon_2, \epsilon_3$ can be evaluated by means of formulae

$$\epsilon_1 = \tfrac{1}{2}(\epsilon_r + \epsilon_z) + \tfrac{1}{2}\sqrt{(\epsilon_r - \epsilon_z)^2 + 4\epsilon_{rz}^2},$$
$$\epsilon_2 = \tfrac{1}{2}(\epsilon_r + \epsilon_z) - \tfrac{1}{2}\sqrt{(\epsilon_r - \epsilon_z)^2 + 4\epsilon_{rz}^2}, \tag{8.13}$$
$$\epsilon_3 = \epsilon_\theta$$

where $\epsilon_r, \epsilon_z, \epsilon_\theta$, and $\epsilon_{rz}$ are related to the velocities by virtue of (8.3).

### 8.3 Determination of stresses

In the present study attention will be confined to the states of stresses corresponding to the equalities (8.6) or (8.7).

We shall solve the system of equations for stresses by the same Lévy method which we employed while examining the equations of plane plastic flow. Let us introduce the function $\chi$, proportional to the sum of the principal stresses

$$2k\chi = \tfrac{1}{2}(\sigma_1 + \sigma_2),$$ (8.14)

and the angle $\vartheta$ between the normal to the plane of maximum shear and the $r$-axis. The sign convention for $\vartheta$ is shown in Fig. 141. Considering a Mohr circle whose radius, according to the yield condition (8.8), is equal to $k$ (Fig. 142), we see that

$$\sigma_r = 2k\chi - k\sin 2\vartheta, \qquad \tau_{rz} = k\cos 2\vartheta,$$
$$\sigma_z = 2k\chi + k\sin 2\vartheta, \qquad \sigma_\theta = k(2\chi \mp 1).$$ (8.15)

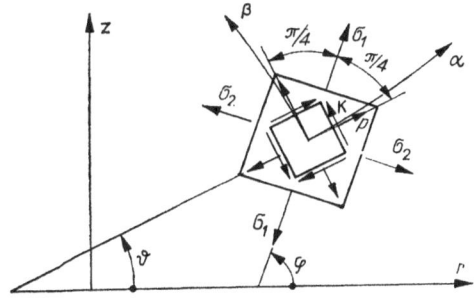

Fig. 141. Directions of principal stresses and sign convention for the angle $\vartheta$

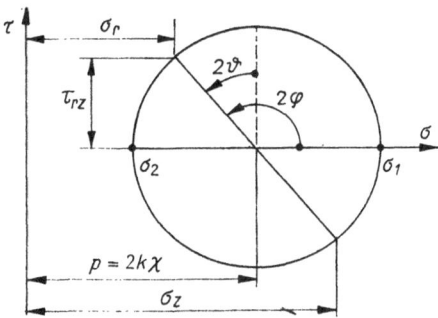

Fig. 142. Auxiliary functions $\chi$ and $\vartheta$ defining the state of stress in the theory of axially symmetric flow

213

The upper minus sign in the expression for $\sigma_\theta$ has been obtained in view of equality (8.6), whereas the lower plus sign refers to equality (8.7).

Thus all the four stress components are expressed by the two new unknowns $\chi$ and $\vartheta$. It can readily be verified that the relations (8.15) identically satisfy the yield condition (8.9), and equality (8.6) or (8.7), whichever is appropriate.

Further, by substituting (8.15) into the equilibrium equations (8.1), one obtains a set of two differential quasi-linear equations in the two unknown functions $\chi$ and $\vartheta$

$$\frac{\partial \chi}{\partial r} - \cos 2\vartheta \frac{\partial \vartheta}{\partial r} - \sin 2\vartheta \frac{\partial \vartheta}{\partial z} = \frac{1}{2r}(\sin 2\vartheta \mp 1),$$

$$\frac{\partial \chi}{\partial z} - \sin 2\vartheta \frac{\partial \vartheta}{\partial r} + \cos 2\vartheta \frac{\partial \vartheta}{\partial z} = -\frac{1}{2r}\cos 2\vartheta. \tag{8.16}$$

This system is hyperbolic for all the possible values of $\chi$ and $\vartheta$; hence it can to be solved by means of the method of characteristics. By the standard procedure (cf. Appendix 2) the differential equations of the characteristics of the first family are found to be

$$\frac{dz}{dr} = \tan\vartheta, \quad d\chi - d\vartheta = \frac{1}{2r}(dz \mp dr), \tag{8.17a}$$

and we shall call them the $\alpha$-*lines.*

The equations of the second family of characteristics are

$$\frac{dz}{dr} = -\cot\vartheta, \quad d\chi + d\vartheta = -\frac{1}{2r}(dz \pm dr), \tag{8.17b}$$

and these characteristics will be referred to as the $\beta$-*lines.*

Equations (8.17) may also be expressed in the different form as

$$d\chi - d\vartheta - (\sin\vartheta \mp \cos\vartheta)\,ds_\alpha/2r = 0, \quad \text{on the } \alpha\text{-line},$$

$$d\chi + d\vartheta + (\cos\vartheta \mp \sin\vartheta)\,ds_\beta/2r = 0, \quad \text{on the } \beta\text{-line}, \tag{8.17'}$$

where $ds_\alpha$ and $ds_\beta$ denote the increments of lengths along the $\alpha$ and the $\beta$-lines, respectively.

It must be remembered that the choice of signs in the equations (8.17) depends upon the specific boundary conditions of the problem.

The characteristics of both families are orthogonal and, as is readily seen, they can be interpreted as the lines of maximum shear, i.e. slip-lines.

Solutions of particular cases may be obtained by finding successive solutions of the boundary value problems for the equations of characteristics (8.17). Integration is conducted numerically after differentials have been replaced by finite differences.

The solution procedures for the basic boundary value problems, viz. the Cauchy problem, for the characteristic problem, and for some of the mixed problems will now be given.

## 1   The Cauchy problem

Let the values of $\chi$ and $\vartheta$ be known along an arc $AB$ of a regular curve in the $rz$-plane (Fig. 143). The functions $\chi$ and $\vartheta$ are continuous along $AB$ and their first derivatives exist. An essential property of the line $AB$

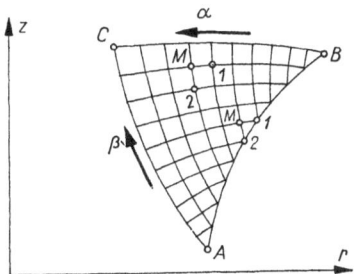

Fig. 143. The Cauchy boundary value problem

is the fact that it intersects each of the characteristics at one point only. These conditions enable us to find a solution in the whole triangular domain $ABC$, formed by the arcs of two characteristics of different families starting from the points $A$ and $B$.

The value of $\chi$ is determined by the stresses acting along $AB$. If the line $AB$ belongs to the free surface of the material, one of the principal stresses in the $rz$-plane, viz. this one which is directed normally to $AB$, must vanish. The other principal stress, oriented tangentially to $AB$ must be equal either to $+2k$ or to $-2k$, depending on the specific problem. In the former case we have $\sigma_1 = 2k$, and $\sigma_2 = 0$, and remembering equality (8.14) we are led to the conclusion that $\chi = 0.5$ along $AB$. Depending on the specific character of the problem the circumferential stress $\sigma_\theta = \sigma_3$ may assume either the value $\sigma_\theta = 0$ or $\sigma_\theta = 2k$, which will be taken into account by fixing the proper sign in the last of the expressions (8.15). If the principal stress parallel to $AB$ is negative, then $\sigma_1 = 0$, and $\sigma_2 =$

$= -2k$, which implies that $\chi = -0.5$ along $AB$. The circumferential stress on $AB$ may also assume two values, either $\sigma_\theta = 0$ or $\sigma_\theta = -2k$.

The value of the angle $\vartheta$ on $AB$ should be determined according to the rule shown in Fig. 141. If $AB$ happens to be a generatrix of the free surface, it constitutes at the same time a trajectory of the principal stress. Once it has been ascertained which of the principal stresses $\sigma_1$ and $\sigma_2$ is algebraically greater, the angle $\vartheta$ may be found at each point of the $AB$-line.

Shield [105] pointed out that if the free edge $AB$ is a straight line through the coordinate origin $O$, then the solution of the Cauchy problem may be reduced to integrating a nonlinear ordinary differential equation. However, attempts to express the solution in a closed form have failed. Numerical results for some values of the angle between the straight boundary $AB$ and the $r$-axis have been tabulated and can be found in [105].

The general case of equations (8.17) is solved by numerical integration. To this end, we take a large enough number of points on the arc $AB$, and we calculate their coordinates $r$ and $z$ and also the corresponding values of the angle $\vartheta$ and the function $\chi$. We shall show how to locate a point $M$ resulting from the intersection of two characteristics passing through the points *1* and *2* on the edge $AB$, and how to compute the values $\vartheta_M$ and $\chi_M$ at $M$. Let us assume that the characteristic joining the points *1* and $M$ corresponds to the $\alpha$-lines family, whereas the characteristic passing through the points *2* and $M$ belongs to the $\beta$-family.

The coordinates $r_M$ and $z_M$ are then found from

$$z_M - z_1 = (r_M - r_1)\tan\vartheta_1,$$
$$z_M - z_2 = -(r_M - r_2)\cot\vartheta_2. \tag{8.18}$$

These equations have been obtained replacing in equations (8.17) the differentials by finite differences. Subscripts *1* and *2* indicate that a given value refers to the point *1* or *2*.

Next, we determine the values $\vartheta_M$ and $\chi_M$ by making use of the following equations

$$\chi_M - \chi_1 - \vartheta_M + \vartheta_1 = (\Delta z_1 - \Delta r_1)/2r_{1M},$$
$$\chi_M - \chi_2 + \vartheta_M - \vartheta_2 = -(\Delta z_2 + \Delta r_2)/2r_{2M}, \tag{8.19}$$

which have been derived, in like manner, from equations (8.17). For sake of definiteness, the upper signs in equations (8.17) have been chosen.

Moreover, the following notation has been introduced in the equations (8.19):

$$r_{1M} = \tfrac{1}{2}(r_1 + r_M), \quad \Delta z_1 = z_M - z_1, \quad \Delta r_1 = r_M - r_1,$$
$$r_{2M} = \tfrac{1}{2}(r_2 + r_M), \quad \Delta z_2 = z_M - z_2, \quad \Delta r_2 = r_M - r_2.$$

To obtain a better approximation we may repeat the computation of the coordinates $r_M$ and $z_M$, and consequently that of $\vartheta_M$ and $\chi_M$, substituting into the recurrent formulae (8.18) the values $\tfrac{1}{2}(\tan\vartheta_1 + \tan\vartheta_M)$ at place of $\tan\vartheta_1$, and $\tfrac{1}{2}(\cot\vartheta_2 + \cot\vartheta_M)$ in lieu of $\cot\vartheta_2$. If it turns out that the second approximation is adequately close to the first one, the calculations can be stopped. If not, we are compelled to keep repeating the above described procedure until there appear two consecutive approximations sufficiently close to each other. In practice, it usually suffices to restrict oneself to the second, or even to the first approximation only.

Knowing the coordinates of the nodes of the characteristic net in a neighbourhood of the line $AB$, we can use them, in the same way as we used the original data at points on $AB$, to locate the new points of the net. Continuing in this manner, we find the state of stress in the whole field within the curvilinear triangle $ABC$.

## 2   The characteristic problem

Along the characteristic arcs $AB$ and $AC$ belonging to different families there are given the values of the coordinates and of the angle $\vartheta$ as well as of the function $\chi$ (Fig. 144). The functions $\vartheta$ and $\chi$ are assumed to be continuous and possessing first derivatives. These data enable us to find

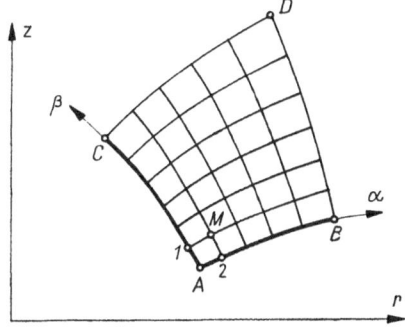

Fig. 144. The characteristic boundary value problem

a solution in the domain $ABDC$, enclosed between the two pairs of characteristics belonging to different families. In order to do so one has to mark off a number of starting points on both arcs $AB$ and $AC$, compute all the relevant quantities, and employ the recurrent formulae (8.18) and (8.19). The process of calculation starts from a pair of points *1* and *2* in the vicinity of the corner $A$.

## 3   *The mixed problem*

The most frequently occurring situation in the mixed problem is as follows: Along a certain line $AB$, defined by an equation $z = z(r)$ and not constituting a characteristic, we know the value of $\vartheta$ (Fig. 145), whereas on

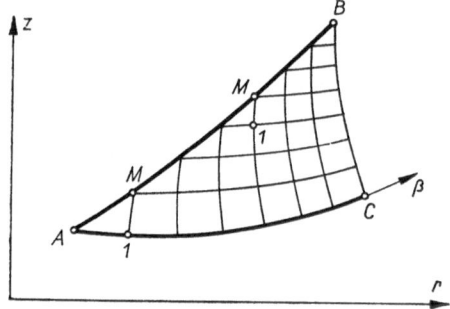

Fig. 145. A mixed boundary value problem

a characteristic $AC$ all the data are known. These starting data enable us to find a solution within a curvilinear triangle $ABC$ created by the two arcs $AB$ and $AC$ and the arc $BC$ of the characteristic of the other family. The numerical computation are performed by means of formulae (8.18) and (8.19), as previously. But now the points on the non-characteristic line $AB$ have to be dealt with in a different manner. Assume that the given characteristic $AC$ belongs to the $\beta$-family. The coordinates of a point $M$ at which a characteristic of the $\alpha$-family through the point *1* intersects the line $AB$ are found from

$$z_M - z_1 = \tfrac{1}{2}(\tan \vartheta_1 + \tan \vartheta_M)\,(r_M - r_1)$$

and from the given equation $z = z(r)$, which must be satisfied by the coordinates $z_M$ and $r_M$. Next we determine the value $\chi_M$ using the first formula (8.19), while $\vartheta_M$ is known by assumption.

When the starting characteristic $AC$ belongs to the $\alpha$-family we proceed in a similar way.

Similarly as in the theory of plane plastic flow, the net of characteristic and the stress distribution can be obtained graphically. A very effective procedure of this kind was proposed by Z. Mróz [91], although the numerical technique described here seems to be more advantageous if a digital computer is used.

## 8.4 Determination of velocities

When determining the velocity components we start from the relations (8.10) and (8.11). Because of equations (8.3) and (8.15), the isotropy condition (8.10) assumes the form

$$\frac{\partial v_r}{\partial z} + \frac{\partial v_z}{\partial r} + \left(\frac{\partial v_r}{\partial r} - \frac{\partial v_z}{\partial z}\right)\cot 2\vartheta = 0. \tag{8.20}$$

This equation, together with the incompressibility condition (8.11)

$$\frac{\partial v_r}{\partial r} + \frac{\partial v_z}{\partial z} = -\frac{v_r}{r}$$

constitutes a set of two equations in two unknowns $v_r$ and $v_z$. It must be noted that the angle $\vartheta$ which appears in equation (8.20) has been already found from the stress solution. The aforementioned set proves to be of the hyperbolic type; hence it can be solved by using the theory of characteristics.

The differential equations for characteristics of the first family $\alpha$ and of the second family $\beta$ are found to be

$$\frac{dz}{dr} = \tan\vartheta, \quad \frac{dz}{dr} = -\cot\vartheta,$$

respectively.

These equations and equations (8.17) for the state of stress are seen to be identical. This leads to the statement that the characteristics for the velocity field and those for the stresses coincide, similarly as they do in the case of plane plastic flow.

Having found the net of characteristics, from the stress solution we may determine the velocities by integrating certain differential equations in velocities that must be satisfied along the characteristics. These equations are found in a well-known manner (Appendix 2) and they take the form

$$\cos\vartheta\,dv_r + \sin\vartheta\,dv_z$$

$$= -\frac{v_r}{2r}(\cos\vartheta\,dr + \sin\vartheta\,dz) \quad \text{on the } \alpha\text{-line},$$

$$\sin\vartheta\,dv_r - \cos\vartheta\,dv_z$$
(8.21)

$$= \frac{v_r}{2r}(-\sin\vartheta\,dr + \cos\vartheta\,dz) \quad \text{on the } \beta\text{-line}.$$

For the purpose of solving a specific problem it appears convenient to utilize the velocity components along the characteristics in lieu of the $v_r$ and $v_z$ components. Let $v_\alpha$ and $v_\beta$ denote these components directed along the $\alpha$ and $\beta$-characteristics, respectively. Relating the components $v_\alpha$ and $v_\beta$ to the components $v_r$ and $v_z$, we have

$$v_\alpha = v_r\cos\vartheta + v_z\sin\vartheta, \quad v_\beta = -v_r\sin\vartheta + v_z\cos\vartheta.$$

By differentiating these relations and substituting into (8.21) we obtain the equations for $v_\alpha$ and $v_\beta$ that must hold along the characteristics,

$$dv_\alpha - v_\beta\,d\vartheta = -(v_\alpha\cot\vartheta - v_\beta)\frac{dz}{2r} \quad \text{along } \alpha\text{-line},$$
(8.22)

$$dv_\beta + v_\alpha\,d\vartheta = (v_\alpha\cot\vartheta - v_\beta)\frac{dz}{2r} \quad \text{along } \beta\text{-line}.$$

In order to determine the velocity components for any specific problem, one has to solve the boundary value problems for equations (8.22). It is to be remembered that the network of characteristics is already known from the stress solution. At this stage, we shall confine ourselves to the description of a solution of the characteristic problem which is of great practical significance.

Let the normal velocity components be known along the arc $AC$ of the $\beta$-line and arc $AB$ of the $\alpha$-line. In other words, $v_\alpha$ is known on $AC$ and $v_\beta$ is known on $AB$ (Fig. 144). These data enable us to find the velocity field in the curvilinear quadrilateral $ABDC$.

First of all we should find the velocity component $v_\alpha$ along $AB$, starting from the point $A$ where both $v_{\alpha A}$ and $v_{\beta A}$ are known. For this purpose, writing the first equation (8.22) in the finite differences form, we obtain the recurrent equation

$$v_{\alpha 2} - v_{\alpha A} - \tfrac{1}{2}(v_{\beta 2} + v_{\beta A})\,(\vartheta_2 - \vartheta_A)$$

$$= -\left[\tfrac{1}{4}(v_{\alpha 2} + v_{\alpha A})\,(\cot\vartheta_2 + \cot\vartheta_A) - \tfrac{1}{2}(v_{\beta 2} + v_{\beta A})\right]\frac{z_2 - z_A}{r_2 + r_A}, \qquad (8.23)$$

where the only unknown is $v_{\alpha 2}$, i.e. the velocity component $v_\alpha$ at a point *2* nearest to the point $A$. Subscripts 2 and $A$ indicate the location of a point at which a quantity is considered. The step-by-step procedure based on equation (8.23) enables us to calculate $v_\alpha$ at a next point, etc., until the terminal point $B$ is reached. At each step the last point for which $v_\alpha$ has been already found plays the role of the starting point $A$. In like manner the component $v_\beta$ along the arc $AC$ can be found.

Let us now show how to determine the velocity components $v_{\alpha M}$ and $v_{\beta M}$ at a point $M$ as soon as the velocities $v_{\alpha 1}, v_{\beta 1}, v_{\alpha 2}, v_{\beta 2}$ at the points *1* and *2* next to the point $M$ are known (Fig. 144). The points *1* and $M$ are on an $\alpha$-line, the points *2* and $M$ are on a $\beta$-line. On account of (8.22), the finite differences equations are

$$v_{\alpha M} - v_{\alpha 1} - \tfrac{1}{2}(v_{\beta M} + v_{\beta 1})\,(\vartheta_M - \vartheta_1)$$

$$= -\left[\tfrac{1}{4}(v_{\alpha M} + v_{\alpha 1})\,(\cot\vartheta_M + \cot\vartheta_1) - \tfrac{1}{2}(v_{\beta M} + v_{\beta 1})\right]\frac{z_M - z_1}{r_M + r_1},$$

$$v_{\beta M} - v_{\beta 2} + \tfrac{1}{2}(v_{\alpha M} + v_{\alpha 2})\,(\vartheta_M - \vartheta_2) \qquad (8.24)$$

$$= \left[\tfrac{1}{4}(v_{\alpha M} + v_{\alpha 2})\,(\cot\vartheta_M + \cot\vartheta_2) - \tfrac{1}{2}(v_{\beta M} + v_{\beta 2})\right]\frac{z_M - z_2}{r_M + r_2}.$$

By solving equations (8.24) with respect to $v_{\alpha M}$ and $v_{\beta M}$, we obtain the recurrent formulae suitable for the consecutive calculation of the velocities at all nodal points of the characteristics network.

Similarly as in the theory of plane plastic flow, the velocity hodograph can be constructed graphically. A procedure of this kind was proposed by Z. Mróz [91] and it proved very useful in solving numerous axially symmetric problems. For lack of space this method will not be described here. All details concerning the graphical method can be found in the original paper [91].

## 8.5    Indentation of a plastic half-space by a flat circular punch

In Chapter 5 the problem of indentation of a plastic half-space by a flat punch was discussed under plane flow conditions. Let us now consider the same problem under conditions of axial symmetry. This problem was investigated by A. Ishlinsky [61] and later treated in greater detail by R. T. Shield [105].

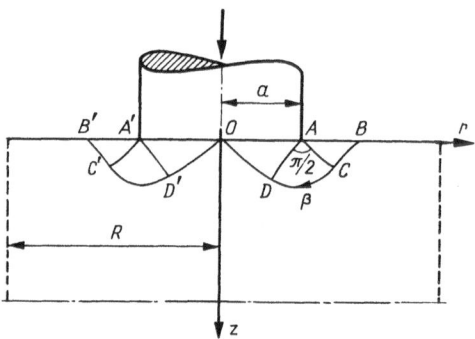

Fig. 146. Indentation of an axially-symmetric punch into half-space

A flat rigid punch with a circular cross-section and having a perfectly smooth surface is pressed into a plastic semi-infinite body $z \geqslant 0$ (Fig. 146). The assumption that the punch surface is smooth means that no friction forces between the punch and the material will be encountered. Similarly as for the plane flow, the present solution is valid merely for the incipient plastic flow, because we shall assume the boundary conditions on the undeformed plane surface of the body. The solution for a more advanced phase of plastic flow, when the material is pushed upwards at the circular edge of the punch to form a "raised lip", is still not known, for reasons of its mathematical complexity.

The treatment of the incipient plastic flow is similar to that under conditions of plane flow. To investigate the plastic stress field in the neighbourhood of the punch, let us start from the free surface $AB$. We assume that on $AB$ $\sigma_1 = \sigma_3 = 0$ and $\sigma_2 = -2k$; hence from (8.14) it follows that $\chi = -0.5$. This is equivalent to the statement that the state of stress is represented in Fig. 140 by a point $P$ at which the extension of the edge $CM$ of the Tresca prism intersects the $\sigma_2$-axis. This assumption will prove correct since the solution obtained will satisfy all the conditions of the problem. Along *AB*, according to the notation shown in Fig. 141, we have $\vartheta = \pi/4$. Since $\sigma_3 = 0$ along *AB*, we assume the lower sign in the expression (8.15) for $\sigma_\theta$ as well as in the equations of characteristics (8.17). Thus the solution reduces to that of the Cauchy problem and the stress field in the triangular domain *ABC* can easily be determined. The characteristic *BC* belongs to the $\beta$-family and *AC* is in the $\alpha$-family.

The point *A* is a singular one and the slip-line net constitutes a centred fan consisting of characteristics of $\alpha$-family passing through *A* and the perpendicular $\beta$-lines. The fan angle at the point *A* equals $\pi/2$. This condition, together with the known values of $\vartheta$ and $\chi$ along *AC*, enable us to determine the state of stress within the curvilinear triangle *ACD*, by solving the characteristic boundary value problem with a singularity at *A*.

Moreover, one can see that in the triangle *ADO* a mixed problem is to be solved, since on the characteristic line *AD* the values of $\chi$ and $\vartheta$ are already known, whereas on the straight line *AE* the angle $\vartheta$ must be equal to $3\pi/4$ on account of the frictionless contact between the punch and the material.

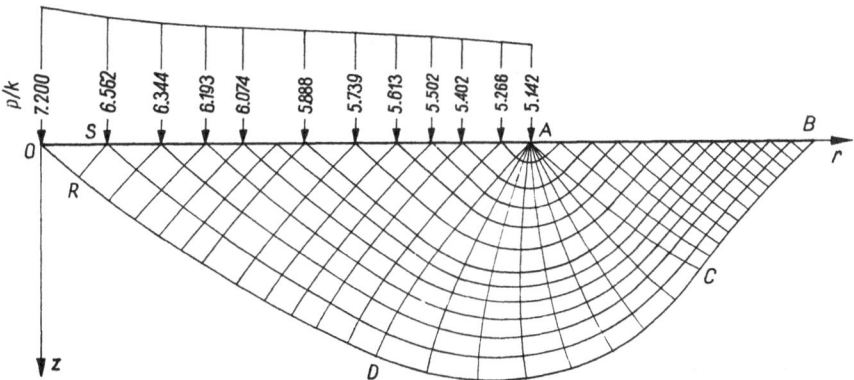

Fig. 147. Slip-line net and pressure distribution in indentation of an axially-symmetric punch (R. T. Shield [105])

The net of characteristics and the distribution of pressure $p(r)$ exerted by the punch on the material are shown in Fig. 147. An average pressure calculated by means of the formula

$$p_0 = \frac{2}{a^2} \int_0^a p(r) r \, dr \tag{a}$$

223

is $p_0 = 2.85 \cdot 2k$. This value is somewhat greater than $p = 2.571 \cdot 2k$ which was obtained for the indentation problem under conditions of plane flow.

It is worth mentioning that E. Levin [75] considered a certain artificial kinematically admissible mechanism of flow which leads to an upper bound on the indentation mean pressure $p_0 = 2.92 \cdot 2k$ very close to Shield's exact value.

Shield [105] demonstrated that the stress field shown in Fig. 147 can be extended into the rigid zone. We shall present such an extension at the end of this section, while considering the compression of a truncated cone.

We shall demonstrate now the technique of calculating the velocities. The normal component of velocity on the contact line $AO$ must be constant and equal to the downward velocity of the punch. The normal velocity component on the characteristic line $BCDO$ must vanish, since the material below this line remains at rest. The characteristic $BCDO$ belongs to the $\beta$-family, thus by the second equation (8.22), taking $v_\alpha = 0$, we get along this characteristic

$$\mathrm{d}v_\beta = -\frac{v_\beta \mathrm{d}r}{2r}.$$

Integrating this we obtain $v_\beta = A\sqrt{r}$ along $BCDO$, where $A$ is a constant. However, $A$ must equal zero since otherwise an infinitely high velocity $v_\beta$ arises at approaching the point $O$ (where $r = 0$). Thus one is bound to conclude that $v_\beta = v_\alpha = 0$ along the whole length of the slip line $BCDO$.

In the curvilinear triangle $ADO$ the boundary values indicate that we have a case of the mixed problem, because $v_\beta = v_\alpha = 0$ on the characteristic $OD$, and the normal velocity component on the straight non-characteristic line $AO$ is equal to the velocity of the indenting punch. However, the numerical calculation is obstructed by the fact that it must start from points lying near the point $O$ where the radius $r$ is minute. In view of equations (8.22), where $r$ appears in the denominator, one would have to assume a very fine mesh of characteristics in the proximity of the point $O$ in order to obtain sufficient accuracy. In [105] this difficulty was overcome by making an assumption that in the neighbourhood of $O$ the distribution of velocities is the same as it would be in the case of compression of a plastic circular cylinder, the closed solution of which is available. Thus the velocities in the region $SOR$ are defined. The remainder of the field of characteristics may be handled by means of the numerical technique, solving first the boundary value problem in the region *RSAD* determined by the values on the edge *AS* and the line *RD*. Next, the values on *AD* and *BCD* make the solution in the region *ABCD* possible. *A* is a singular point of the velocity field.

The calculation of velocities is a time-consuming process, therefore a digital computer was partly employed to obtain the results in [105]. The smaller the radius *r*, the more does the accuracy of calculations depend

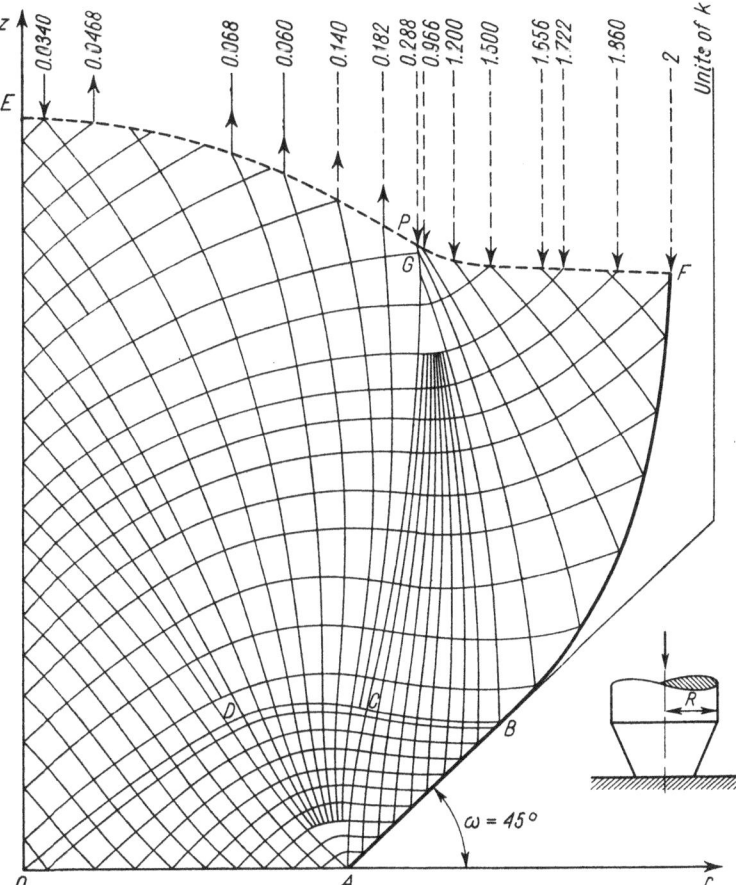

Fig. 148. Compression of a truncated cone. The deforming region is bounded by slip-line *BCDO*. The slip-line net above *BCDO* represents extension of the stress field into rigid region (after L. Dietrich and K. Turski [25])

225

upon the mesh of the slip-line net. For this reason, when calculating the velocity distribution in the region *RSAD*, the mesh of the net was made much finer than it is visualized in Fig. 147. Moreover, a numerical check of inequalities (8.12b) was made. To make sure that these inequalities are satisfied throughout the whole region, an adequately great number of points in the field was investigated.

For practical purposes we may draw the conclusion, without having to calculate numerically the velocity field, that the plastically raised region of the free surface of the material at the edge of the punch has a width equal to *AB*.

The slip-line fields for an allied problem of compression of a truncated cone are given in the author's work [130] and in the paper by L. Dietrich and K. Turski [25]. An example of such a solution for a frictionless contact is shown in Fig. 148. The deforming region is bounded by the slip-line *BCDO*. The extension of the field into the rigid remainder of the cone above this line has been constructed according to the procedure used by Shield [105] who presented the complete solution to the indentation problem (Fig. 146). This procedure parallels that used by Bishop for the corresponding indentation problem under conditions of plane flow (see Fig. 70).

The slip-line mesh within the deforming region has been computed starting from the stress free edge *AB*, where the state of stress is obviously the same as that in the foregoing problem of indentation. Thus we have $\chi = -0.5$ and $\vartheta = \omega + \pi/4$ on *AB*. Now, proceeding according to the scheme described above, we find the slip-line net in the region *ABCDO*. We begin the construction of the extension of the field into the rigid region above *BCDO* by assuming that the state of stress reaches the yield criterion there. First we solve the mixed boundary value problem determined by the already known data on *BCDO* and by the symmetry condition that the slip-lines cut the axis *OE* at angles equal to $\pi/4$.[1]

---

[1] In solving the mixed problem, a certain difficulty arises in the vicinity of the z-axis, where $r = 0$, because the relations (8.17) to be satisfied along the characteristics become undefined. This difficulty can be overcome by determining the limits of the right-hand sides of these equations as $r \to 0$. Let us write the relations (8.17) in the slightly different form

$$d\chi - d\vartheta = \frac{1}{2r}(\tan\vartheta - 1)dr \text{ along } \alpha\text{-lines},$$

$$d\chi + d\vartheta = -\frac{1}{2r}(-\cot\vartheta + 1)dr \text{ along } \beta\text{-lines}.$$

We find the field to the right of *BG* by solving an inverse Cauchy problem determined by the data already known along *BG* and by the assumption that it terminates in a stress-free surface *BF*. At *G* the $\alpha$-lines begin to intersect one another and the line of stress discontinuity marked as broken has to be introduced. The hypothetical stress-free boundary becomes at the point *F* parallel to the axis of symmetry. Now, starting from *F* we begin to construct the line of stress discontinuity *FPE* determined by the condition that the field above it should be a uniaxial compression or tension parallel to the *z*-axis. The stresses above *FPE* were found to be below the yield magnitude as indicated in the figure.

It is interesting to note that the extension of the stress field furnishes a safe estimate of the radius *R* of the cylindrical part of the cone (see Fig. 148). For instance, from the extension given by Shield [105] for the indentation problem, which may be treated as the limit case ($\omega = 0$) of the compression of a cone, we obtain the required ratio $R^*/a = 3.20$ (comp. Fig. 146). For smaller values of the radius *R* the exact solution is not known and the slip-line field shown in Fig. 147 is kinematically admissible only. Thus it gives merely an upper estimate of the indentation force. A lower bound may be obtained for instance from the statically admissible stress field shown in Fig. 149. For a given ratio $R/a < R^*/a$ we find an angle $\omega$ for which the extended stress field of the type presented in Fig. 148 will be entirely inscribed within the actual contour of the indented cylinder. The lower bound corresponding to this field may be found from Fig. 150, where the load factor $f = p_0/2k$ [$p_0$ is the mean pressure calculated from (a)] for a cone is given.

In Fig. 150 the load factor for the cone is compared to the load factor $f = (1 + \frac{1}{2}\pi - \omega)$ obtained for the compression of a truncated wedge under conditions of plane flow. The largest difference is 9.8 percent for $\omega = 0$.

---

Thus on the *z*-axis, where $r = 0$ and $\vartheta = \frac{1}{4}\pi$, we have an indefinite form 0/0. By utilizing de l'Hospital's rule we find that the limit is equal to $d\vartheta$ for the right-hand side of the first relation, and $-d\vartheta$ for the second relation. Thus, when approaching the *z*-axis we can replace the original equations by their approximate form

$d\chi - 2d\vartheta = 0$ along $\alpha$-lines,

$d\chi + 2d\vartheta = 0$ along $\beta$-lines.

These equations can be integrated. Thus finally we have

$\chi - 2\vartheta = $ const along $\alpha$-lines,

$\chi + 2\vartheta = $ const along $\beta$-lines.

227

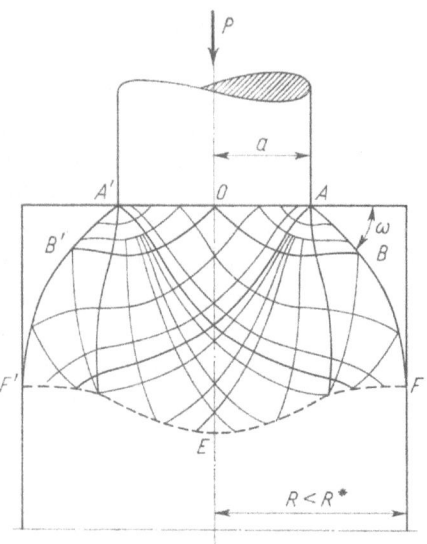

Fig. 149. Statically admissible stress field for indentation of a plastic cylinder by a rigid punch

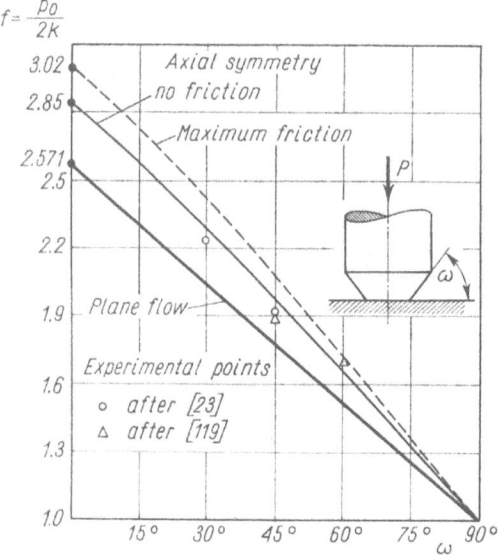

Fig. 150. Relation between the load factor and the angle $\omega$ of compressed truncated cone and wedge

Numerous other comparisons of this kind indicate that good estimates of forces for axially symmetric problems can be obtained from analogous solutions of plane flow.

The experimental results [23, 119] shown in Fig. 150 demonstrate that the actual forces needed to deform the cone are very close to theoretical estimates, although L. Dietrich [23] has shown that the actual distribution of contact stresses is different than that predicted by the theory.

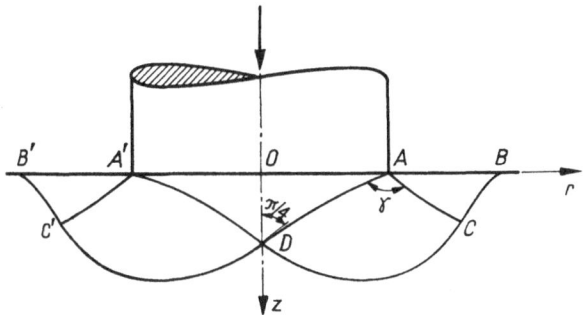

Fig. 151. Indentation of a rough punch into a half-space

If the surface of the punch is rough, the solution takes the form shown schematically in Fig. 151 [34]. We begin the procedure by solving the Cauchy problem defined by the data on $AB$, similarly as in the case of a smooth punch. However, now the angle $\gamma$ of the slip-line fan is determined by the condition that the slip-line $AD$ must cut the $z$-axis at the angle of 45°. The region $ADA'$ is rigid and moves downwards together with the punch. An analogous problem for a truncated cone was examined in the author's work [24] and in the paper by N. P. Suh *et al.* [119].

Indentation of a circular punch with the central hole into a semi-infinite plastic body was investigated by V. A. Zhalin *et al.* [138].

### 8.6 Indentation of a rigid cone. Rockwell hardness test

Let us examine now the problem of indentation of a plastic half-space by a rigid cone of the total angle $2\delta$ (Fig. 152), The solution to this problem, which is closely related to the Rockwell hardness test, was given by F. J. Lockett [79].

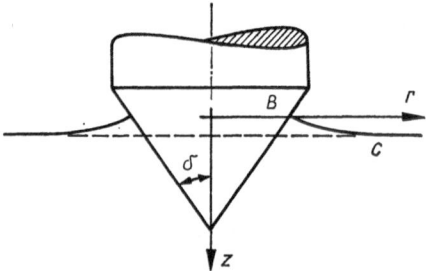

Fig. 152. Indentation of a plastic half-space by a rigid cone

Let the surface of the cone be perfectly smooth so that no friction forces can appear between the indenter and the material. The material penetrated by the cone is displaced and forms an axially symmetric surface whose generatrix is to be determined. This problem is much more difficult to solve than that of the wedge indentation (Section 5.6), where the boundary of the displaced material was straight. But it is evident that also in the present case the deformation remains geometrically similar at each stage of penetration. Thus the generatrix preserves geometric similarity throughout the process. This peculiar feature of the problem significantly facilitates the solution, although it still cannot be obtained in a direct way and a laborious iteration procedure has to be used.

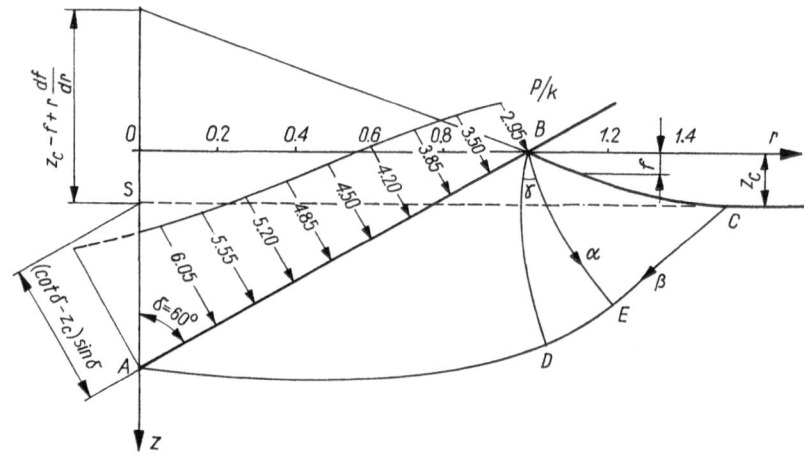

Fig. 153. Slip-line field and pressure distribution for indentation of a half-space by a cone (F. J. Lockett [79])

The net of characteristics for the particular case $\delta = 60°$ is shown in Fig. 153. The $r$-axis of the coordinate system passes through the point $B$ at which the surface of the displaced material cuts the surface of the indentor. Without loss of generality one can assume that the radius $OB$ is equal to unity. The unknown shape of the displaced surface of the material $BC$ will be described by a function $z = f(r)$ which is still to be found. The condition of volume preservation will be satisfied if the volume of the material displaced above the original surface is equal to the volume of the impression below it. Thus we can write

$$r_c^2 z_c - 2 \int_1^{r_c} rf(r) \, dr = \tfrac{1}{3} \cot \delta, \tag{8.25}$$

where $r_C$ and $z_C$ are the coordinates of the point $C$.

On the stress free boundary $BC$ there exist the same conditions as in the case of indentation by a circular punch. Thus we have there $\sigma_\theta = 0$, $\sigma_1 = 0$ and $\sigma_2 = -2k$, or in different terms

$$\chi = -0.5, \quad \vartheta = \frac{\pi}{4} + \arctan \frac{df}{dr} \text{ on } BC. \tag{8.26}$$

Thus the characteristics $BE$ and $BD$ belong to the $\alpha$-lines family, and $CEDA$ belongs to the $\beta$-lines family.

The principal directions are known along the contact line $AB$ since frictionless contact has been assumed. Thus we have

$$\vartheta = \frac{\pi}{4} + \delta \text{ along } AB. \tag{8.27}$$

These data are not yet sufficient for the determination of the stress field within the region $BCEDA$ since the shape of the edge is not known. Thus even in this sense the problem is statically indeterminate.

Consider now the conditions which must be satisfied by the velocity components on the boundaries of the deforming region. Along the characteristic $CEDA$ these conditions are the same as in the case of indentation by a circular punch. Thus we have $v_\alpha = 0$ along $CEDA$ since the material below it remains at rest and the normal velocity component must be continuous across this line. For the same reasons as in the punch problem the other component $v_\beta$ must also be zero on $CEDA$. Thus the first condition imposed on velocities is

$$v_\alpha = v_\beta = 0 \text{ along } CEDA. \tag{8.28}$$

231

To make the configuration preserve geometrical similarity throughout the process, the velocity distribution on the contact line $AB$ and on the free edge $BC$ must be such that each instantaneous configuration is obtained by enlarging the previous one with respect to a fixed point. In our case this fixed point is $S$, where the axis of the cone intersects the original surface of the material. The similarity requires only that a point situated on the surface is in the new configuration also situated on the surface. Its position on the new surface not necessarily must be similar to the previous one. Hence only the velocity component normal to the free surface must have a strictly defined value. The similarity will be preserved if the normal velocity component on $BC$ is distributed proportionally to the distance of the tangent to $BC$ from the point $S$.

For all the points of the contact line $AB$ this distance is $(\cot \delta - z_C) \sin \delta$ (remember that $\overline{OB} = 1$ is assumed). For each point of the free surface $BC$ the distance from $S$ to the tangent line is $[z_C - f + r(\mathrm{d}f/\mathrm{d}r)]\cos \varepsilon$, where $\varepsilon$ is the angle at which the tangent is inclined to the $r$-axis. Bearing in mind that $\cos \varepsilon = [1 + (\mathrm{d}f/\mathrm{d}r)^2]^{-1/2}$, we finally arrive at the condition

$$v_\alpha + v_\beta = \varkappa \frac{z_C - f + r(\mathrm{d}f/\mathrm{d}r)}{\sqrt{1 + (\mathrm{d}f/\mathrm{d}r)^2}} \quad \text{along } BC. \tag{8.29}$$

Similarly we obtain the condition

$$v_\alpha - v_\beta = \varkappa(\cot \delta - z_C)\sin \delta \quad \text{on } AB. \tag{8.30}$$

In (8.29) and (8.30) $\varkappa$ stands for a proportionality constant. Without loss of generality one can assume that $\varkappa = 1$.

In order to satisfy both conditions (8.28) and (8.29) at $C$ we must assume that $\mathrm{d}f/\mathrm{d}r = 0$ at that point. Hence the contour of displaced material is tangent to the original surface at $C$. Comparison of (8.30) and (8.28) leads to the conclusion that a velocity singularity occurs at the vertex $A$. This is kinematically admissible, because the basic equations for velocities (8.22) also display a singularity on the symmetry axis $r = 0$.

Let us pass to the determination of the slip-line field. On the contact line $AB$ the value of the angle $\vartheta$ is known. If the function $\chi$ were known there, the problem could be solved in the standard manner and the shape of the free surface $BC$ easily determined. Conditions (8.28) on $CEDA$ and (8.30) on $AB$ would be sufficient to calculate the velocity field, satisfying automatically condition (8.29) on $BC$. Unfortunately, the function $\chi$ is not known along $AB$. Thus the problem has to be solved by means of an iteration procedure. We may, for instance, assume some distribution of $\chi$ along $AB$, then find the shape of the edge $BC$ and the velocities on it. They obviously will not satisfy condition (8.29). Thus one has to correct the distribution of $\chi$ and to compute once more the velocity distribution on $BC$, and so forth until condition (8.29) will be fulfilled.

In [79] another iteration procedure was used. In each step a shape $z = f(r)$ of the free edge $BC$ was assumed, and then the velocity distribution on the contact line $AB$ was examined. If condition (8.30) on $AB$ is satisfied, we stop the computations. It is worth mentioning that conditions (8.28) and (8.29) enable us to calculate simultaneously the velocity field while computing the slip-line net. This remarkably facilitates computations on a digital computer. In [79] all computations were made on a digital computer, and six to seven iterations were calculated until a satisfactory result was obtained.

The slip-line net calculated for $\delta = 60°$ and shown in Fig. 153 contains a centred fan $EBD$ of angle $\gamma$. The value of this angle decreases with the decreasing of the angle of the cone $\delta$ and for the limit value $\delta^*$ $\approx 52° \, 30'$ it becomes zero. Thus this type of solution is valid for $\delta \geqslant 52° \, 30'$. For smaller values of $\delta$ the solution is not known.

The equation $z = f(r)$ of the free edge $BC$ was found numerically for various values of $\delta$. For the most important case $\delta = 60°$ the coordinates of that edge are as follows:

| $r$ | 1.000 | 1.057 | 1.113 | 1.170 | 1.227 | 1.283 | 1.340 | 1.397 | 1.454 | 1.510 |
|---|---|---|---|---|---|---|---|---|---|---|
| $z$ | 0.000 | 0.031 | 0.055 | 0.075 | 0.091 | 0.105 | 0.117 | 0.126 | 0.133 | 0.137 |

The radius at which the point $C$ is situated is $r_C = 1.510$. The force needed to make the impression of the radius $r_B = 1.000$ is $P/k\pi = 4.28$.

The extension of the slip-line field into the rigid region has been given by Dao-Duy-Tien [20] (Fig. 154). This extension is similar to that shown in Fig. 148. But now a centred fan $CAH$ had to be introduced in order to satisfy the conditions on the symmetry axis $AQ$. This extension is of particular practical significance, because it allows us to theoretically estimate the minimum dimension of specimens for the Rockwell hardness test. It is clearly seen that the width of the tested element should not be smaller than $2c$, where $c = 2.448a$. Since the radius of the impression $a$ is directly related to the hardness of the material, we may easily estimate the minimum width $d = 2c$ as related to the hardness (Fig. 155). It is interesting to note that the empirical recommendations for the width are very close

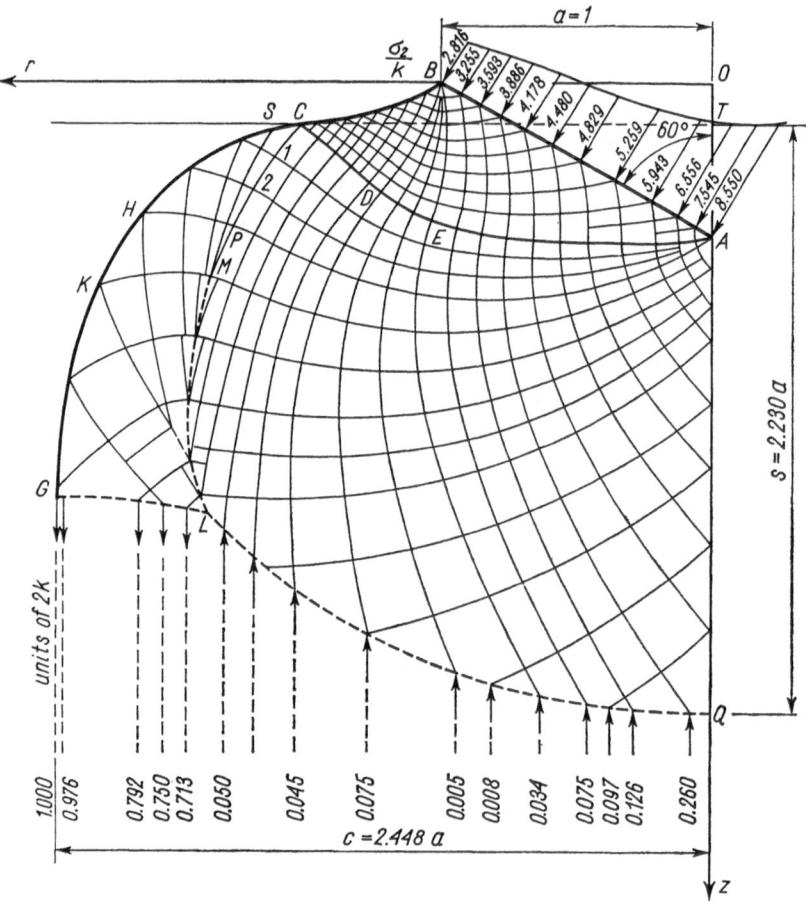

Fig. 154. Extension of the slip-line field into rigid region for the cone indentation problem (after Dao-Duy-Tien [20])

to those resulting from the plasticity solution. However, the stress field shown in Fig. 154 cannot be utilized for an estimation of the minimum thickness of the tested specimen, because the tensile stresses appearing in the column of the material supporting the extended field cannot be transmitted onto the support of the testing apparatus. Thus our stress field is valid only for infinitely long elements.

The above theoretical analysis of the Rockwell hardness test, treated as a problem in plasticity, is fairly well confirmed by experimental investi-

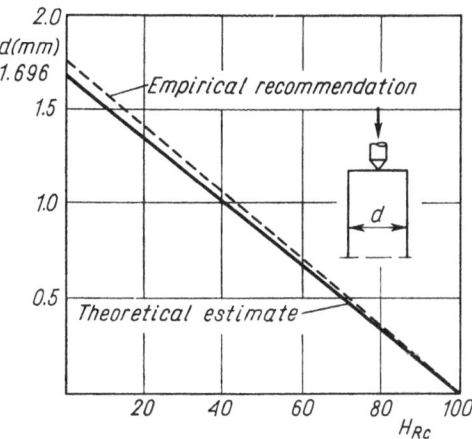

Fig. 155. The relationship between the required specimen diameter and the Rockwell hardness

gations. For instance, Dao-Duy-Tien [20] found that the experimentally determined minimum diameter of cylindrical specimens indented by a cone is only $1 \div 12$ percent greater than the theoretical one.

D. S. Dugdale [33] investigated experimentally indentations of cones into specimens of various materials. He found that the elevation of the lip around the cone $z_C$ is close to the computed value. But the actual radius $r_C$ of displaced material is larger than that resulting from the slip-line solution. Also the actual indentation force was found to be greater up to 12 percent than its theoretical estimate. These discrepancies are attributed to the influence of friction, neglected in the theoretical analysis and also to the strain-hardening of the material.

More recent investigations [2] are also confirming the practical significance of the theoretical solution, although in some cases the actual deformation mode is different than that predicted by the theory.

## 8.7   Compression and press forging of axially symmetric elements

Let us examine the compression of a short cylinder between two rigid parallel plates. The slip-line solution to this problem was given by K. Kwaszczyńska and Z. Mróz [73] for various diameter–thickness ratios and different friction conditions. Figure 156 presents the static field for the ratio $2R/H = 3$.

235

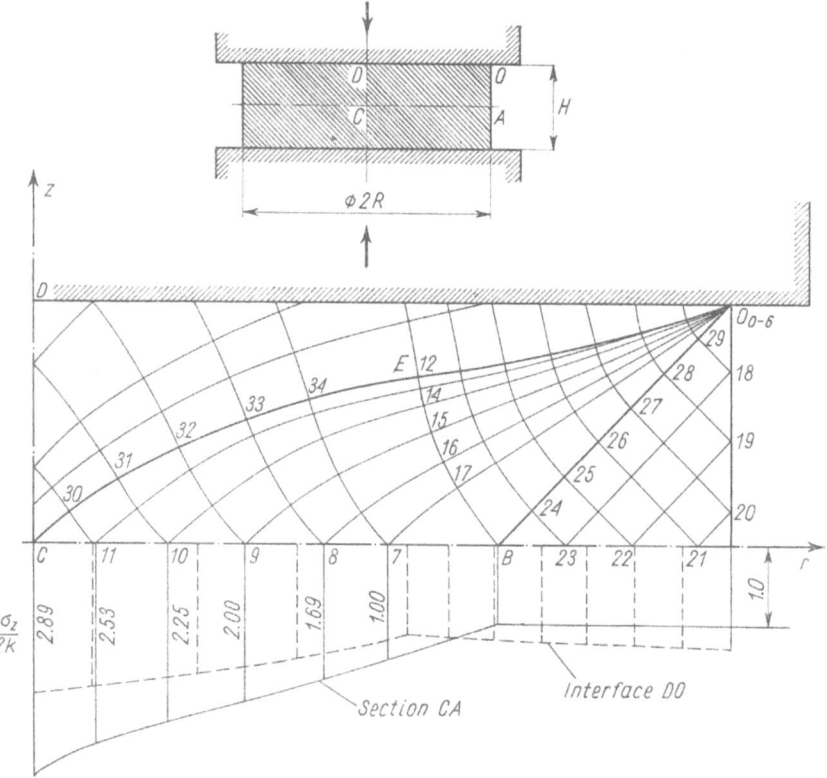

Fig. 156. Slip-line field for the problem of compression of a short plastic cylinder between rigid plates (after K. Kwaszczyńska and Z. Mróz [73])

It is readily seen that the circumferential strain rate $\epsilon_\theta = \epsilon_3$ must be positive. Thus the state of stress in the compressed cylinder corresponds to the edge $CM$ of the Tresca prism (Fig. 140) where $\epsilon_3 \geqslant 0$. The boundary condition $\sigma_r = \sigma_1 = 0$ on the stress free surface $OA$ will be satisfied if we assume the uniaxial compression conditions $\sigma_z = -2k$, $\sigma_r = \sigma_\theta = 0$ in the triangle $AOB$. Now we can determine the slip-line net numerically by using the standard procedure (note the similarity of this problem to that shown in Fig. 85 for plane plastic flow). The angle of the centred fan $BOE$ at $O$ is determined by the condition that the slip-line $OEC$ must pass through the centre $C$ of the cylinder. The deforming region is bounded by the slip-line $OEC$, above which the material remains rigid. The stress field has been extended into this region in order to estimate the stress distribution on the contact surface $OD$. The distribution so obtained is shown in the figure together with the distribution along the section $AC$. Our stress field will be statically admissible if friction forces can develop on the plate-cylinder interface $OD$. The necessary coefficient of friction is determined by calculating the ratio $\tau_{rz}/\sigma_z$ on $OD$, it was found that we must have $\mu \geqslant 0.344$.

In [73] also the velocity field for this problem was examined. Note that the two principal strain rates $\epsilon_1$ and $\epsilon_2$ in the meridian plane are not always of opposite signs, as required by the inequalities (8.12b) corresponding to the states of stress represented by the edge $CM$ in Fig. 140, but are both negative in the region $ABO$. Thus the velocity field in this region corresponds to the edge $DN$ of the Tresca prism, where the two principal stresses $\sigma_1$ and $\sigma_2$ are identical.

Since no velocity field can be associated with our slip-line field, the latter is statically admissible only, and thus it gives only a lower bound on the mean contact pressure equal to $p^- = 1.27 \cdot 2k$. In [73] also an upper bound $p^+ = 1.36 \cdot 2k$ was evaluated numerically on the basis of the velocity hodograph. Thus the difference between the two bounds is small. Similar results were obtained for other $2R/H$ ratios. Hence it is seen that for practical purposes the estimate of the compression force obtained from a slip-line solution is sufficiently accurate.

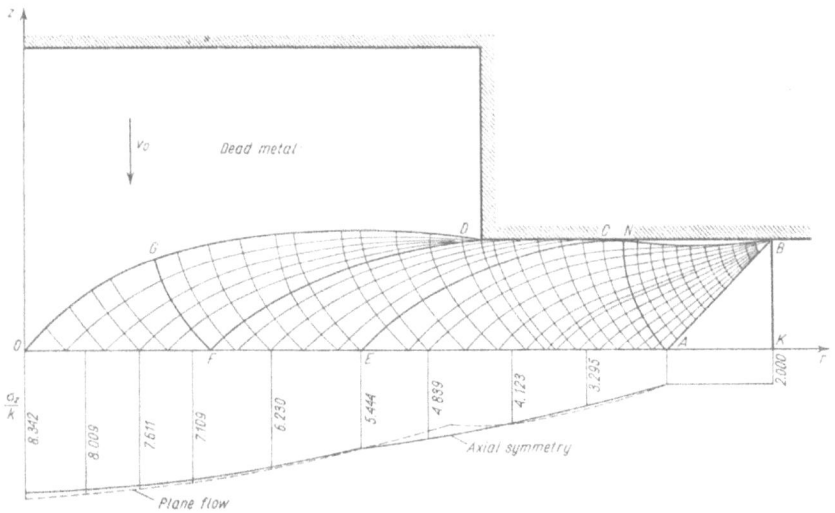

Fig. 157. Axially symmetric closed die forging. Slip-line field

Consider now a simple case of axially symmetric closed-die forging. The material is squeezed into a flash through the space between the closing dies (comp. Fig. 135).

The slip-line field (Fig. 157) is similar to that for the compression of a short cylinder. We begin from the triangle *ABK*, where the state of stress is identical to that in the region *AOB* in Fig. 156. The centred fan is bounded by the slip-line *BNC*, which at *C* becomes tangent to the perfectly rough die surface *BD*. The narrow portion of the material above *BNC* remains in the rigid state. Now, solving in the standard manner a sequence of boundary value problems, we arrive at the slip-line *DF*. The point *D* is evidently a singular one. Starting from it we can compute the centred fan of lines *FDG* and then the triangle *FGO*.

The distribution of the stress $\sigma_z$ along the central section is shown in the figure. For ready comparison there is also presented the distribution of $\sigma_z$ calculated for the analogous problem under conditions of plane flow. As in the previous cases, the difference between the two distributions is very small. Thus we have confirmed once more the conclusion that plane flow solutions may be used for the estimation of forces in axially symmetric problems.

### 8.8 Special problems in axially symmetric flow

There exist a number of particular axially symmetric processes, as for instance ironing of thin-walled cups or extrusion of thin-walled tubes, which may be treated approximately as problems of plane plastic flow. When the wall thickness of the cup or the tube is sufficiently small as compared with the diameter, the circumferential strain is negligible compared with the strain components resulting from the reduction in thickness. Thus we may assume that $\epsilon = v_r/r \approx 0$ and solve the problem according to the theory of plane flow described in Chapter 4. Such an approach has been proposed by R. Hill (see [50]), who investigated the ironing of a thin-walled cup forced by means of a punch through a die. The wall thickness of the cup was reduced while its internal diameter did not change. This approach was applied later by other authors for the analysis of other problems [7, 104].

An interesting analysis of the possible error of such an approximation was given by A. J. M. Spencer [118]. He considered a problem of ex-

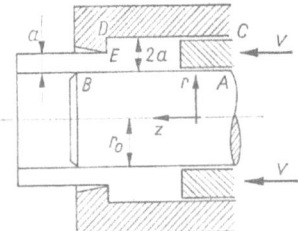

Fig. 158. Extrusion of a tube

trusion of a tube through a die (Fig. 158). The internal diameter of the tube did not change while the thickness was reduced to one-half of the original value. The surfaces of the mandrel *AB* and of the container *CD* were perfectly smooth. Let the ratio $\delta = a/r_0$ be small enough for the magnitude $\delta^2$ to be neglegible in comparison to unity.

The solution given by Spencer was obtained by means of a perturbation method. A known slip-line solution corresponding to the conditions of plane flow (Fig. 159a) was perturbed. The dead material in the region *EDH* was at rest, and deformation took place only in the centred fan *HEK*. Details of rather complex computations will be found by the reader in the original work [118]. We shall shortly outline only the final results which are of practical interest.

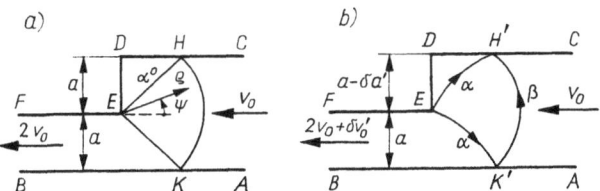

Fig. 159. Extrusion of a tube; a) slip-line field for a plane flow problem, b) perturbed slip-line field for an axially symmetric problem (A. J. M. Spencer [118])

The perturbed slip-line field is shown in Fig. 159b. The equations of slip-lines bounding the region *EK'H'* are as follows:

$$EK': \quad \psi = -\frac{\pi}{4} + \frac{a}{r_0}\left(\frac{1}{2+\pi/2} - \frac{1}{4}\right)\left(2 - \frac{\varrho}{\sqrt{2a}}\right),$$

$$K'H': \quad \varrho = \sqrt{2}a - a\frac{a}{r_0}\left(\frac{\sin\psi}{2} - \frac{\sqrt{2}\psi}{2+\pi/2}\right),$$

$$EH': \quad \psi = \frac{\pi}{4} + \frac{a}{r_0}\left(\frac{1}{2+\pi/2} - \frac{1}{4}\right)\left(2 - \frac{\varrho}{\sqrt{2}a}\right).$$

The coefficients of the ratio $a/r_0$ are small (for example, $[(2+\pi/2)^{-1} - \frac{1}{4}]$ $\approx 0.0301$). The perturbed slip-line field only slightly differs from the field for plane flow (Fig. 159a), even if $\delta = a/r_0$ is as large as 0.25.

The axial force $P$ applied to the punch is equal to the resultant in the $z$-direction of the tractions on $EH'$ or $K'H'$,

$$P = 2\pi r_0 k a\left[2 + \pi + \frac{a}{r_0}\left(\frac{3\pi}{2} + 4\right)\right].$$

The mean extrusion pressure is $p = P/[4a\pi(r_0 + a)]$, or up to order $\delta k$,

$$p = k\left[1 + \frac{\pi}{2} + \frac{a}{r_0}\left(\frac{3\pi}{2} + 4\right)\right]. \tag{8.31}$$

Thus the mean pressure is greater than the pressure $p = k(1 + \pi/2)$ resulting from the plane flow solution. But the reduction of area is also greater. For $a/r_0 = 0.1$ the difference between the estimates of the pressure obtained from the axially symmetric solution and that of the plane flow is 6.5 percent.

# Plane stress

## 9.1 General relations

A state of plane stress exists in a thin sheet of material strained by forces acting in its median plane. If we choose this plane for the $xy$-plane, thereby directing the $z$-axis normally, then the stress components $\sigma_z$, $\tau_{zx}$ and $\tau_{zy}$ vanish. The distribution of the components $\sigma_x$, $\sigma_y$ and $\tau_{xy}$ across the thickness is generally non-uniform, but their variations appear insignificant. Hence they are assumed to be distributed uniformly. The thickness usually changes during the process but as long as the derivative $dh/ds$, where $s$ stands for the distance between two closely located points of the median surface, is small the stress state may be looked upon as two-dimensional.

The general theory of plane stress has not found much application in problems of metal working. However, it constitutes the basis of the theory of drawing and stretch forming of thin shells (see Chapter 11). Of great practical significance are also the axially-symmetric problems, considered further in this chapter.

At a generic point of the plastic region under consideration, one has to determine the three components of stress, namely: $\sigma_x$, $\sigma_y$, and $\tau_{xy}$, the two components of velocity $v_x$ and $v_y$, and the thickness $h$. Similarly as in the plane flow theory, we shall restrict ourselves to a rigid/perfectly-plastic material, i.e. to one showing no strain-hardening. However, as will be seen further, under conditions of axi-symmetrical plane stress it will be possible to obtain solutions where strain-hardening is taken into account.

The stress components must satisfy the following equations of equilibrium

$$\frac{\partial}{\partial x}(h\sigma_x) + \frac{\partial}{\partial y}(h\tau_{xy}) = 0,$$

$$\frac{\partial}{\partial y}(h\sigma_y) + \frac{\partial}{\partial x}(h\tau_{xy}) = 0.$$

(9.1)

The incompressibility condition furnishes the relation

$$\epsilon_x + \epsilon_y + \epsilon_h = 0,$$

where $\epsilon_x = \partial v_x / \partial x$ and $\epsilon_y = \partial v_y / \partial y$. The strain rate $\epsilon_h$ in the $z$-direction may be derived from the expression

$$dh = \frac{\partial h}{\partial x} dx + \frac{\partial h}{\partial y} dy + \frac{\partial h}{\partial t} dt$$

for the increment of the thickness; hence

$$\epsilon_h = \frac{1}{h} \frac{dh}{dt} = \frac{1}{h} \left( \frac{\partial h}{\partial t} + v_x \frac{\partial h}{\partial x} + v_y \frac{\partial h}{\partial y} \right).$$

The expression in brackets represents the material rate of change of the thickness. The incompressibility condition may be finally written down as

$$\frac{\partial h}{\partial t} + \frac{\partial}{\partial x} (h v_x) + \frac{\partial}{\partial y} (h v_y) = 0. \tag{9.2}$$

In contrast to the state of plane flow, the theory of plane stress assumes different forms depending which of the two basic yield conditions is employed.

The Huber–Mises yield condition is now given by [comp. (3.9)]

$$\sigma_x^2 - \sigma_x \sigma_y + \sigma_y^2 + 3\tau_{xy}^2 = 3k^2. \tag{9.3}$$

This relation is obtained from the general form (3.5) for $\sigma_z = \tau_{zx} = \tau_{zy} = 0$.

Relations between strain rates and stresses are given by the associated flow law (see (3.16)) which, after elimination of the parameter $\lambda$, become

$$\frac{1}{2\sigma_x - \sigma_y} \frac{\partial v_x}{\partial x} = \frac{1}{2\sigma_y - \sigma_x} \frac{\partial v_y}{\partial y} = \frac{1}{6\tau_{xy}} \left( \frac{\partial v_x}{\partial y} + \frac{\partial v_y}{\partial x} \right). \tag{9.4}$$

Concerning the Tresca yield condition, two cases must be distinguished. If the principal stresses have opposite signs, i.e. $\sigma_1 \sigma_2 \leqslant 0$, (or, put differently $\sigma_x \sigma_y \leqslant \tau_{xy}^2$), the Tresca yield condition assumes the form

$$(\sigma_x - \sigma_y)^2 + 4\tau_{xy}^2 = 4k^2 \quad \text{for } \sigma_x \sigma_y \leqslant \tau_{xy}^2. \tag{9.5}$$

This relation can directly be derived from the Mohr circle for stresses. According to the Tresca yield condition a material begins to flow plastically at the moment when the maximum shearing stress reaches the value $k$, or in other terms, when the diameter of the largest Mohr circle equals $2k$. In the present case ($\sigma_1 \sigma_2 \leqslant 0$) the largest circle has the diameter equal to the distance between the points $\sigma_1$ and $\sigma_2$ on the $\sigma$-axis, for the third principal stress $\sigma_3 = \sigma_z = 0$ is situated between $\sigma_1$ and $\sigma_2$. Hence the basic form of the Tresca yield condition for $\sigma_1 \sigma_2 \leqslant 0$ can be written as

$$\sigma_1 - \sigma_2 = 2k \quad (\sigma_1 > \sigma_2). \tag{9.6}$$

Now elementary considerations lead to the expression (9.5).

Expressions (9.5) and (9.6) are the same as the expressions (4.7) and (4.6) for the plane plastic flow. In the stress space $\sigma_x$, $\sigma_y$, $\tau_{xy}$ condition (9.5) is represented by the cylindrical part of the surface shown in Fig. 26. This cylindrical part coincides with the surface for the plane flow represented in Fig. 43.

The associated flow law may be written in the form

$$\frac{1}{\sigma_x - \sigma_y} \frac{\partial v_x}{\partial x} = \frac{1}{\sigma_y - \sigma_x} \frac{\partial v_y}{\partial y} = \frac{1}{4\tau_{xy}} \left( \frac{\partial v_x}{\partial y} + \frac{\partial v_y}{\partial x} \right). \tag{9.7}$$

These relations are the same as relations (4.8) for the plane plastic flow. This is obvious because the yield conditions (9.5) and (4.7) have the same form in both cases.

When the principal stresses have the same sign, thus when $\sigma_1 \sigma_2 \geqslant 0$, or in the equivalent form, when $\sigma_x \sigma_y \geqslant \tau_{xy}^2$, the largest Mohr circle has the diameter equal to $|\sigma_1|$ if we adopt the subscript convention yielding $|\sigma_1| > |\sigma_2|$. Indeed, this is so because $\sigma_2$ lies between $\sigma_1$ and $\sigma_3 = \sigma_z = 0$. This largest circle must have its diameter equal to $\sigma_{p1}$. Hence from elementary considerations we obtain

$$(\sigma_x - \sigma_y)^2 + 4\tau_{xy}^2 = 4[\sigma_{p1} - \tfrac{1}{2}|\sigma_x + \sigma_y|]^2 \quad \text{for } \sigma_x \sigma_y \geqslant \tau_{xy}^2. \tag{9.8}$$

The Tresca yield condition for the plane state of stress is represented in the plane of the principal stresses $\sigma_1$, $\sigma_2$ by a hexagon $ABCDEF$ (Fig. 163), obtained by intersecting the Tresca prism (Fig. 24) with the plane $\sigma_3 = 0$, or by intersecting the surface shown in Fig. 26 with the plane $\tau_{xy} = 0$. The case $\sigma_1 \sigma_2 \geqslant 0$ corresponds to the edges $AF$, $FE$, $DC$, and $CB$ of the hexagon. According to the associated flow law (3.13) the vectors representing strain rates in the stress space must be orthogonal to the yield surface. Thus we obtain directly from Fig. 163 that

$$\epsilon_1 = 0 \text{ on the edges } EF \text{ and } CB,$$
$$\epsilon_2 = 0 \text{ on the edges } AF \text{ and } CD. \tag{9.9}$$

The corresponding relations for the velocity components $v_x$ and $v_y$ will be not given here, because they do not seem to be of much use in problems of plastic forming.

The equations of equilibrium (9.1), the incompressibility condition (9.2) together with the assumed yield condition and the associated flow law, constitute a set of six equations in six unknowns. It can be seen that the stress state and the velocity field are coupled since the variable thickness $h$ appears in the equilibrium equations.

No method of solving such system has been developed as yet, hence we are forced to make some simplifications. A general method of solution was given by Sokolovsky [109], who assumed $h = \text{const}$, which means that the thickness remains unchanged during the process of plastic flow and is the same at each point. Hill [50] presented a solution by assuming that the thickness remains the same during the process, but it has different predetermined values at each point of the domain considered, i.e. $h = h(x, y)$. Clearly, neither of these assumptions is compatible with the incompressibility condition (9.2).

It is, however, worth emphasizing that under condition of axial symmetry it is possible to obtain solutions which take into account both strain-hardening of the material and the variation of thickness occurring during the deformation process. We shall discuss this problem in this chapter.

## 9.2   Solution of plane stress equations for the Huber–Mises yield condition

Assuming that the thickness $h(x, y)$ is known, the problem is reduced to determining the three stress components $\sigma_x$, $\sigma_y$, and $\tau_{xy}$, and the two components of the velocity $v_x$ and $v_y$ at any generic point of the region under consideration. We have a system of five equations made up of the equilibrium equations (9.1), the Huber–Mises yield condition (9.3), and the relations of the flow law (9.4). A remarkable feature of this system is the fact that, exactly as in the case of plane plastic flow, the velocities appear neither in the equilibrium equations nor in the yield condition. These three equations in three unknowns $\sigma_x$, $\sigma_y$, $\tau_{xy}$ may be solved independently and then one should check if the solution satisfies all the kinematical conditions of the problem.

Let us introduce a new function $\omega$, by means of which the sum of, and the difference between the principal stresses can be expressed

$$\tfrac{1}{2}(\sigma_1 + \sigma_2) = \sqrt{3}\,k\cos\omega, \qquad \tfrac{1}{2}(\sigma_1 - \sigma_2) = k\sin\omega. \tag{9.10}$$

Making use of Mohr's circle (compare Fig. 42), we fignd the stress components $\sigma_x$, $\sigma_y$, $\tau_{xy}$ to be

$$\sigma_x = k\left(\sqrt{3}\cos\omega + \sin\omega\cos 2\varphi\right),$$
$$\sigma_y = k\left(\sqrt{3}\cos\omega - \sin\omega\cos 2\varphi\right), \qquad \tau_{xy} = k\sin\omega\sin 2\varphi, \qquad (9.11)$$

where $\varphi$ denotes an angle between the direction of the larger principal stress and the $x$-axis.

It is readily verified that equations (9.11) are compatible with the yield condition (9.3) for arbitrary values of $\omega$ and $\varphi$. Substituting (9.11) into the equilibrium equations (9.1) yields a system of two partial differential quasi-linear equations in the two unknown functions $\omega$ and $\varphi$,

$$\cos\omega\sin 2\varphi\,\frac{\partial\omega}{\partial x} + 2\sin\omega\cos 2\varphi\,\frac{\partial\varphi}{\partial x} -$$

$$- \left(\sqrt{3}\sin\omega + \cos\omega\cos 2\varphi\right)\frac{\partial\omega}{\partial y} + 2\sin\omega\sin 2\varphi\,\frac{\partial\varphi}{\partial y}$$

$$= -\sin\omega\sin 2\varphi\,\frac{\partial\ln h}{\partial x} - \left(\sqrt{3}\cos\omega - \sin\omega\cos 2\varphi\right)\frac{\partial\ln h}{\partial y},$$

$$- \left(\sqrt{3}\sin\omega - \cos\omega\cos 2\varphi\right)\frac{\partial\omega}{\partial x} - 2\sin\omega\sin 2\varphi\,\frac{\partial\varphi}{\partial x} + \qquad (9.12)$$

$$+ \cos\omega\sin 2\varphi\,\frac{\partial\omega}{\partial y} + 2\sin\omega\cos 2\varphi\,\frac{\partial\varphi}{\partial y}$$

$$= -\left(\sqrt{3}\cos\omega + \sin\omega\cos 2\varphi\right)\frac{\partial\ln h}{\partial x} - \sin\omega\sin 2\varphi\,\frac{\partial\ln h}{\partial y}.$$

We shall solve this system by means of the theory of characteristics. We start with an analysis of the Cauchy problem, i.e. we assume that along a certain line $y = y(x)$ the values of $\omega$ and $\varphi$ are given. As already known, these data are sufficient to determine the functions $\omega$ and $\varphi$ in a neighbourhood of that line, provided that the partial derivatives $\partial\omega/\partial x$, ... ..., $\partial\varphi/\partial y$ can be determined. Clearly, this can be done by supplementing equations (9.12) with the following obvious equalities:

$$d\omega = \frac{\partial\omega}{\partial x}\,dx + \frac{\partial\omega}{\partial y}\,dy, \qquad d\varphi = \frac{\partial\varphi}{\partial x}\,dx + \frac{\partial\varphi}{\partial y}\,dy$$

which must be satisfied along the line $y = y(x)$.

If the characteristic determinant formed by the coefficients of the derivatives is equal to zero, then the derivatives along the line $y = y(x)$ are not defined and the Cauchy problem cannot be solved (see Appendix 2). Then the line $y = y(x)$ is said to be a *characteristic* of the system (9.12). When solving this determinant equated to zero with respect to $dy/dx$, we obtain the differential equations of characteristics

$$\frac{dy}{dx} = \frac{\sqrt{3}\sin\omega\sin 2\varphi \pm \sqrt{3-4\cos^2\omega}}{\sqrt{3}\sin\omega\cos 2\varphi - \cos\omega}, \tag{9.13}$$

the positive and negative signs in the numerator corresponding to the first and to the second family of characteristics. The differential relations that must be satisfied along these characteristics may be obtained by replacing in the characteristic determinant an arbitrary column by the column of free terms containing no derivatives, and equating to zero the determinant thus obtained. Finally these relations assume the form

$$2\left(d\varphi \mp \frac{\sqrt{3-4\cos^2\omega}}{2\sin\omega}\,d\omega\right)$$

$$= \left[(1-\sqrt{3}\cot\omega\cos 2\varphi)\frac{\partial\ln h}{\partial y} + \sqrt{3}\cot\omega\sin 2\varphi\frac{\partial\ln h}{\partial x}\right]dx -$$

$$- \left[\sqrt{3}\cot\omega\sin 2\varphi\frac{\partial\ln h}{\partial y} + (1+\sqrt{3}\cot\omega\cos 2\varphi)\frac{\partial\ln h}{\partial x}\right]dy, \tag{9.14}$$

the upper, negative sign on the left-hand side referring to the first family, the lower, positive sign, to the second family of characteristics.

From the equations (9.13) it follows that the characteristics are not always real; hence the system (9.12) need not necessarily be hyperbolic. This occurs only when $(3-4\cos^2\omega)$ is positive, which takes place when $\pi/6 < \omega < 5\pi/6$ or $7\pi/6 < \omega < 11\pi/6$. For the remaining values of $\omega$, this system is elliptic and it cannot be solved by means of the theory of characteristics. This range can be shown in the principal stress plane $\sigma_1\sigma_2$. The Huber–Mises ellipse is obtained by intersecting the cylinder in Fig. 23 with the plane $\sigma_3 = 0$ (Fig. 160), or the ellipsoid for plane stress (Fig. 25) with the plane $\tau_{xy} = 0$. Let us now find the points corresponding to the limiting values of the function $\omega$. To this end, let us solve equations (9.1 ) with respect to the principal stresses. We get

$$\sigma_1 = 2k\cos(\omega - \pi/6),$$
$$\sigma_2 = 2k\cos(\omega + \pi/6). \tag{9.15}$$

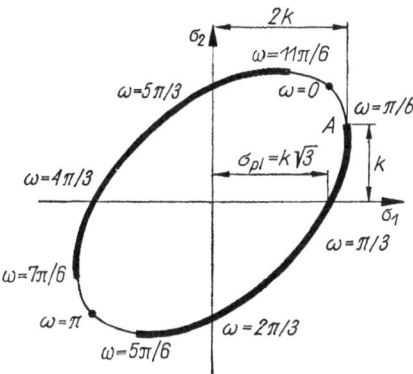

Fig. 160. The Huber–Mises ellipse for the plane state of stress

Substituting the limiting values of $\omega$ into (9.15), we find the range on the ellipse where the system (9.12) is hyperbolic. This is shown in Fig. 160 by heavy lines. It can be seen that the remaining range in which the system is elliptic appears to be relatively narrow. When the principal stresses bear opposite signs, one can be certain that the system is hyperbolic. However, if the signs are the same, we must check whether the state of stress belongs to the hyperbolic range.

It can be seen from (9.13) that for the particular case $(3-4\cos^2\omega) = 0$ the system (9.12) has one family of characteristics. Thus for the values $\omega = \pi/6;\ 5\pi/6,\ 7\pi/6,\ 11\pi/6$ it is parabolic.

From now on we shall confine ourselves to equations within the hyperbolic range. A solution outside this range proves to be exceedingly difficult to tackle.

In order to simplify the equations of characteristics, let us introduce some auxiliary notation due to V. V. Sokolovsky [110]:

$$2\psi = \pi - \arccos \frac{\cot\omega}{\sqrt{3}},$$

$$\chi = -\frac{1}{2}\int_{\pi/6}^{\omega} \frac{\sqrt{3-4\cos^2\omega}}{\sin\omega}\,d\omega.$$

(9.16)

The equations of characteristics assume now the form

$$\frac{dy}{dx} = \tan(\varphi \pm \psi),$$

(9.17₁)

$$2(\mathrm{d}\varphi \pm \mathrm{d}\chi)$$

$$= \left[(1 - \sqrt{3}\cot\omega \cos 2\varphi)\frac{\partial \ln h}{\partial y} + \sqrt{3}\cot\omega \sin 2\varphi \frac{\partial \ln h}{\partial x}\right]\mathrm{d}x$$

$$- \left[\sqrt{3}\cot\omega \sin 2\varphi \frac{\partial \ln h}{\partial y} + (1 + \sqrt{3}\cot\omega \cos 2\varphi)\frac{\partial \ln h}{\partial x}\right]\mathrm{d}y, \quad (9.17_2)$$

where, as previously, the upper (positive) signs correspond to the first family, while the lower (negative) signs are appropriate for the second family of characteristics.

Assuming $h = $ const, i.e. that the distribution of thickness is uniform, we obtain the well-known classic equations of characteristics given by V. V. Sokolovsky

$$\frac{\mathrm{d}y}{\mathrm{d}x} = \tan(\varphi + \psi), \quad \chi + \varphi = \text{const}, \quad (9.18a)$$

for the first family, and

$$\frac{\mathrm{d}y}{\mathrm{d}x} = \tan(\varphi - \psi), \quad \chi - \varphi = \text{const}, \quad (9.18b)$$

for the second family.

In contrast to plane plastic flow, the characteristics do not constitute an orthogonal net. They intersect each other at an angle $2\psi$ which, in the general case, varies from point to point.

It must be strongly emphasized that the characteristics given by the differential equations (9.17) or (9.18) cannot be interpreted physically. In particular, they are not to be looked upon as slip-lines as it was in the case of plane flow.

For practical purposes we quote here from [110] a diagrammatic representation of the function $\chi = \chi(\omega)$ by means of the following table.

| $\omega$ | $\pi/6$ | 0.611 | 0.785 | 0.873 | 1.047 | 1.222 | 1.396 | $\pi/2$ |
|---|---|---|---|---|---|---|---|---|
| $-\chi$ | 0 | 0.030 | 0.139 | 0.201 | 0.339 | 0.485 | 0.633 | $\pi/4$ |
| $\omega$ | $\pi/2$ | 1.745 | 1.920 | 2.094 | 2.269 | 2.357 | 2.571 | $5\pi/6$ |
| $-\chi$ | $\pi/4$ | 0.936 | 1.085 | 1.230 | 1.370 | 1.432 | 1.541 | $\pi/2$ |

To obtain a solution of a specific problem one must successively solve the boundary value problems for the equations (9.17) or (9.18); these may be Cauchy problem, characteristic problems or various types of the mixed problem. The computations are based on Massau's method after the differentials have been replaced by finite differences. We shall discuss the details of the numerical procedure in Chapter 11 devoted to the theory of drawing and stretchforming of shells, from which the plane state of stress discussed here may be obtained as a particular case.

### 9.3 Velocity field associated with the Huber–Mises yield condition

The equations (9.4) may be expressed in the form

$$(2\sigma_y - \sigma_x)\frac{\partial v_x}{\partial x} - (2\sigma_x - \sigma_y)\frac{\partial v_y}{\partial y} = 0,$$

$$6\tau_{xy}\frac{\partial v_x}{\partial x} - (2\sigma_x - \sigma_y)\frac{\partial v_x}{\partial y} - (2\sigma_x - \sigma_y)\frac{\partial v_y}{\partial x} = 0.$$

$$(9.19)$$

Since the stress state is known, this system contains two unknowns $v_x$ and $v_y$. It has two families of real characteristics. If the stresses are expressed by (9.11), the differential equations of characteristics of the system (9.19) are found to be

$$\frac{dy}{dx} = \tan(\varphi \pm \psi),$$

where $\psi$ is defined by the relations (9.16). This leads to the significant conclusion that the characteristics for velocities and the characteristics for stresses coincide. The following differential relations must be satisfied along those characteristics

$$dv_x + dv_y \tan(\varphi + \psi) = 0,$$

$$dv_x + dv_y \tan(\varphi - \psi) = 0,$$

$$(9.20)$$

for the first and for the second family, respectively. To solve a specific problem, one has to find the solutions for the boundary value problems for the equations (9.20). There is no need to numerically determine the net of characteristics for the velocities as it coincides with that for stresses.

Since the equations (9.20) have been derived in a formal way, they explain nothing about the properties of characteristics. But we can investigate these properties by discussing the basic system of equations (9.19). Let us consider a certain line $L$ by assuming a local system of coordinates $x, y$ at an arbitrary point on it with the $x$-axis directed tangentially to $L$. Let us assume that the velocity components $v_x$ and $v_y$ along

$L$ are known, and hence the derivatives $\partial v_x/\partial x$ and $\partial v_y/\partial x$ along $L$ are also known. The remaining two derivatives $\partial v_x/\partial y$ and $\partial v_y/\partial y$, necessary to solve the Cauchy boundary value problem, are unknown, for we do not know what the variations of the velocities $v_x$ and $v_y$ in the neighbourhood of the $L$-line are going to be. However, these derivatives may be determined from (9.19), unless it so happens that we have simultaneously

$$2\sigma_x - \sigma_y = 0 \quad \text{and} \quad \partial v_x/\partial x = 0.$$

The first relation is satisfied along $L$ if it represents a characteristic of the stress field. To prove this, let us assume that along $L$ the stress components $\sigma_x$, $\sigma_y$ and $\tau_{xy}$ are given. Hence the derivatives $\partial \sigma_x/\partial x$, $\partial \sigma_y/\partial x$, and $\partial \tau_{xy}/\partial x$ along $L$ are also known. Then the derivatives $\partial \sigma_y/\partial y$ and $\partial \tau_{xy}/\partial y$ can be determined from the equilibrium equations (9.1). The sixth derivative $\partial \sigma_x/\partial y$ may be evaluated by means of the yield condition (9.3) differentiated with respect to $y$

$$(2\sigma_x - \sigma_y)\frac{\partial \sigma_x}{\partial y} + (2\sigma_y - \sigma_x)\frac{\partial \sigma_y}{\partial y} + 6\tau_{xy}\frac{\partial \tau_{xy}}{\partial y} = 0,$$

provided that $2\sigma_x - \sigma_y \neq 0$. The vanishing of the term $2\sigma_x - \sigma_y$ along $L$ means that $L$ constitutes a characteristic for the stress, and hence also for the velocities.

The second condition, $\partial v_x/\partial x = 0$, shows that along the characteristic the velocity of relative elongation vanishes. We have shown in Chapter 4 that the characteristics under conditions of plane plastic flow also exhibit this property.

### 9.4   Velocity discontinuities

Apart from discontinuities in velocities analogous to those discussed in Chapter 4, devoted to plane flow, there may occur under certain conditions of plane stress new types of discontinuities. These appear to be due to the localized waisting of the material which will be further referred to as necking. This type of discontinuity was investigated by Hill [51].

The mathematical idealization of necking may be obtained by assuming that along a line $L$ there occurs a sudden reduction of thickness of the material. Such a line may be looked upon as the limiting case of a localized waisting having a small width $b \to 0$ (Fig. 161). Such necking

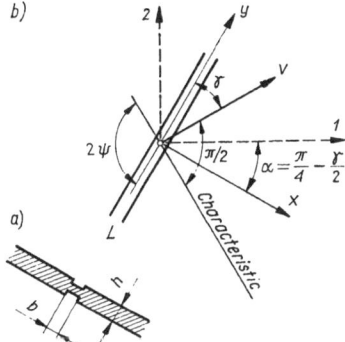

Fig. 161. Idealization of necking in a sheet of metal (R. Hill [51])

is caused by the discontinuity of the normal component of velocity on the line $L$. The discontinuity is connected with the state of stress in the neck.

Let us assume that the relative velocity of displacement of the rigid parts of the material on either side of $L$ is represented by a vector $\mathbf{v}$, inclined to $L$ at an angle $\gamma$ (Fig. 161b). Hence both the normal and the tangential velocity components are discontinuous.

Adopting at a generic point a local coordinate system $x, y$ as shown in Fig. 161b, we have the following components of the strain rate tensor

$$\epsilon_x = \frac{v_x}{b} = \frac{v \sin\gamma}{b}, \quad \epsilon_y = 0, \quad \epsilon_{xy} = \frac{v_y}{2b} = \frac{v \cos\gamma}{2b}.$$

The strain rate $\epsilon_y$ must vanish, because the process of plastic flow is considered to be restricted to the region of necking, while adjacent parts of the material move as rigid bodies. The condition $\epsilon_y = 0$ indicates that the line $L$ constitutes a characteristic for velocities. The other line of vanishing strain rate is clearly the normal to the vector of the velocity discontinuity $\mathbf{v}$. Hence this line constitutes the second characteristic. The velocity component directed along this second characteristic varies continuously across $L$. The angle of inclination of the characteristics is known to be equal to $2\psi$, hence the angle $\gamma$ between the vector $\mathbf{v}$ and the neck is found to be $\gamma = 2\psi - \pi/2$.

If $\gamma = 0$, the vector of velocity discontinuity is parallel to the line $L$ and there occurs merely a relative slip of the material situated on either side of $L$. In this case $v_x = 0$ and no localized necking takes place, which

251

shows that the line $L$ may be looked upon as a line of discontinuity of the same type as that discussed in the theory of plane flow.

The other limiting case is defined by $\gamma = \pi/2$, which means that across the line $L$ only the normal velocity component suffers a jump. The characteristics of both families coincide, because the angle of mutual inclination is found to be $2\psi = \gamma + \pi/2 = \pi$. We have now $v_y = 0$, and $\epsilon_y = \epsilon_{xy} = 0$. The localized necking along $L$ is caused by the strain rate $\epsilon_x = v/b$ only. The stress state in the neck corresponds to the parabolic points on the yield ellipse (Fig. 160), e.g. where $\sigma_1 = 2k$, $\sigma_2 = k$.

In the general case the values of the principal strain rates can be obtained by means of the known transformation formulae or from Mohr's representation, bearing in mind that the angle $\alpha$ between the principal axis $1$ and the $x$-asis is equal to $\alpha = \pi/4 - \gamma/2$. The principal strain rates are found to be

$$\epsilon_1 = \frac{v}{2b}(1+\sin\gamma), \quad \epsilon_2 = -\frac{v}{2b}(1-\sin\gamma), \quad \epsilon_3 = \epsilon_z = -\frac{v}{b}\sin\gamma.$$

The third principal strain rate $\epsilon_3 = \epsilon_z$ in the normal direction to the $xy$-plane has been obtained on account of the incompressibility of the material.

We determine the stress state in the neck from the yield condition

$$\sigma_1^2 - \sigma_1\sigma_2 + \sigma_2^2 = 3k^2,$$

and the relations between the strain rates and the stress components

$$\frac{\epsilon_1}{2\sigma_1 - \sigma_2} = \frac{\epsilon_2}{2\sigma_2 - \sigma_1}.$$

The principal stresses are found to be

$$\sigma_1 = k\frac{3\sin\gamma + 1}{\sqrt{1 + 3\sin^2\gamma}},$$

$$\sigma_2 = k\frac{3\sin\gamma - 1}{\sqrt{1 + 3\sin^2\gamma}}.$$

The angle $\gamma$ is determined by

$$\sin\gamma = \frac{\sigma_1 + \sigma_2}{3(\sigma_1 - \sigma_2)}.$$

In the particular case of a strip subjected to uniaxial tension we have (Fig. 162): $\sigma_1 = \sqrt{3}k$, $\sigma_2 = 0$ and $\gamma = \arcsin 1/3 = 19° 28'$. The inclina-

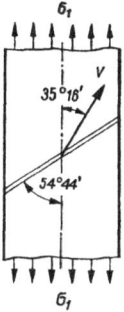

Fig. 162. Oblique necking of a strip in tension (R. Hill [51])

tion of the neck with respect to the axis of the strip is $\psi = \pi/4 + \gamma/2$ $= 54° 44'$, whereas the vector of discontinuity in velocity will be inclined at $\alpha = \pi/4 - \gamma/2 = 35° 16'$.

## 9.5 Solution of plane stress equations under the Tresca yield condition

The geometrical representation of Tresca's yield condition in the case of plane stress is a hexagon in the $\sigma_1 \sigma_2$-plane (Fig. 163).

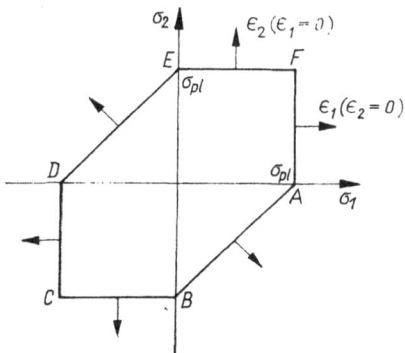

Fig. 163. Tresca hexagon for the plane state of stress

Let us consider first the case where the principal stresses have opposite signs, which corresponds to the segments $AB$ and $DE$ in Fig. 163. The yield condition is then expressed by equation (9.5) which is the same

253

as that for plane flow. To satisfy the yield condition identically, let us express the stress components in terms of a new function $\chi$ and the angle $\varphi$ between the direction of the greater principal stress and the x-axis

$$\sigma_x = 2k\chi + k\cos 2\varphi,$$
$$\sigma_y = 2k\chi - k\cos 2\varphi, \tag{9.21}$$
$$\tau_{xy} = k\sin 2\varphi.$$

Substituting (9.21) in the equilibrium equations (9.1) we obtain a set of two partial differential, quasi-linear equations in the two unknown functions $\chi$ and $\varphi$:

$$\cos 2\varphi \frac{\partial \varphi}{\partial x} + \sin 2\varphi \frac{\partial \varphi}{\partial y} + \frac{\partial \chi}{\partial y}$$
$$= -\frac{1}{2}\sin 2\varphi \frac{\partial \ln h}{\partial x} - \frac{1}{2}(2\chi - \cos 2\varphi)\frac{\partial \ln h}{\partial y},$$
$$-\sin 2\varphi \frac{\partial \varphi}{\partial x} + \cos 2\varphi \frac{\partial \varphi}{\partial y} + \frac{\partial \chi}{\partial x} \tag{9.22}$$
$$= -\frac{1}{2}(2\chi + \cos 2\varphi)\frac{\partial \ln h}{\partial x} - \frac{1}{2}\sin 2\varphi \frac{\partial \ln h}{\partial y}.$$

This system is always hyperbolic; hence it possesses two families of real characteristics. Proceeding in the same manner as previously, we obtain the following differential equations of characteristics

$$\frac{dy}{dx} = \tan\left(\varphi \pm \frac{\pi}{4}\right),$$
$$d\varphi \pm d\chi = \left[\chi \sin 2\varphi \frac{\partial \ln h}{\partial x} + \frac{1}{2}(1 - 2\chi\cos 2\varphi)\frac{\partial \ln h}{\partial y}\right]dx - \tag{9.23}$$
$$- \left[\frac{1}{2}(1 + 2\chi\cos 2\varphi)\frac{\partial \ln h}{\partial x} + \chi \sin 2\varphi \frac{\partial \ln h}{\partial y}\right]dy,$$

the upper, positive sign, and the lower, negative sign corresponding to the first and to the second family of characteristics, respectively.

Assuming a uniform distribution of thickness, we arrive at the classic differential equations of characteristics, for the plane state of stress

$$\frac{dy}{dx} = \tan\left(\varphi \pm \frac{\pi}{4}\right), \quad \chi \pm \varphi = \text{const.} \tag{9.24}$$

These are seen to be identical with the respective equations for plane plastic flow. However, it should be emphasized that in plane flow they bear a general meaning, i.e. they are valid under both the Tresca and the Huber–Mises yield conditions, and hold for an arbitrary state of stress. But now they are valid merely for the Tresca yield condition in the case of a precisely defined state of stress, viz. in the range $\sigma_1 \sigma_2 \leqslant 0$.

There exists an important physical interpretation of the characteristics given by (9.23) or in the special case by (9.24). They are inclined at the angles $\pm \pi/4$ to the principal directions at a generic point, hence they may be considered as the lines of maximum shear stress or, to put it differently, as slip-lines.

It can readily be verified that the flow law (9.7) associated with the Tresca yield condition (9.5) is identical with that for plane plastic flow [compare relations (4.8) in Chapter 4]. Thus the velocity analysis given in Chapter 4 applies also in the case of plane stress considered here. An important feature of the velocity field is the fact that its characteristics coincide with the characteristics for stresses.

Let us discuss now the case where the principal stresses have the same sign, in other words when $\sigma_1 \sigma_2 \geqslant 0$ or $\sigma_x \sigma_y \geqslant \tau_{xy}^2$. These stress states are represented by segments $BC$, $CD$, $EF$ and $AF$ (Fig. 163). A solution of the stress equations for $h = \text{const}$ was given by V. V. Sokolovsky [110]. We shall present it in a slightly different manner, assuming that $h = h(x, y)$ is known. The yield condition (9.8) is identically satisfied if we introduce a new function $\lambda$ related to the principal stresses by

$$\tfrac{1}{2}(\sigma_1 - \sigma_2) = \lambda \sigma_{\text{pl}}, \qquad \tfrac{1}{2}|\sigma_1 + \sigma_2| = (1 - \lambda)\sigma_{\text{pl}}. \tag{9.25}$$

This simply means that one of the principal stresses is equal to the yield point in uniaxial tension, which can readily be seen by adding these two equations. This is consistent with Fig. 163. Making use of Mohr's circle, we find the stress components to be

$$\begin{aligned} \sigma_x &= \sigma_{\text{pl}}[\varkappa(1-\lambda) + \lambda \cos 2\varphi], \\ \sigma_y &= \sigma_{\text{pl}}[\varkappa(1-\lambda) - \lambda \cos 2\varphi], \end{aligned} \qquad \tau_{xy} = \sigma_{\text{pl}} \lambda \sin 2\varphi, \tag{9.26}$$

where $\varkappa = +1$ for $\sigma_1 > 0$, $\sigma_2 > 0$, and $\varkappa = -1$ for $\sigma_1 < 0$, $\sigma_2 < 0$. $\varphi$ is the angle between the direction of the larger principal stress and the $x$-axis.

Substituting (9.26) in the equilibrium equations (9.1), we obtain after rearrangements the equation

$$2\lambda\sin2\varphi\frac{\partial\varphi}{\partial x}-2\lambda(\varkappa+\cos2\varphi)\frac{\partial\varphi}{\partial y}$$

$$= (\varkappa+\cos2\varphi)\frac{\partial\ln h}{\partial x}+\sin2\varphi\frac{\partial\ln h}{\partial y}, \tag{9.27}$$

at the left-hand side of which there appear the derivatives of one unknown function $\varphi$ only. It is known that such an equation has one family of real characteristics (see Appendix 2). The differential equation of characteristics has the form

$$\frac{dy}{dx} = -\frac{\varkappa+\cos2\varphi}{\sin2\varphi}. \tag{9.28}$$

We shall not give here the differential equations that must be satisfied along these characteristics neither shall we exhibit the second equation, resulting from substituting (9.26) in the equilibrium equations, and containing the derivatives of the angle $\varphi$ as well as the function $\lambda$. These equations are rather cumbersome and their significance for solving specific problems is limited.

Nevertheless even the incomplete analysis offered here justifies the conclusion that a problem of the parabolic type is involved, because the equations have one family of real characteristics expressed by the differential equation (9.28). These characteristics, as in the case of the principal stress components with opposite signs, may be referred to as slip-lines. Let us discuss this problem in more detail. If the principal stresses have opposite signs, the planes of maximum shear stress are normal to the median plane of a sheet (Fig. 164a), and inclined at 45° to the principal

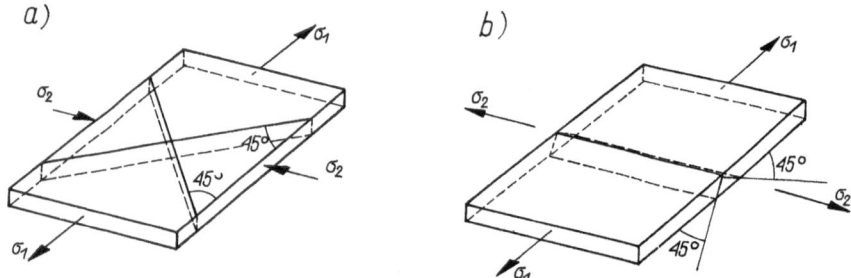

Fig. 164. Planes of maximum shear stresses; a) principal stresses with opposite signs, b) principal stresses with identical signs

directions. The intersections of these planes with the surface of the sheet constitute the slip lines which coincide with the characteristics of the basic system of equations (9.22). There are two planes of this kind at a generic point, giving rise to two slip-lines. However, if the signs of the principal stresses are identical (Fig. 164b), the planes of maximum shear stress intersect the surface of the material at an angle of 45° and, although there exist two planes of this kind at a generic point, Fig. 164b clearly shows that they intersect the surface along a single, common straight line. Here, the single family of slip-lines coincides with the analytically obtained family of characteristics of the differential equation (9.27).

## 9.6 Plane stress problems under axial symmetry

The theory of axi-symmetrical plane stress problems has important practical applications. It leads also to a better understanding of the general plane stress theory, since it permits to obtain solutions in which both the variation in the wall-thickness as the result of the deformation process, and strain-hardening in its general form are taken into account.

Let us now replace the Cartesian system of coordinates $x$, $y$ by the polar coordinates $r, \theta$. Because of the assumed axial symmetry, all the magnitudes will be independent of $\theta$, thus being functions of the radius $r$ and the time $t$ only. The stress state is defined by two stress components: The radial stress $\sigma_r$ and the circumferential stress $\sigma_\theta$. Only such problems will be considered for which $\tau_{r\theta} = 0$. Thus the radial and circumferential directions coincide with the directions of the principal stresses $\sigma_r$ and $\sigma_\theta$.

The equilibrium conditions reduce to one equation

$$\frac{d}{dr}(\sigma_r rh) - \sigma_\theta h = 0. \tag{9.29}$$

The incompressibility condition

$$\epsilon_r + \epsilon_\theta + \epsilon_h = 0$$

must hold with the strain rates $\epsilon_r$ and $\epsilon_\theta$ expressed in terms of the radial velocity $v$ as follows:

$$\epsilon_r = \frac{\partial v}{\partial r}, \quad \epsilon_\theta = \frac{v}{r}. \tag{9.30}$$

Due to axial symmetry, the velocity field is uniquely determined by the radial component $v_r = v(r, t)$. The strain rate $\epsilon_h$ across the thickness will be determined by making use of the formula for a thickness increment

$$\mathrm{d}h = \frac{\partial h}{\partial r}\,\mathrm{d}r + \frac{\partial h}{\partial t}\,\mathrm{d}t.$$

We get

$$\epsilon_h = \frac{1}{h}\left(\frac{\partial h}{\partial t} + v\frac{\partial h}{\partial r}\right).$$

Substitution of this expression and of expressions (9.30) into the incompressibility condition leads to its final form

$$\frac{1}{h}\left(\frac{\partial h}{\partial t} + v\frac{\partial h}{\partial r}\right) + \frac{\partial v}{\partial r} + \frac{v}{r} = 0. \tag{9.31}$$

The theory of axially symmetric plane stress problems assumes different forms depending on which of the two basic yield conditions is employed.

The Huber–Mises yield condition is given by

$$\sigma_r^2 - \sigma_r\sigma_\theta + \sigma_\theta^2 = 3k^2. \tag{9.32}$$

The flow rule associated with (9.32) takes the form

$$\frac{\epsilon_r}{2\sigma_r - \sigma_\theta} = \frac{\epsilon_\theta}{2\sigma_\theta - \sigma_r}. \tag{9.33}$$

Substituting (9.30) into (9.33), we obtain the following relation

$$\frac{1}{2\sigma_r - \sigma_\theta}\frac{\partial v}{\partial r} = \frac{1}{2\sigma_\theta - \sigma_r}\frac{v}{r} \tag{9.34}$$

between the velocity $v$ and the stress components.

As for the Tresca yield condition, two cases must be distinguished. If the stresses $\sigma_r$ and $\sigma_\theta$ have opposite signs, i.e. if $\sigma_r\sigma_\theta \leqslant 0$, the Tresca yield condition assumes the form

$$|\sigma_r - \sigma_\theta| = 2k. \tag{9.35}$$

Since the stress components $\sigma_r$ and $\sigma_\theta$ prove to be principal stresses, the various plane states of stress under axial symmetry will be represented by the hexagon shown in Fig. 163. The yield condition (9.35), holding for the case $\sigma_r\sigma_\theta \leqslant 0$, corresponds to the edges $AB$ and $DE$ of the hexagon.

According to the associated flow rule a vector of strain rate must be perpendicular to the yield surface. Thus from Fig. 163 we immediately obtain that for the edges $AB$ and $DE$

$$\epsilon_r = -\epsilon_\theta. \tag{9.36}$$

Thus to ensure that the incompressibility condition is satisfied, the strain rate component $\epsilon_h$ must vanish, i.e. $\epsilon_h = 0$. This means that the thickness of the sheet does not vary during the process of deformation.

In the other instance of the Tresca yield condition, when the signs of both principal stresses are the same, or in other words when $\sigma_r \sigma_\theta \geqslant 0$, we have

$$|\sigma_r| = \sigma_{p1} \quad \text{for} \quad |\sigma_r| > |\sigma_\theta|,$$
$$|\sigma_\theta| = \sigma_{p1} \quad \text{for} \quad |\sigma_\theta| > |\sigma_r|. \tag{9.37}$$

The associated flow law may be obtained from Fig. 163. Thus we have

$$\epsilon_\theta = 0 \text{ for } |\sigma_r| = \sigma_{p1} \text{ and } \epsilon_r = 0 \text{ for } |\sigma_\theta| = \sigma_{p1}. \tag{9.38}$$

One of the strain rate components must obviously vanish, because the states of stress considered correspond to the edges $BC$, $CD$, $EF$ and $AF$ of the Tresca hexagon (Fig. 163). The strain rate vector which is perpendicular to an edge is parallel to the respective axis of the coordinate system.

In the present study the Huber–Mises yield condition (9.32) will be used along with the relation (9.34).

To solve a specific problem, a set of four equations must be tackled: The equilibrium equation (9.29), the incompressibility condition (9.31), the yield condition (9.32), and the flow law (9.34). The set contains four unknowns $\sigma_r$, $\sigma_\theta$, $v$ and $h$.

## 9.7  Plastic deformation of flat rings

The analysis of large plastic deformations of flat rings presented here finds numerous practical applications, which will be considered in the following sections. First the problem will be solved neglecting the strain-hardening of the material. In the next section a simple method of solution accounting for the strain-hardening phenomenon is presented. In both cases the Huber–Mises yield condition and the associated flow law will be assumed, since they do well agree with the real behaviour of metals.

The Tresca yield condition, and particularly the flow law associated to it, lead to practically unacceptable results concerning mainly the variation of the wall-thickness of the deformed ring. In some works, for example in Hill's book [50], the Tresca yield condition is used together with the flow law (9.34) associated to the Huber–Mises condition. Such an approach gives results close to those obtained for the Huber–Mises yield condition, and it is mathematically slightly simpler than that presented below.

Consider a flat ring having a uniform initial thickness $h_0 = $ const, and initial outer and inner radii denoted by $R_0$ and $r_0$, respectively. The ring can be loaded by uniformly distributed tractions at the outer or inner rim. Four of the possible combinations of the loading mode are shown in Fig. 165. Let us denote by $b$ the time-dependent radius of the stress free (inner or outer) edge. If we assume the radial velocity of that edge to be equal to unity, i.e. $v = 1$, the magnitude of the radius $b$ may be looked

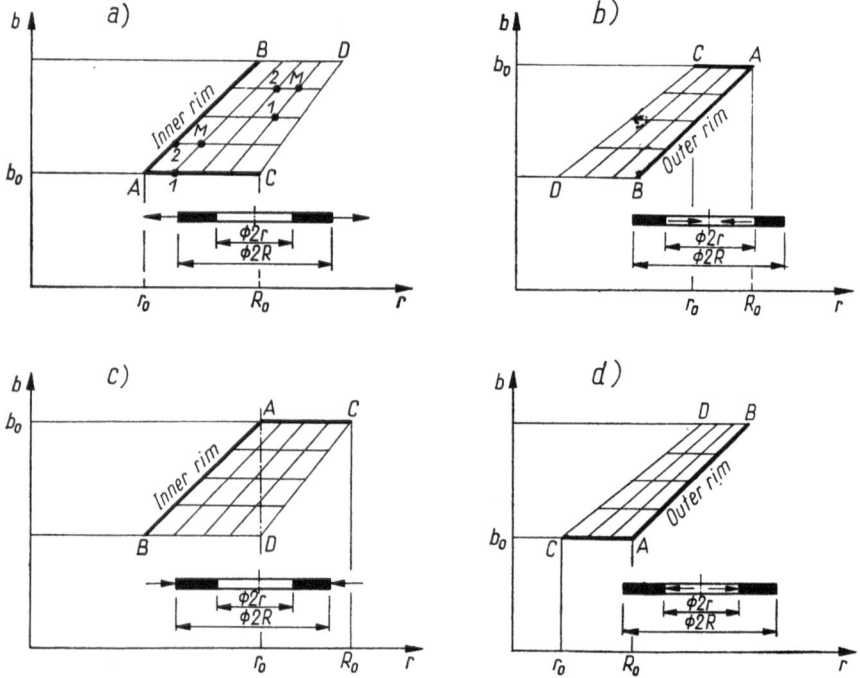

Fig. 165. Characteristic boundary and initial value problems for various modes of deformation of flat rings [126]

upon as a measure of time [50]. On account of this the incompressibility condition (9.31) may be written down in the slightly different form

$$\frac{1}{h}\left(\frac{\partial h}{\partial b}+v\frac{\partial h}{\partial r}\right)+\frac{\partial v}{\partial r}+\frac{v}{r}=0. \tag{9.39}$$

Eliminating the derivative $\partial v/\partial r$ by means of equation (9.34), we obtain the equation

$$\frac{\partial h}{\partial b}+v\frac{\partial h}{\partial r}=\frac{hv}{r}\cdot\frac{\sigma_r+\sigma_\theta}{\sigma_r-2\sigma_\theta},$$

which now contains the derivatives of one function $h$ only. Hence the first family of characteristics is given by the differential equations (see Appendix 2)

$$dr-vdb=0,\qquad \frac{dh}{dr}=\frac{h}{r}\frac{\sigma_r+\sigma_\theta}{\sigma_r-2\sigma_\theta}, \tag{9.40}$$

whereas the differential equations of the second family of characteristics

$$b=\text{const},\qquad \frac{dv}{dr}=-\frac{v}{r}\frac{2\sigma_r-\sigma_\theta}{2\sigma_\theta-\sigma_r}. \tag{9.41}$$

follow directly from equation (9.34). Along the characteristics of this family we have the following differential equation

$$\frac{dh}{dr}=-\frac{h}{r}\frac{\sigma_r-\sigma_\theta}{\sigma_r}-\frac{h}{\sigma_r}\frac{d\sigma_r}{dr}, \tag{9.42}$$

derived from the equation of equilibrium (9.29). These characteristics represent various stages of the advancing process defined by an instantaneous magnitude of the free edge radius $b$. The stresses along the radius for each instant must obviously satisfy the equilibrium condition.

The set of equations (9.29), (9.34) and (9.39) possesses, as we have shown above, two families of characteristics; hence it proves to be of the hyperbolic type. The equations of characteristics (9.40)–(9.42) must be supplemented by the Huber–Mises yield condition (9.32). This condition is seen to be identically satisfied if the stress components $\sigma_r$ and $\sigma_\theta$ are expressed in terms of an auxiliary function $\omega$, similarly as the principal stress components were expressed by (9.15) in the general plane stress theory. Thus we arrive at the relations

$$\sigma_r=2k\cos(\omega-\tfrac{1}{6}\pi),$$
$$\sigma_\theta=2k\cos(\omega+\tfrac{1}{6}\pi). \tag{9.43}$$

Substituting these relations into the differential equations of characteristics, we obtain the final form of the equations for the first family of characteristics

$$dr - v\,db = 0, \qquad \frac{dh}{dr} = -\frac{h}{r}\lambda(\omega), \tag{9.44}$$

where

$$\lambda(\omega) = \frac{\cos(\omega - \tfrac{1}{6}\pi) + \cos(\omega + \tfrac{1}{6}\pi)}{2\cos(\omega + \tfrac{1}{6}\pi) - \cos(\omega - \tfrac{1}{6}\pi)},$$

and for the second family, where $b = $ const, we have

$$\frac{dv}{dr} = \frac{v}{r}\varphi(\omega), \tag{9.45a}$$

where

$$\varphi(\omega) = \frac{2\cos(\omega - \tfrac{1}{6}\pi) - \cos(\omega + \tfrac{1}{6}\pi)}{2\cos(\omega + \tfrac{1}{6}\pi) - \cos(\omega - \tfrac{1}{6}\pi)},$$

and

$$\frac{dh}{dr} = -\frac{h}{r}\left[1 - \frac{\cos(\omega + \tfrac{1}{6}\pi)}{\cos(\omega - \tfrac{1}{6}\pi)}\right] + h\tan(\omega - \tfrac{1}{6}\pi)\frac{d\omega}{dr}. \tag{9.45b}$$

Thus the problem has been reduced to the integration of the ordinary differential equations (9.44) and (9.45). The integration is performed by solving the respective boundary and initial value problems for these equations.

Let us consider these problems in the plane of the independent variables $r, b$. All cases shown in Fig. 165 reduce to the characteristic problem, i.e. to the problem in which all the magnitudes that matter are known in advance along the two starting characteristics belonging to different families [126]. It is readily seen that characteristics of the first family (9.44) represent the paths of particular particles in the $r, b$-plane, and the magnitudes found along them describe the stress and deformation history which a particle undergoes. One of those characteristics, shown in Fig. 165 as $AB$, corresponds to the stress free rim of the ring. Hence we have along it $\sigma_r = 0$. Hence on account of (9.43) we have along $AB$ $\omega = 2\pi/3$ or $\omega = 5\pi/3$, depending on whether the circumferential stress component $\sigma_\theta$ is compressive or tensile, respectively. The velocity of the stress free rim is also known in advance since we assumed that $v = 1$ holds there. Furthermore, the differential equations (9.44) for the thickness $h$ may be integrated along this characteristic, because the value of $\omega$ remains constant on *AB*. For $\omega = 2\pi/3$ and $\omega = 5\pi/3$ we obtain after integration is done the same relation between the thickness of the ring at the free rim and its changing radius $b$, namely $h = h_0/\sqrt{b}$, assuming without loss of generality that the initial radius of the free rim is equal to unity, i.e. that $b_0 = 1$. Thus all the magnitudes, namely $\omega$, $v$ and $h$ are known along the characteristic *AB*, representing the stress-free rim (outer or inner), depending on the loading mode. Note that according to equation (9.44) the characteristic *AB* is in any case rectilinear since we have assumed $v = 1$ at the free rim.

The characteristics of the second family (9.45), i.e. the lines $b = $ const, represent the various stages of the process, corresponding to the instantaneous values of the radius $b$ of the free rim. The magnitudes $h$, $\sigma_r$ and $\sigma_\theta$ found along these characteristics show the variation of the thickness and stress components along the radius for a fixed $b$. At the beginning of the process, i.e. for $b = b_0 = 1$, the thickness of the ring is constant and equal to $h_0$. In other words, along the characteristic $b = b_0$ we have $h = h_0 = $ const. This characteristic is denoted in Fig. 165 by *AC*. At the point *A*, corresponding to the non-loaded rim, we must evidently have $\omega = 2\pi/3$ or $\omega = 5\pi/3$. It follows that equation (9.45b) is suitable for numerical integration, for example by means of the Runge–Kutta method, thus yielding the values of the function $\omega$ along *AC*. Similarly the velocity distribution along *AC* can be found by integrating equation (9.45a) in which the function $\varphi(\omega)$ is already known, and the initial condition is obviously $v = 1$ at the point *A*. Thus all sought for magnitudes $\omega$, $v$, and $h$ along the characteristic *AC* of the second family can be found.

These data together with the previously found data along *AB*, constitute the characteristic boundary value problem, and hence they permit to find all the sought for functions inside the entire quadrangle *ABDC* in the $r$, $b$-plane. It is seen that there is no limitation on the position of the point *B* on the characteristic line *AB*. Hence the solution can be extended arbitrarily far. In other words, it can be obtained for arbitrarily large deformations. It retains of course a physical sense only until the decohesion of the material or loss of stability occurs. Note that characteristics of the first family (9.44) are curvilinear, except for the starting line *AB*.

The equations of characteristics are integrated numerically, replacing differentials by finite differences. Let us consider three neighbouring points *1, 2,* and *M*, where the sector *1–M* constitutes a segment of the characteristic of the first family, while *2–M* represents a segment of the characteristic of the second family (Fig. 165a). If all functions are given at the points *1* and *2*, the position of the third point $M$ in the $r$, $b$-plane and all magnitudes at this point may be found by means of the recurrence formulae. The coordinate $b_M$ of the point $M$ is equal to the coordinate of the point 2, i.e. $b_M = b_2$, because the two points lie on the same characteristic $b$ = const. The abscissa $r_M$ can be found from the equation $dr - v\,db = 0$ of the characteristic of the first family joining both points *1* and $M$

$$r_M = r_1 + v_1(b_M - b_1).$$
(9.46)

The suffix notations 1 and $M$ indicate that the term has the value corresponding to the point *1* or $M$.

The wall-thickness $h_M$ at the point $M$ is obtained from the relation [compare (9.44)]

$$h_M = h_1 - \frac{h_1}{r_1}(r_M - r_1)\,\lambda(\omega)_1.$$
(9.47)

Now the value $\omega_M$ can be found from the formula

$$\omega_M = \omega_2 + \frac{h_M - h_2}{h_2 \tan(\omega - \frac{1}{6}\pi)_2} + \frac{r_M - r_2}{r_2 \tan(\omega - \frac{1}{6}\pi)_2}\left[1 - \frac{\cos(\omega + \frac{1}{6}\pi)}{\cos(\omega - \frac{1}{6}\pi)}\right]_2,$$
(9.48)

which is a consequence of equation (9.45b). The velocity $v_M$ can be obtained from equation (9.45a) in the form of the finite differences formula

$$v_M = v_2 + (r_M - r_2)\frac{v_2}{r_2}\varphi(\omega)_2.$$
(9.49)

The so obtained magnitudes at the point $M$ constitute the first approximation of the solution at that point. If a more accurate analysis is needed, the calculations should be repeated taking in the foregoing recurrence formulae the mean values of all magnitudes. Thus the second approximation of the coordinate $r'_M$ is computed from the equation

$$r'_M = r_1 + \tfrac{1}{2}(v_1 + v_M)\,(b_M - b_1),$$
(9.46a)

instead of (9.46). The term $v_M$ represents the value of the velocity at $M$ obtained from the first approximation. Instead of (9.47) we have now

$$h'_M = h_1 - \frac{h_1 + h_M}{r_1 + r_M}\cdot\frac{1}{2}[\lambda(\omega)_1 + \lambda(\omega)_M]\,(r_M - r_1),$$
(9.47a)

where $h_M$ and $\lambda(\omega)_M$ represent the first approximations obtained previously.

Similarly the formula for the second approximation of $\omega_M$ takes the form

$$\omega'_M = \omega_2 + \frac{4(h_M - h_2)}{(h_2 + h_M)\,[\tan(\omega - \frac{1}{6}\pi)_2 + \tan(\omega - \frac{1}{6}\pi)_M]} +$$
$$+ \frac{2(r_M - r_2)}{(r_M + r_2)\,[\tan(\omega - \frac{1}{6}\pi)_2 + \tan(\omega - \frac{1}{6}\pi)_M]} \times$$
$$\times \left[ 2 - \frac{\cos(\omega + \frac{1}{6}\pi)_2}{\cos(\omega - \frac{1}{6}\pi)_2} - \frac{\cos(\omega + \frac{1}{6}\pi)_M}{\cos(\omega - \frac{1}{6}\pi)_M} \right]. \tag{9.48a}$$

Instead of (9.49) we have for the second approximation

$$v'_M = v_2 + (r_M - r_2)\,\frac{v_2 + v_M}{r_2 + r_M} \cdot \frac{1}{2}\,[\varphi(\omega)_2 + \varphi(\omega)_M]. \tag{9.49a}$$

If necessary, a third and further approximations can be computed in the same manner, but usually the second approximation gives satisfactory results.

To begin the numerical computations the starting characteristics $AB$ and $AC$ should be divided into an appropriate number of segments. Then at the nodal points of those segments all necessary values should be determined. The first three points permitting to start computations according to the described procedure are those in the vicinity of the point $A$.

As an example let us consider the process of expanding a flat ring

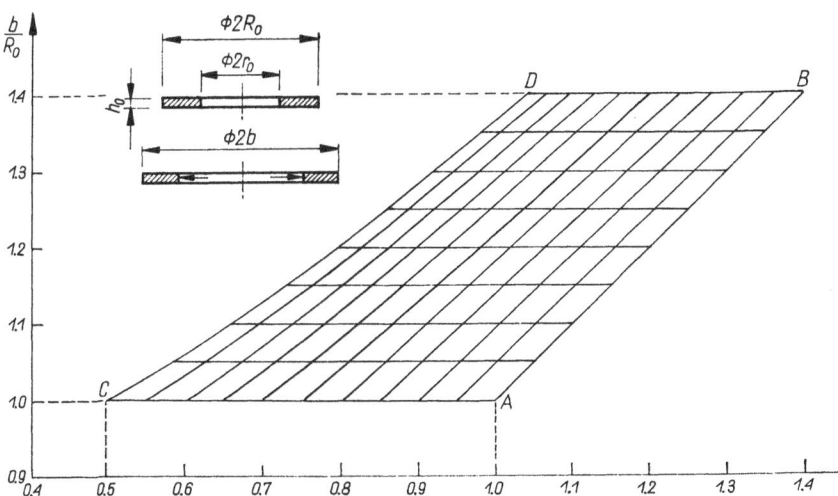

Fig. 166. The mesh of characteristics for the process of expanding a flat ring [126]

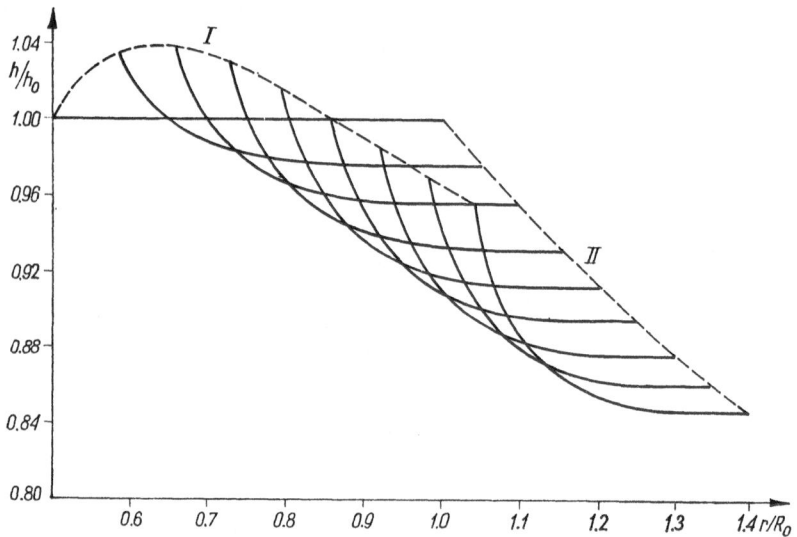

Fig. 167. Variation in wall thickness in the process of expanding a flat ring [126]

loaded by a pressure applied inside the hole as shown schematically in
Fig. 165d. Along the characteristics *AB* of the first family, which corresponds to the stress-free outer edge, we have $\omega = 5\pi/3$, $h = h_0/\sqrt{b}$ and
$v = 1$. Along the characteristic *AC* of the second family there is $h = h_0$
= const, where $h_0$ is the sheet thickness before deformation. The calculated
mesh of characteristics is shown in Fig. 166. Figure 167 shows how the wall-
thickness varies along the radius during some stages of the expansion
process. In the vicinity of the hole appears a specific thickening of the sheet.

### 9.8   Strain-hardening solutions of
### axially symmetric plane stress problems

In the axially symmetric problems considered in the previous section the
strain-hardening of the material can be accounted for in a relatively simple
way. This can be done following the author's method [126] which will be
presented in this section. Assume a general strain-hardening hypothesis,
according to which the yield surface is simultaneously expanded and trans-
lated in the stress space. This comprises, as the particular cases, the isotropic and the kinematic strain-hardening hypotheses. For the initial yield

surface we take the Huber–Mises yield ellipse. Instead of (9.32) we have now

$$(\sigma_r - a_r)^2 - (\sigma_r - a_r)(\sigma_\theta - a_\theta) + (\sigma_\theta - a_\theta)^2 = 3k^2, \tag{9.50}$$

where $a_r$, $a_\theta$, and $k$ are certain functions of strains,

$$a_r = a_r(\varepsilon_r, \varepsilon_\theta), \quad a_\theta = a_\theta(\varepsilon_r, \varepsilon_\theta), \quad k = k(\varepsilon_i). \tag{9.50a}$$

Among the particular cases covered by the condition (9.50) are the kinematic hardening rule, which we obtain by putting $k = k_0 = \text{const}$, and the isotropic hardening rule, for which $a_r = a_\theta = 0$. Assuming the linear hardening law, we have for the kinematic hardening rule

$$a_r = c(2\varepsilon_r + \varepsilon_\theta), \quad a_\theta = c(2\varepsilon_\theta + \varepsilon_r). \tag{9.51}$$

The flow law associated with the yield condition (9.50) takes the following form

$$\frac{\partial v}{\partial r} = \frac{v}{r} \frac{2(\sigma_r - a_r) - (\sigma_\theta - a_\theta)}{2(\sigma_\theta - a_\theta) - (\sigma_r - a_r)}. \tag{9.52}$$

The set of four equations (9.50), (9.52), (9.39), and (9.29), in the four unknowns $\sigma_r$, $\sigma_\theta$, $h$ and $v$, is the basis for the calculation of the distribution of stress and strain. This set will be solved by means of the method of successive approximations aided by the observation that the strain-hardening of the material has merely a small influence upon the variation of the thickness of the sheet. Hence as the first approximation a solution without hardening may be taken, and then further approximations can be computed. Having found the solution for the non-hardening material according to the procedure described in the foregoing section, and particularly the distribution of the wall-thickness along the radius, we can calculate the strain increments

$$d\varepsilon_\theta = \frac{dr}{r},$$

$$d\varepsilon_h = \frac{dh}{h}, \tag{9.53}$$

$$d\varepsilon_r = -d\varepsilon_\theta - d\varepsilon_r$$

by dividing the whole deformation process into a sufficiently great number of small steps. The strains themselves can be obtained by summing up the strain increments. The equivalent strain may be found from the formula

$$\varepsilon_i = \int \sqrt{\tfrac{1}{2}(d\varepsilon_\theta^2 + d\varepsilon_h^2 + d\varepsilon_r^2)}. \tag{9.54}$$

The integration should be done for each particle from the beginning of the process, or in other words, along the trajectory of the particle in the $r, b$-plane. The values $\varepsilon_r$, $\varepsilon_\theta$ and $\varepsilon_l$ so obtained permit to obtain for each instant and each radius the magnitudes of $a_r$, $a_\theta$, and $k$ by means of the above relations. Now the procedure of the second approximation can be started by substituting into the yield condition (9.50) and into the relation (9.52) the just now obtained magnitudes $a_r$, $a_\theta$, and $k$ as known functions of the radius $r$ and of the time measured as the actual value of the radius $b$ of the stress-free rim. Thus the problem reduces to the solution for a ring with a known non-homogeneity of the yield locus.

The stress components will be expressed in terms of the auxiliary function

$$\sigma_r = a_r + 2k\cos(\omega - \tfrac{1}{6}\pi),$$
$$\sigma_\theta = a_\theta + 2k\cos(\omega + \tfrac{1}{6}\pi). \tag{9.55}$$

These expressions satisfy identically the yield condition (9.50) for any arbitrary value of $\omega$.

Similarly as in the case of a non-hardening material, the set of governing equations is of the hyperbolic type and it has, therefore, two families of characteristics. The first family of characteristics is determined by the equations

$$dr - v\, db = 0, \qquad \frac{dh}{dr} = -\frac{h}{r}\lambda(\omega). \tag{9.56}$$

The equations of the second family $b = $ const have the form

$$\frac{dv}{dr} = \frac{v}{r}\varphi(\omega), \tag{9.57a}$$

$$\frac{dh}{dr} = -\frac{h}{r}\left[1 - \frac{a_\theta + 2k\cos(\omega + \tfrac{1}{6}\pi)}{a_r + 2k\cos(\omega - \tfrac{1}{6}\pi)}\right] -$$
$$-\frac{h}{a_r + 2k\cos(\omega - \tfrac{1}{6}\pi)}\left[\frac{da_r}{dr} + 2\cos(\omega - \tfrac{1}{6}\pi)\frac{dk}{dr} - \right.$$
$$\left. - 2k\sin(\omega - \tfrac{1}{6}\pi)\frac{d\omega}{dr}\right], \tag{9.57b}$$

where $\lambda(\omega)$ and $\varphi(\omega)$ denote the same auxiliary functions as in equations (9.44) and (9.45).

The boundary value problems and the numerical technique will not be discussed here since they are virtually the same as in the previous case of a non-hardening material. Having calculated the second approximation of the variation of the thickness $h$, we may find the new, corrected magnitudes of $a_r$, $a_\theta$, and $k$, and then start to compute the third approximation, and so on. Concrete examples show that the method is quickly converging. The difference between the second and third approximations does not exceed 3 percent.

As an example let us consider the process of expanding a flat ring loaded by tensile tractions at the outer edge (Fig. 165a). This problem finds important practical applications, since it is directly connected with processes such as the forming of flanges by means of a flat punch (Fig. 168),

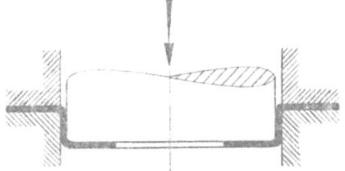

Fig. 168. Forming of a flange by means of a flat punch

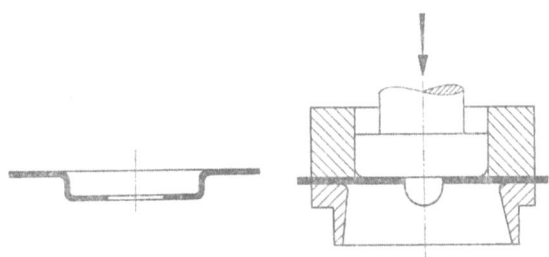

Fig. 169. Siebel and Pomp formability test

and with Siebel's and Pomp's formability test (Fig. 169). In such cases we must of course stop computations at the vertical line passing through the point $C$ in the $r$, $b$-plane. Numerical calculations have been performed under the isotropic strain-hardening hypothesis. The linear relation for the pure shear in the form $\tau = k_0 + c\varepsilon$ has been assumed, where the tangential modulus $c$ is constant and $\varepsilon$ represents half the shear angle $\gamma$. Such a linear relation corresponds to the generalized relation $k = k_0 + c\varepsilon_i$, allowing to find the value of $k$ in each of the successive approximations. The problem

269

has been solved for $c = k_0$ approximately corresponding to mild steel. The net of characteristics obtained for a strain-hardening material is shown in Fig. 170. The difference between this net and the net resulting from the solution for a non-hardening material is insignificant. The two cases differ also very little in the variation of the thickness. Figure 171 shows the solution for the thickness taking into account the strain-hardening. Continuous lines represent the distribution of the thickness along the radius for certain

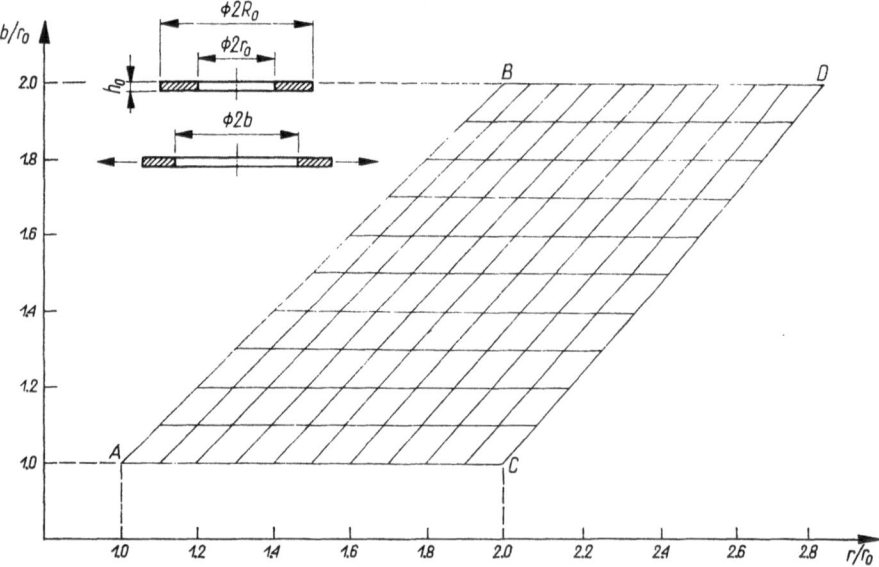

Fig. 170. The mesh of characteristics for the process of expanding of a flat ring loaded by tensile tractions at the outer edge [126]

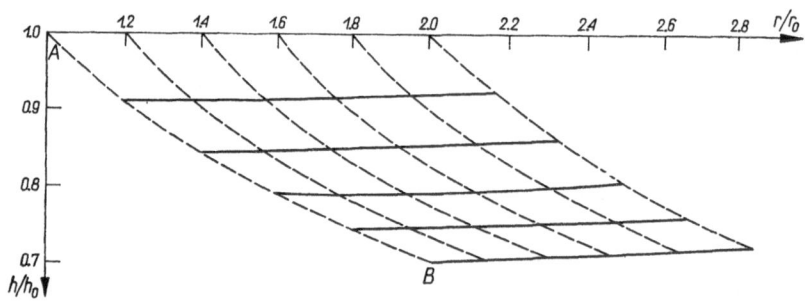

Fig. 171. Variation in wall thickness in the process of expanding a flat ring [126]

270

values of the radius *b* of the expanding hole. Dashed lines show the history of the wall-thickness variation for certain sections of the ring. In particular, the line *AB* represents the thickness variation of the inner edge at various instants. It is seen that the greatest reduction of the thickness takes place at the edge of the hole. At every instant the distribution of the thickness along the radius differs little from the uniform distribution. Figure 172

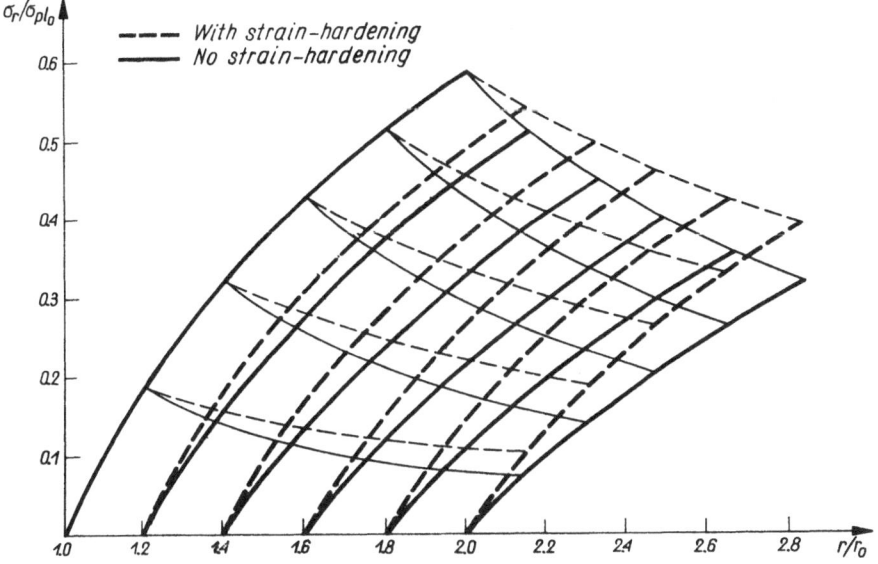

Fig. 172. Radial stresses in the expanded ring

shows the values of the radial stress $\sigma_r$. Continuous lines have been obtained from the solution with no strain-hardening, while dashed lines from the solution with strain-hardening. Heavy lines represent the stress variation along the radius for certain values of the radius of the hole, and thin lines, the history of the radial stress for selected elements of the ring.

## 9.9 Drawing of cups from circular blanks

We shall analyse presently the stress and strain distribution in a circular blank without any hole from which a cylindrical cup is drawn (Fig. 173). The process may be considered as a particular case of deformation of

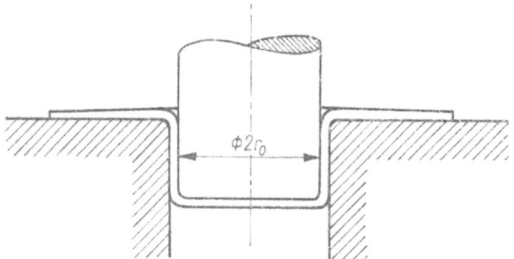

Fig. 173. Drawing of a cup from a circular blank

a ring as shown in Fig. 165b. The computations should be stopped at the vertical line drawn from the point $C$ in the $r, b$-plane, because this line corresponds to the instants at which a material particle reaches the rim of the hole in the die. This problem has been solved by R. Hill [50] for the Tresca yield criterion and the Lévy–Mises relation (9.34). A certain variant of the same problem has been considered by V. V. Sokolovsky [112] who used the Huber–Mises yield condition and the associated flow rule (9.34). From his work originates Fig. 174 showing the variation of the wall-

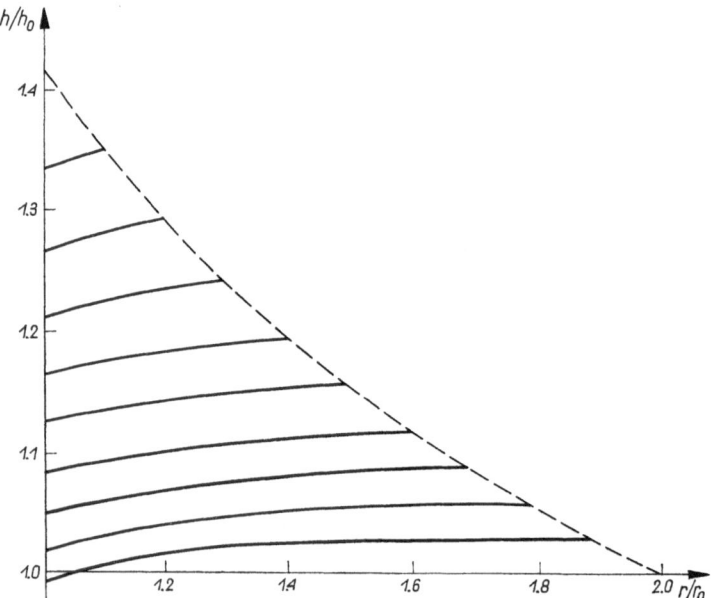

Fig. 174. Variations in wall thickness at various stages of drawing of a cup (V. V. Sokolovsky [112])

thickness at various stages of the drawing. This solution has been obtained for a non-hardening material.

In [112] the same problem has also been solved assuming a linearized yield condition and the associated flow rule. The results obtained, particularly about the variation of the wall-thickness, are completely different from those shown in Fig. 174. First of all, the rise of the thickness towards the hole is obtained, which is not confirmed experimentally. Such a result indicates once more that flow rules associated with linearized yield conditions give unrealistic deformation patterns.

A much more difficult problem, where the blank of the outer radius $R_0$ has a central hole of radius $r_0$, has been solved by Z. Marciniak [84]. The blank, resting on the flat die with a hole of a radius $a$ ($R_0 > a > r_0$), is pushed downwards by a punch as shown in Fig. 175. The problem consists in the analysis what will be the final shape of the drawpiece depending on the ratios of the inner and outer radii of the blank to the radius of the punch.

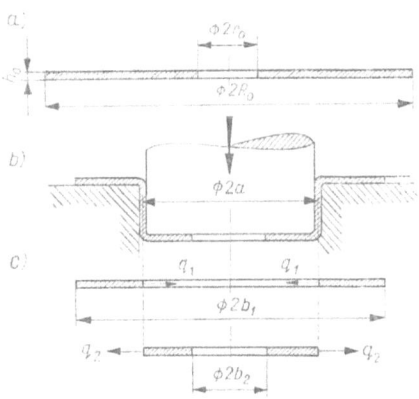

Fig. 175. Drawing of a drawpiece from a circular blank with a central hole

Assume that plastic deformation takes place in the two flat portions only, namely in the flange resting on the die plate and in the bottom part adjacent to the flat foot of the punch. As shown in Fig. 175c, the flange is pulled towards the centre by means of tractions $q_1$ per unit length of circumference, applied at the radius $a$, and the bottom part is extended outwards by means of tractions $q_2$ per unit length, also applied at the radius $a$. The two tractions are equal to the product of the radial stress $\sigma_r$ and the wall-thickness $h$. If the friction forces on the interface between the cylindrical part of the drawpiece and the surface of die hole, and bending effects at the rounded edges of the die hole and the punch are disregarded, the equilibrium condition requires that the equality

$$q_1 = q_2 \qquad\qquad (a)$$

is satisfied at every instant.

The method of solution consists in analysing separately the flange deformation beginning from the initial value of the stress free edge radius $b_1 = R_0$, up to the value $b_1 = a$ of that radius, and the deformation of the bottom part beginning from $b_2 = r_0$ and up to $b_2 = a$. It is readily seen that the process of flange deformation reduces to the problem on Fig. 165b. Computations for an arbitrary $b_1$ should be stopped at the vertical straight line $r = a$ in the $r, b$-plane. This line corresponds to the instants when a flange particle reaches the hole of the die. Similarly deformation of the bottom part of the cup reduces to the scheme on Fig. 165a. Also in this case computations are carried up to the vertical line $r = a$ for any arbitrary value of $b_2$. Solving the two problems in the manner described in the foregoing sections, we find the radial stress $\sigma_r$ and the value of the thickness $h$ at the radius $a$. Then the traction $q_1 = q_1(b_1)$ required to deform plastically the flange of an arbitrary radius $b_1$ may be determined from the relation $q = \sigma_r h$. Similarly the friction $q_2 = q_2(b_2)$, necessary to cause a plastic deformation of the bottom part at an arbitrary instant defined by the value of the hole radius $b_2$, can be found.

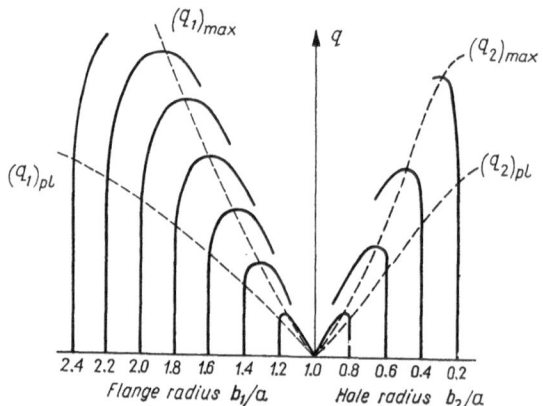

Fig. 176. Tractions at various stages of drawing of a drawpiece from a blank with a hole (after Z. Marciniak [84])

Numerical computations have been performed for mild steel (0.05% C), assuming the hypothesis of isotropic hardening and an actual stress-strain relation given by a diagram obtained from the pure tension test. Figure 176 shows the diagrams of $q_1(b_1)$ and $q_2(b_2)$ obtained for various ratios $R_0/a$ and $r_0/a$ of initial dimensions. If the tractions are small, the material remains rigid, which is represented by initial vertical segments of diagrams. The flange becomes plastic when the tractions reach the value $(q_1)_{pl}$. As the process of deformation is continued, the value of tractions increases due to the strain-hardening of the material until the maximum value $(q_1)_{max}$ is reached. Further deformation of the flange is accompanied by decreasing tractions. Similarly the bottom part reaches the plastic state for $(q_2)_{pl}$. The magnitude $(q_2)_{max}$ represents the maximum value of tractions.

The curves in Fig. 176 and the condition of equilibrium (a) suffice to solve our problem. The course of the process depends on the correlation of four tractions $(q_1)_{pl}$, $(q_1)_{max}$, $(q_2)_{pl}$ and $(q_2)_{max}$ if the ratios $R_0/a$ and $r_0/a$ are fixed. It is most convenient to present the results in the plane of two variables $R_0/a$ and $r_0/a$ (Fig. 177). We can distinguish five regions in that plane, in each of them the deformation of the blank being different. These regions are separated by the lines satisfying the equalities

$$(q_1)_{pl} = (q_2)_{pl}, \quad (q_1)_{max} = (q_2)_{pl},$$

$$(q_1)_{pl} = (q_2)_{max}, \quad (q_1)_{max} = (q_2)_{max}.$$

Different modes of deformation correspond to each of the five regions.

Region I is defined by $(q_1)_{max} < (q_2)_{pl}$. Here the maximum value of the traction $(q_1)_{max}$, occurring during blank deformation, is smaller than the traction $(q_2)_{pl}$ necessary to begin expansion of the bottom part. Hence the bottom remains rigid and the diameter $2r_0$ of the hole does not change.

Region II corresponds to $(q_1)_{pl} < (q_2)_{pl} < (q_1)_{max} < (q_2)_{max}$. Initially, only deformation of the flange takes place, until the traction $q_2$ reaches the value $(q_2)_{pl}$. Then the flange and the bottom begin to deform plastically together. The outer radius of the flange diminishes and simultaneously the hole in the bottom part expands. This stage goes on up to the moment, when the traction $q_1$ reaches the value $(q_1)_{max}$. Further motion of the punch causes deformation of the flange only, the bottom part being unloaded.

Region III is given by $(q_1)_{pl} < (q_2)_{pl} < (q_2)_{max} < (q_1)_{max}$. After an initial plastic flow of the flange, followed by simultaneous flow of both the flange and the bottom, the traction $q_2$ reaches the value $(q_2)_{max}$. From

275

this moment on the flange deformation is stopped, while the bottom part undergoes further deformation.

Region IV is described by $(q_2)_{pl} < (q_1)_{pl} < (q_2)_{max} < (q_1)_{max}$. The course of deformation is similar to that in region III. Now, however, in the initial stage the bottom part only is deformed, and then follows a simultaneous flow of the two parts.

Region V is the one where $(q_2)_{max} < (q_1)_{pl}$. In such a case the maximum value of tractions needed to deform the bottom is smaller than tractions required to start deformation of the flange. Hence the outer radius of the flange remains unchanged.

The four possible types of drawpieces corresponding to different values of the ratios $R_0/a$ and $r_0/a$ are shown in Fig. 177.

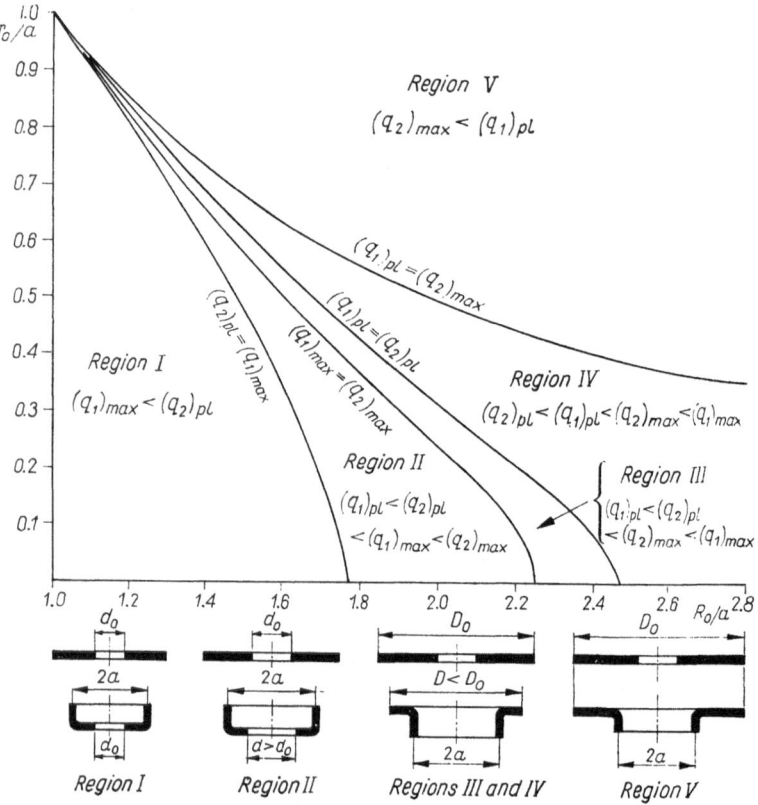

Fig. 177. Regions of various deformation modes of a blank with a central hole (after Z. Marciniak [84])

# Axially symmetric problems of plastic forming of shells under conditions of plane stress

## 10.1 Introduction

In this chapter we present solutions to processes of plastic forming of axially symmetric shells where only one of the surfaces of the shell is in contact with the die. Under such conditions the plane state of stress may be assumed, provided the wall thickness is small when compared with other characteristic dimensions, and particularly with the radii of curvature. The contact pressure $p$ between the surface of the shell and the die is in such a case very small when compared with internal tractions and thus it may be disregarded in the yield condition. Since the wall thickness is small, we may neglect variation of stresses across the thickness. Thus the membrane state of stress may be assumed.

All solutions presented in this chapter have been obtained for the Huber–Mises yield condition and the associated flow rule. These solutions would be radically simplified if the flow law associated with the Tresca yield condition were used. However, deformation patterns so obtained depart very far from the real ones. Some remarks concerning this problem are given at the end of Section 10.2.

Assume an orthogonal system of coordinates $\alpha, \theta$ with the $\alpha = $ const-lines coinciding with the parallel circles, and the $\theta = $ const-lines forming meridians on the surface of the shell (Fig. 178). Let us denote the normal stress acting on the meridian section by $\sigma_\theta$, and the stress on the parallel section by $\sigma_r$. On account of the axial symmetry assumed we have obviously $\tau_{r\theta} = 0$. Consider an infinitesimal element of the shell bounded by two pairs of the coordinate lines $\alpha = $ const, $\alpha + d\alpha = $ const and $\theta = $ const, $\theta + d\theta = $ const (Fig. 179). It is readily seen that the $\alpha$-coordinate represents the angle between the normal to the shell and the axis of sym-

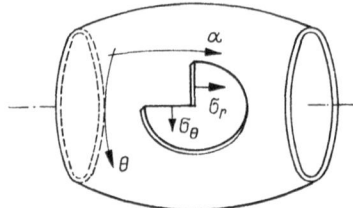

Fig. 178. Coordinate axes for the analysis of plastic deformation of shells

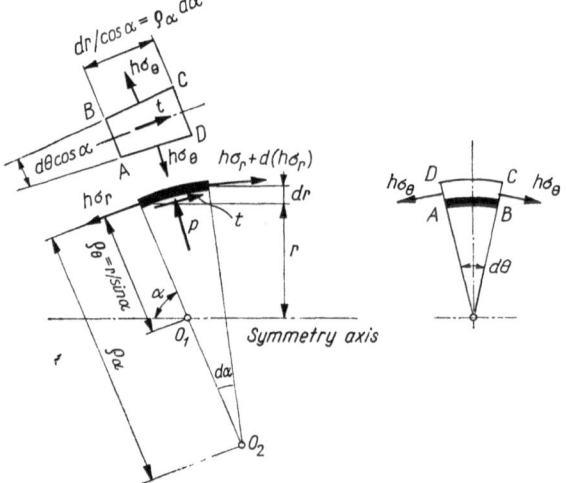

Fig. 179. Stresses acting on an element of the shell

metry of the shell, and the $\theta$-coordinate represents the angle between the meridian plane passing through a generic point and an arbitrarily chosen meridian plane $\theta = 0$. The shell is of double curvature, the radius of curvature in the meridional direction $\varrho_\alpha$ coincides with the radius of curvature of the generatrix. The other radius in the direction of the parallel $\varrho_\theta$ is equal to the length of the generatrix of a cone formed by normals to the surface of the shell drawn from one parallel circle. In practical applications it is more convenient to use the radius $r$ of the parallel circle as the coordinate, instead of the coordinate $\alpha$.

A small element cut out from the shell is loaded by tractions of the magnitudes $h\sigma_\theta$ and $h\sigma_r$ (Fig. 178). These tractions are directed tangentially to the median surface of the shell. The wall thickness of the shell is denoted by $h$. In the direction normal to the shell the pressure of the as yet unknown magnitude $p$ is applied over the entire surface of the element $dF = r\,d\theta\,dr/\cos\alpha$. This pressure, which is necessary to keep the element in equilibrium, represents reactive forces on the sheet-die interface. Assuming that the relative motion of the material particles can take place along the meridians of the die only, we can take into account forces $t$ per unit area tangent to the die surface and directed along meridians, i.e. along the lines $\theta = \mathrm{const}$. These forces are due to the friction of the material sliding over the die surface. Assuming the Coulomb friction we have $t = \mu p$, where $\mu$ is the friction coefficient. The direction of the forces $t$ is opposite to that of the vector of the relative velocity of motion and should be defined for each specific problem. The stress components $\sigma_r$, $\sigma_\theta$ and the external agencies $p, t$ depend for reasons of symmetry on the coordinate $r$ only. Also the wall thickness, changing its magnitude due to the plastic deformations, depends on $r$ only. Thus $h$ is also one of the unknown functions.

Consider now the condition of equilibrium of the shell element. The resultant force acting on its edge $AB$ of length $r\,d\theta$ is $h\sigma_r r\,d\theta$, and on the opposite edge $CD$ of length $(r+dr)d\theta$ there acts the force $[h\sigma_r + d(h\sigma_r)] \times \times (r+dr)d\theta$. The forces applied on the edges $AD$ and $BC$ are of the same magnitude $h\sigma_\theta\,dr/\cos\alpha$. Projecting all these forces on the direction tangent to the meridian and taking into account the resultant of friction forces acting on our element, we obtain

$$[h\sigma_r + d(h\sigma_r)]\,(r+dr)\,d\theta - h\sigma_r r\,d\theta - h\sigma_\theta\,dr\,d\theta + tr\,d\theta\,dr/\cos\alpha = 0.$$

The third term represents the projection of both forces acting on the edges $AD$ and $BC$. The resultant of these forces is $h\sigma_\theta\,d\theta\,dr/\cos\alpha$ and it is directed orthogonally to the symmetry axis of the shell. Thus it makes the angle $\alpha$ with the normal to the shell surface. The projection of this resultant force on the direction tangent to the meridian is obtained from the expression $h\sigma_\theta(dr/\cos\alpha)\,d\theta\cos\alpha$.

Finally the equation of equilibrium, the small magnitudes of higher orders being rejected, takes the form

$$\frac{d(hr\sigma_r)}{dr} - h\sigma_\theta + \frac{tr}{\cos\alpha} = 0. \tag{10.1}$$

Passing to the condition on the projections on the direction normal to the shell, we obtain

$$h\sigma_r r\,d\theta\,d\alpha + h\sigma_\theta\,d\theta\,\frac{dr}{\cos\alpha}\sin\alpha - pr\,d\theta\,dr/\cos\alpha = 0.$$

279

The small terms of higher orders have been neglected here by assuming that the traction $h\sigma_r$ is applied on the edge $CD$, i.e. just as on the edge $AB$. Bearing in mind that $dr/\cos\alpha = \varrho_\alpha d\alpha$ and $r/\sin\alpha = \varrho_\theta$, we arrive at the well-known form of the second equation of equilibrium

$$\frac{\sigma_r}{\varrho_\alpha} + \frac{\sigma_\theta}{\varrho_\theta} = \frac{p}{h}. \tag{10.2}$$

This equation does not contain derivatives and it will be used in specific solutions to calculate the magnitude of the pressure $p$ after the stress components have been found. Also the force $t$ in equation (10.1) can be expressed in terms of stresses by means of (10.2) since $t = \mu p$.

## 10.2  The steady-state forming operations

Among the steady-state axially symmetric problems are numerous practically important processes of plastic forming of shells, such as drawing and sinking of tubes, various types of redrawing of cups, etc. We shall begin by deriving some basic equations for the case of tube drawing (Fig. 180), and it will be seen that they preserve their validity for other

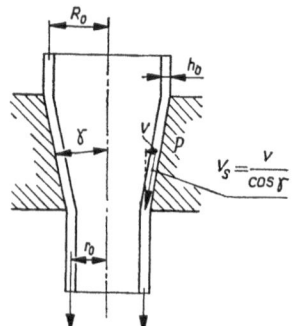

Fig. 180. Drawing of a tube

forming operations by means of various conical tools. The problem of tube drawing has been solved by A. A. Ilyushin [58] for a non-hardening material, and by H. W. Swift [120] who accounted for the isotropic hardening, adopting, however, a rather crude assumption that the equivalent strain is equal to the absolute value of circumferential strain. Such an assumption radically simplifies the solution, because the circumferential strain does not depend on the state of stress and for any arbitrary radius $r$ it may be found from the formula $\varepsilon_\theta = \ln(r/R_0)$. Hence the distribution of the yield point in pure shear $k$ along the generatrix can be found in advance, and then the problem with so obtained non-homogeneity may be solved. Such a procedure gives satisfactory results in the case of tube drawing, differing from the exact solutions by not more than 3 percent. But in other cases, in particular for the non-stationary processes, the method does not work so well, giving considerable errors exceeding 20 percent. Because of this the method of successive approximations described in the foregoing chapter for the true plane state of stress will also be used here [124, 126]. This method is based upon the observation that in processes of the type discussed here, the variation of the wall thickness along the tool generatrix depends only slightly on the strain-hardening of the material. Consequently as the first approximation the solution with no hardening will be taken. It permits us to find the deformation path of each particle and to determine strains at each point by integration along that path. Then the hardening parameters along the generatrix for the assumed strain-hardening hypothesis and the given stress-strain diagram can be found. The non-homogeneity so obtained enables us to begin computations of the next approximation. The procedure should be repeated until two successive approximations give results sufficiently close to each other. Similarly as in the previous cases, the method is rapidly converging. The difference between the second and third approximations usually does not exceed 3 percent.

Four unknowns must be determined at every point: The stress along the generatrix $\sigma_r$, the circumferential stress $\sigma_\theta$, the wall-thickness $h$, and the radial component of flow velocity of the material $v$. The radii of curvature of the conical part of the tube are $\varrho_\alpha = \infty$ and $\varrho_\theta = r/\cos\gamma$. Hence equation (10.2) yields the equation $p = (\sigma_\theta h\cos\gamma)/r$, which together with the equation $t = \mu p$ leads to the elimination of the force $t$ from equation (10.1). Defining the sign of the term representing friction forces in the equation of equilibrium, one must bear in mind that the stress component $\sigma_\theta$ is negative (compressive stress) and, therefore, this term must have a sign opposite to that indicated by the actual direction of the friction force $t$. Remembering that $\gamma = \frac{1}{2}\pi - \alpha$, we arrive at the equation of equilibrium

$$\frac{\mathrm{d}}{\mathrm{d}r}(\sigma_r rh) - \sigma_\theta h(1 + \mu\cot\gamma) = 0. \tag{10.3}$$

The yield condition will be assumed in the general form

$$(\sigma_r - a_r)^2 - (\sigma_r - a_r)(\sigma_\theta - a_\theta) + (\sigma_\theta - a_\theta)^2 = 3k^2, \tag{10.4}$$

where $a_r$, $a_\theta$, and $k$ are certain functions of the state of strain. Condition (10.4) is identical with condition (9.50). Taking $k = k_0 = $ const, we obtain the hypothesis of kinematic hardening. Isotropic hardening is obtained if $a_\theta = a_r = 0$ is taken. Computing the first approximation, we assume that there is no strain-hardening at all. Thus we put $k = k_0 = $ const and $a_r = a_\theta = 0$. All relations given in this section retain their validity also for the first approximation. Computing subsequent approximations, we find $a_r$, $a_\theta$, and $k$ in the same way as in Section 9.8. Also in the present case the strain increments can be obtained from the equations

$$d\varepsilon_\theta = \frac{dr}{r}, \quad d\varepsilon_h = \frac{dh}{h}, \quad d\varepsilon_r = -d\varepsilon_\theta - d\varepsilon_h. \tag{10.5}$$

The components of the strain rate tensor at a generic point $P$ (Fig. 180) depend on the flow velocity $v_s = v/\sin\gamma$

$$\epsilon_r = \frac{dv_s}{ds}, \quad \epsilon_\theta = \frac{v_s \sin\gamma}{r}.$$

Bearing in mind that $ds = dr/\sin\gamma$, we obtain

$$\epsilon_r = \frac{dv}{dr}, \quad \epsilon_\theta = \frac{v}{r}. \tag{10.6}$$

Equations (10.6) have a form similar to (9.30).

The relation between the radial component of the flow velocity $v$ and stress components is analogous to (9.52)

$$\frac{dv}{dr} = \frac{v}{r} \frac{2(\sigma_r - a_r) - (\sigma_\theta - a_\theta)}{2(\sigma_\theta - a_\theta) - (\sigma_r - a_r)}. \tag{10.7}$$

However now we have the ordinary derivative $dv/dr$ instead of the partial derivative $\partial v/\partial r$ since the process is stationary and the velocity depends on the radius $r$ only, being independent of time.

The last equation is the incompressibility condition which states that the amount of the material flowing through each cross-section must be the same. In other words, $d(rhv/\sin\gamma)/dr = 0$, or in equivalent form

$$\frac{dv}{dr} + \frac{v}{r} + \frac{v}{h}\frac{dh}{dr} = 0. \tag{10.8}$$

Equations (10.3), (10.4), (10.7), and (10.8) constitute a set of four equations in four unknown functions. We can eliminate the velocity $v$ from these equations by substituting (10.7) into (10.8). Hence we obtain

$$\frac{dh}{dr} = -\frac{h}{r}\frac{(\sigma_r - a_r) + (\sigma_\theta - a_\theta)}{2(\sigma_\theta - a_\theta) - (\sigma_r - a_r)}. \tag{10.9}$$

Next, by means of (10.9), we can eliminate $h$ from the equation of equilibrium (10.3). The equation

$$\frac{d\sigma_r}{dr} + \frac{\sigma_r}{r}\frac{(\sigma_\theta - a_\theta) - 2(\sigma_r - a_r)}{2(\sigma_\theta - a_\theta) - (\sigma_r - a_r)} - \frac{\sigma_\theta}{r}(1 + \mu\cot\gamma) = 0 \tag{10.10}$$

thus obtained constitutes, together with the yield condition (10.4), a system of two equations in two unknowns $\sigma_r$ and $\sigma_\theta$. Let us express the stress components in terms of a new function $\omega$ as follows:

$$\begin{aligned}
\sigma_r &= a_r + 2k\cos(\omega - \tfrac{1}{6}\pi),\\
\sigma_\theta &= a_r + 2k\cos(\omega + \tfrac{1}{6}\pi).
\end{aligned} \tag{10.11}$$

These expressions satisfy identically the yield condition (10.4). On substituting them into (10.10) we obtain finally a differential equation with one unknown function $\omega$ only

$$\begin{aligned}
\frac{d\omega}{dr} = {} & \frac{a_r + 2k\cos(\omega - \tfrac{1}{6}\pi)}{2kr}\frac{\cos(\omega + \tfrac{1}{6}\pi) - 2\cos(\omega - \tfrac{1}{6}\pi)}{\sin(\omega - \tfrac{1}{6}\pi)\,[2\cos(\omega + \tfrac{1}{6}\pi) - \cos(\omega - \tfrac{1}{6}\pi)]} - \\
& -\frac{a_\theta + 2k\cos(\omega + \tfrac{1}{6}\pi)}{2kr\sin(\omega - \tfrac{1}{6}\pi)}(1 + \mu\cot\gamma) + \\
& +\frac{1}{2k\sin(\omega - \tfrac{1}{6}\pi)}\frac{da_r}{dr} + \cot(\omega - \tfrac{1}{6}\pi)\frac{d\ln k}{dr}.
\end{aligned} \tag{10.12}$$

A solution of this equation may be obtained numerically for instance by means of the Runge–Kutta method, starting from the radius at which, if there is no back-pull, we have $\sigma_r = 0$, and consequently $\omega = 2\pi/3$. If the back-pull is applied, its value is known and consequently the stress components at this radius are also known. Having found $\omega$, we obtain in the same way the thickness $h$ from the equation

$$\frac{dh}{dr} = \frac{h}{r}\frac{\cos(\omega - \tfrac{1}{6}\pi) + \cos(\omega + \tfrac{1}{6}\pi)}{\cos(\omega - \tfrac{1}{6}\pi) - 2\cos(\omega + \tfrac{1}{6}\pi)}, \tag{10.13}$$

obtained by substituting (10.11) into (10.9). The known value of the thickness before entry into the die constitutes the initial condition.

Similarly the radial component $v$ of the flow velocity can be found from equation (10.7), which on substituting (10.11) takes the form

$$\frac{dv}{dr} = \frac{v}{r} \frac{2\cos(\omega - \frac{1}{6}\pi) - \cos(\omega + \frac{1}{6}\pi)}{2\cos(\omega + \frac{1}{6}\pi) - \cos(\omega - \frac{1}{6}\pi)}. \tag{10.14}$$

However, in practical applications, the computation of the velocity $v$ is not necessary, and it usually is omitted, for the state of strain is uniquely determined from the equations (10.5) using the distribution of the wall thickness $h$.

The equations just derived enable us to solve numerous stationary problems of plastic forming of tubes, in which the tube is in contact with the tool on one surface only (inner or outer). Examples of such problems are the drawing and sinking by means of a conical die and expansion on a conical mandrel. The so-called *back-pull* can easily be taken into account. The only difference between these cases is in the presence of other boundary conditions.

A numerical example concerning the case of drawing (Fig. 180) has been computed in two variants: For the isotropic and for the kinematic strain-hardening hypothesis. For the two hypotheses the same linear strain-hardening law for pure shear $\tau = k_0 + c\varepsilon$ was assumed, where the tangent modulus $c$ is constant and $\varepsilon$ is half the shear angle $\gamma$. In order to show the influence of strain-hardening in the unperturbated state, the friction forces on the tube-die interface were disregarded, although it is not difficult to introduce these forces.

Assuming the theory of isotropic strain-hardening, we put $a_r = a_\theta = 0$. The hardening law assumed here corresponds to the relation $k = k_0 + c\varepsilon_i$, enabling us to find the value of $k$ in each of the consecutive approximations. Computations have been performed for $c/k_0 = 1$, which corresponds approximately to the properties of mild steel.

For the kinematic strain-hardening hypothesis we have $k = k_0 = $ const, and thus the last term in (10.12) containing the derivative of $k$ vanishes. On account on the linear hardening assumed above the parameters $a_r$ and $a_\theta$ are given by

$$a_r = c(2\varepsilon_r + \varepsilon_\theta), \qquad a_\theta = c(2\varepsilon_\theta + \varepsilon_r),$$

[compare condition (3.12)].

For the two variants of strain-hardening the method of successive approximations proved quickly converging. The approximation following the solution with no hardening differs from the next one by less than 3 percent. Further approximations give no practically noticeable differences.

For the computing of the function $\omega$ from (10.12) we have in each approximation the boundary condition $\sigma_r = 0$, or $\omega = 2\pi/3$ for $r = R_0$, for on this radius always exists the state of uniaxial compression by the stresses $\sigma_\theta = -\sigma_{p1}$. Hence on this radius we have $\varepsilon_\theta = -2\varepsilon_r$ and therefore $a_r = 0$. The equation (10.13) for the wall thickness $h$ should be integrated

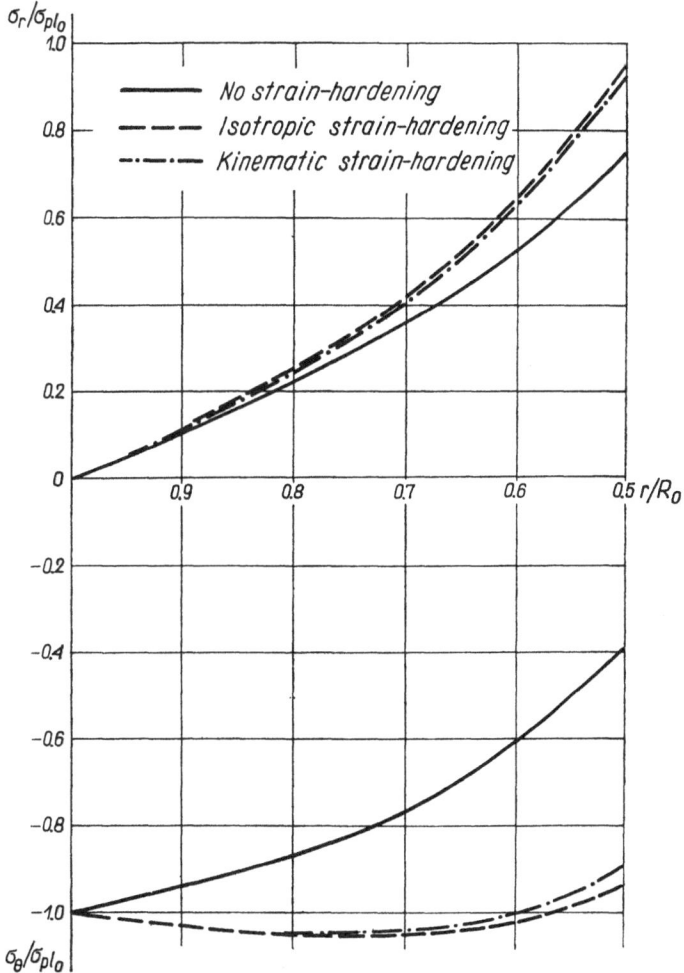

Fig. 181. Stresses in a tube during the drawing process

with the boundary condition $h = h_0$ for $r = R_0$. Computing the radial velocity $v$ from (10.14) we may assume $v = v_0$ for $r = R_0$.

Figure 181 represents diagrams of the stress components $\sigma_r$ and $\sigma_\theta$. The difference between the influences of the two types of hardening appears insignificant. The difference in the effects of the drawing force is somewhat larger. This force equals $P = 2\pi r_0 h_1 (\sigma_r)_1 \cos\gamma$, where the suffix 1 denotes the values attained by the wall thickness and stress $\sigma_r$ in the exit section, i.e. for $r = r_0$. In the case of isotropic hardening the force $P$ amounts to $P = 1.31 P_0$, while the kinematic hardening leads to the value $P = 1.19 P_0$, where $P_0$ is the drawing force for the non-hardening material. The variation of the wall thickness along the die is shown in Fig. 182.

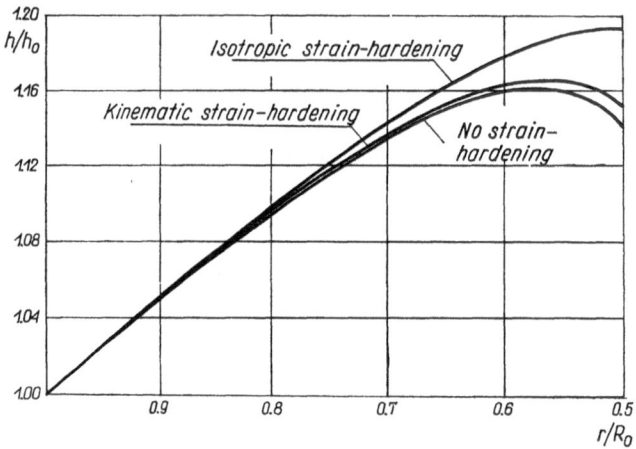

Fig. 182. Variation in wall thickness of the tube along the die

Note that in computations only boundary conditions for $r = R_0$ were employed. Hence each of the radii $r$ may be taken in Fig. 182 as the radius at the exit. In other words, the figure represents also the relation between the wall thickness of the drawn tube and the ratio of the two radii of the die, i.e. $r_0/R_0$. It is interesting to note that the assumption of kinematic hardening leads to a thickness distribution very close to that obtained with no hardening at all. The cause of this is explained to some extent by Fig. 183 in which the curve $AB$ represents the loading path for a material with no hardening, and the curve $AC$—for a material with kinematic hardening. The curve $OO'$ is the path of the central point of the ellipse of plasticity shifted in the stress plane. It is seen that in both cases

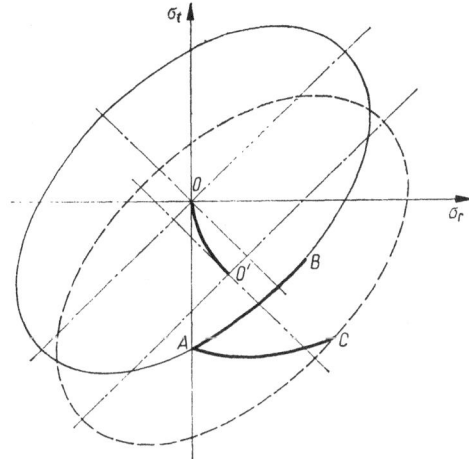

Fig. 183. Loading path in the tube drawing; *AB*—for non-hardening material, *AC*—for material with kinematical hardening

we are on the same segment of the ellipse, fixed or shifted, hence the strain incrèments, which according to the associated flow rule are ortho-gonal to the ellipse, are very close to each other in both cases. For the isotropic strain-hardening hypothesis the initial ellipse expands and thus its curvature decreases. The normal directions differ now from those ob-tained from the solution with no strain-hardening evidently more than from those obtained from the solution with the kinematic hypothesis. Figure 184 presents the loading paths and the strain rate vectors cor-responding to both hypotheses.

The analysis of steady state processes including tube drawing is sig-nificantly simplified if a linearized yield condition and the associated flow rule are assumed [96]. However, as shown by V. V. Sokolovsky [112, 114], such an assumption can lead to the distribution of the wall thickness completely different from that obtained for the flow law associated with the Huber–Mises yield condition. If, for instance, the flow law associated with the Tresca yield condition is employed, the wall thickness of the drawn tube does not change at all, which is in contradiction with ex-perimental observations. More detailed discussion of this problem is given by M. Singh [108]. As a general rule the flow laws associated with the linear yield conditions should be avoided in the analysis of processes, in

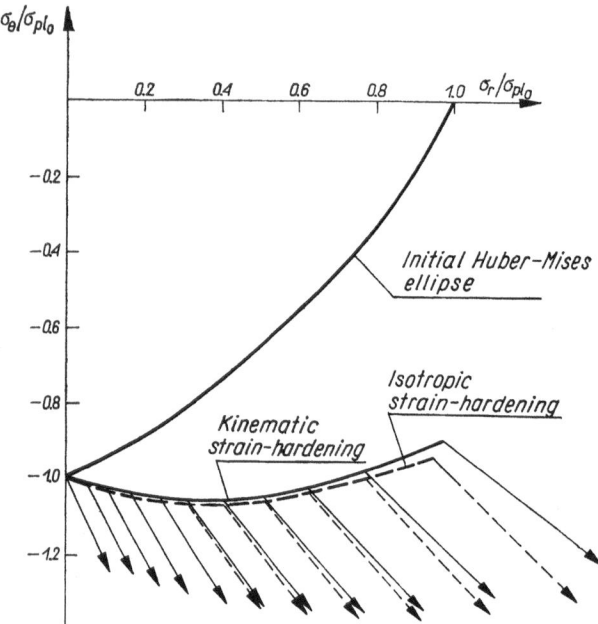

Fig. 184. Strain increment vectors in tube drawing process

which large deformations take place. Such flow law may, however, be used with no danger of a significant error in the analysis of plastic forming operations accompanied by small deformations. This is so for instance in the case of the stretch-forming of thin-walled shells of arbitrary double curvature described in Chapter 11.

### 10.3   Dynamic solution to the tube drawing

Consider now the process of tube drawing, with the inertia forces being taken into account.

The equation of equilibrium (10.3) must now be complemented by the inertia term

$$\varrho\left(\frac{\partial v}{\partial t}+v\frac{\partial v}{\partial r}\right),\tag{i}$$

where $\varrho$ stands for the density of the material. We have assumed that the velocity $v$ depends on time. Thus we have $v = v(r, t)$, which means that the process will be considered as a non-steady-state one. All the

288

magnitudes that matter will therefore depend on the radius $r$ and the time $t$. The governing system of equations has an insufficient number of characteristics and its solution is connected with serious difficulties. For that reason we shall consider only the case of a steady-state drawing, assuming that the drawing velocity $v^*$ at the exit from the die is constant.

In the expression (i) the term $\partial v/\partial t$ vanishes and the equation of motion assumes the form

$$\frac{d(\sigma_r hr)}{dr} - \sigma_\theta h - \varrho \frac{hr}{\sin \gamma} v \frac{dv}{dr} = 0. \tag{ii}$$

The friction forces between the tube and the die have been deliberately neglected in this equation to make the effect of the dynamic forces more visible. Note, however, that these forces can easily be included into considerations in the same way as it was done in the foregoing section.

If non-hardening material is assumed, the Huber–Mises yield condition has the form

$$\sigma_r^2 - \sigma_r \sigma_\theta + \sigma_\theta^2 = 3k^2, \tag{iii}$$

and the associated flow law is expressed by two equations: the relation between the velocity $v$ and the stresses

$$\frac{dv}{dr} = \frac{v}{r} \frac{2\sigma_r - \sigma_\theta}{2\sigma_\theta - \sigma_r}, \tag{iv}$$

and the incompressibility condition [see (10.8)]

$$\frac{dv}{dr} + \frac{v}{r} + \frac{v}{h} \frac{dh}{dr} = 0. \tag{v}$$

The velocity $v$ can be eliminated from the two latter equations. In this way we obtain

$$\frac{dh}{dr} = \frac{h}{r} \frac{\sigma_r + \sigma_\theta}{\sigma_r - 2\sigma_\theta}. \tag{vi}$$

The yield condition (iii) will be identically satisfied if the stresses are expressed by means of a new function $\omega$ [compare (9.43)]

$$\sigma_r = 2k\cos(\omega - \tfrac{1}{6}\pi), \qquad \sigma_\theta = 2k\cos(\omega + \tfrac{1}{6}\pi). \tag{vii}$$

Now the thickness $h$ can be eliminated from the equation of motion. By substituting (vii), we finally arrive at the equation

$$\frac{d\omega}{dr} = \frac{1}{r} \frac{3 + \dfrac{\varrho}{k} \dfrac{v^2}{\sin \gamma} [2\cos(\omega - \tfrac{1}{6}\pi) - \cos(\omega + \tfrac{1}{6}\pi)]}{2\sin(\omega - \tfrac{1}{6}\pi) [\cos(\omega - \tfrac{1}{6}\pi) - 2\cos(\omega + \tfrac{1}{6}\pi)]}, \tag{viii}$$

and two remaining equations

$$\frac{dv}{dr} = \frac{v}{r} \frac{2\cos(\omega - \frac{1}{6}\pi) - \cos(\omega + \frac{1}{6}\pi)}{2\cos(\omega + \frac{1}{6}\pi) - \cos(\omega - \frac{1}{6}\pi)},$$

$$\frac{dh}{dr} = -\frac{h}{r} \frac{\cos(\omega - \frac{1}{6}\pi) + \cos(\omega + \frac{1}{6}\pi)}{2\cos(\omega + \frac{1}{6}\pi) - \cos(\omega - \frac{1}{6}\pi)}. \qquad \text{(ix)}$$

Thus the problem has been reduced to the integration of a system of equations

$$\omega' = F_1(\omega, v, r), \qquad v' = F_2(v, \omega, r), \qquad h' = F_3(h, \omega, r),$$

with the boundary conditions

$$\left. \begin{array}{l} \omega = \frac{2}{3}\pi \\ h = h_0 \end{array} \right\} \text{ for } r = R_0 \quad \text{and} \quad v = v^* \text{ for } r = r_0.$$

The latter condition is not convenient for numerical computations, thus we replace it without loss of generality by the condition $v = v_0$ for $r = R_0$. Such a condition not only does simplify computations but it also allows us to stop the computations at any arbitrary radius $r_0$ which may be treated as the radius at the exit. Thus the solution will be valid for an arbitrary reduction ratio $(R_0 - r_0)/R_0$. Thus the boundary conditions will be assumed in the form

$$\omega = \omega_0 = \tfrac{2}{3}\pi, \qquad h = h_0, \qquad v = v_0 \text{ for } r = R_0. \qquad \text{(x)}$$

Solution for a given velocity $v^*$ at the exit may be obtained by interpolation.

The system of equations (viii) and (ix) will be solved by means of the Pickard method of successive approximations, by assuming as the first approximation of the solution the boundary values of the sought for functions. Therefore we have

$$\omega_{(1)}(r) = \omega_0,$$

$$h_{(1)}(r) = h_0,$$

$$v_{(1)}(r) = v_0.$$

The second approximation will be computed by integration of the equations

$$\omega'_{(2)} = F_1(\omega_0, v_0, r),$$

$$v'_{(2)} = F_2(v_0, \omega_0, r),$$

$$h'_{(2)} = F_3(h_0, \omega_0, r).$$

Similarly the $n$th approximation is computed by solving the equations

$$\omega'_{(n)} = F_1(\omega_{(n-1)}, v_{(n-1)}, r),$$
$$v'_{(n)} = F_2(v_{(n-1)}, \omega_{(n-1)}, r),$$
$$h'_{(n)} = F_3(h_{(n-1)}, \omega_{(n-1)}, r).$$

Each approximation must of course satisfy the boundary conditions (x). Computations can be done by means of any finite differences method.

The system of governing equations (viii), (ix) differs from the system of equations (10.12), (10.13), (10.14) only by the term containing $v^2$ in the numerator on the right-hand side of (viii). Thus the significance of the inertia term can easily be examined.

It can easily be verified that the expression $\beta = 2\cos(\omega - \tfrac{1}{6}\pi) - \cos(\omega + \tfrac{1}{6}\pi)$ varies within the range $\sqrt{3}/2 \leqslant \beta \leqslant \sqrt{3}$, since we have $0 \geqslant \sigma_0 \geqslant -\sqrt{3}\,k$. Thus the magnitude of the term $\varrho v^2/k\sin\gamma$ will indicate how the inertia effects influence the solution. Thus it is clearly seen that the dynamic effects might be of significant magnitude as compared with the number 3 appearing in the numerator of (viii) only when the drawing velocity is of the order $v^* = 10^3 \div 10^4$ cm/sec ($6000 \div 6000$ m/min). In industrial processes we have drawing velocities not exceeding $v^* = 100$ m/min. Thus for practical purposes the problem of drawing can be treated as a quasi-static one.

Figure 185 shows the distribution of the wall thickness of a copper tube drawn with various velocities. The computations have been carried out according to the procedure described above.

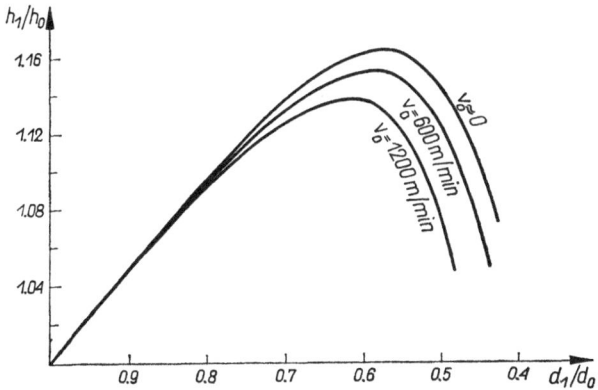

Fig. 185. Variation in wall thickness of the tube along the die for various drawing velocities

### 10.4   Non-stationary forming operations

In the previous chapter a number of non-stationary deformations of flat rings, and certain plastic forming processes in which the deforming part of the drawpiece is flat were considered. In this section we shall present solutions for non-stationary axially symmetric problems of plastic forming of thin shells, in which the quasi-plane state of stress may be assumed. To this group of problems belong such processes as expansion and sinking of the end of a tube, the final stage of tube drawing, etc. The method of solution is the same as in the case of deformation of flat rings. But the boundary value problems are usually of a different type. The basic equations will be for definiteness derived for the case of expanding of the end of a tube (Fig. 186a), although they preserve their validity also in other cases, as for instance in the case shown in Fig. 186b.

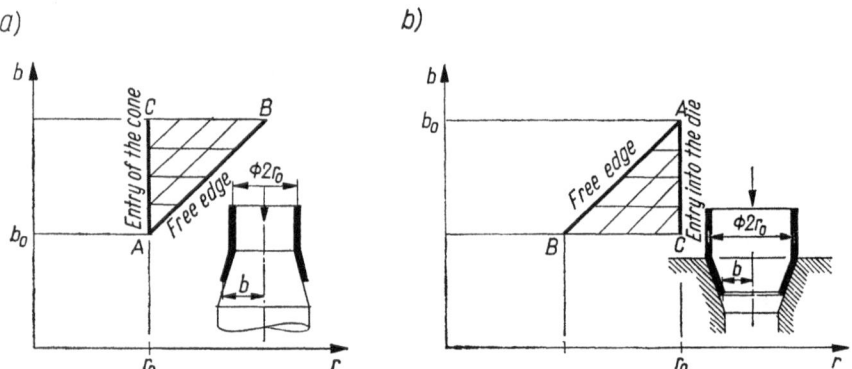

Fig. 186. Basic types of boundary value problems for non-stationary operations of plastic forming of tubes

We have to find at every point the four unknowns: the stress component along the generatrix $\sigma_r$, the circumferential stress $\sigma_\theta$, the wall thickness $h$ and the radial component of the flow velocity $v$. All the unknowns depend not only on the radius $r$, but also on the time $t$. As a measure of time, the radius $b$ of the free edge of the tube will be taken, assuming that the radial velocity component is equal to unity ($v = 1$) on this edge, similarly as in the case of flat rings.

The equation of equilibrium (10.3) and the yield condition (10.4) do not change. The relation between the flow velocity and the radial component of velocity takes now the form

$$\frac{\partial v}{\partial r} = \frac{v}{r}\frac{2(\sigma_r - a_r) - (\sigma_\theta - a_\theta)}{2(\sigma_\theta - a_\theta) - (\sigma_r - a_r)}, \tag{10.15}$$

which is identical with that for flat rings [compare (9.52)]. Also the incompressibility condition

$$\frac{1}{h}\left(\frac{\partial h}{\partial b} + v\frac{\partial h}{\partial r}\right) + \frac{\partial v}{\partial r} + \frac{v}{r} = 0 \tag{10.16}$$

preserves the previous form (9.39).

The method of solution is the same as in the foregoing cases and consists in computing the successive approximations, taking as the first approximation the solution for a perfectly plastic (non-hardening) material. Every subsequent approximation is obtained by solving a problem with a known non-homogeneity found from the previous approximation. Each approximation is solved numerically by the method of characteristics. The equations of characteristics are obtained in the same manner as in the case of flat rings.

For the first family of characteristics we have

$$dr - v\,db = 0, \tag{10.17}$$

$$\frac{dh}{dr} = -\frac{h}{r}\lambda(\omega),$$

where

$$\lambda(\omega) = \frac{\cos(\omega - \frac{1}{6}\pi) + \cos(\omega + \frac{1}{6}\pi)}{2\cos(\omega + \frac{1}{6}\pi) - \cos(\omega - \frac{1}{6}\pi)}.$$

The second family of characteristics is determined by the equations

$$b = \text{const}, \tag{10.18a}$$

$$\frac{dv}{dr} = \frac{v}{r}\varphi(\omega),$$

where

$$\varphi(\omega) = \frac{2\cos(\omega - \frac{1}{6}\pi) - \cos(\omega + \frac{1}{6}\pi)}{2\cos(\omega + \frac{1}{6}\pi) - \cos(\omega - \frac{1}{6}\pi)},$$

and

$$\frac{dh}{dr} = -\frac{h}{r}\left[1 - \frac{a_\theta + 2k\cos(\omega + \tfrac{1}{6}\pi)}{a_r + 2k\cos(\omega - \tfrac{1}{6}\pi)}(1 + \mu\cot\gamma)\right] - \frac{h}{a_r + 2k\cos(\omega - \tfrac{1}{6}\pi)} \times$$

$$\times \left[\frac{da_r}{dr} + 2\cos(\omega - \tfrac{1}{6}\pi)\frac{dk}{dr} - 2k\sin(\omega - \tfrac{1}{6}\pi)\frac{d\omega}{dr}\right]. \qquad (10.18\text{b})$$

In these equations $\omega$ has the same meaning as in Section 10.2 [see expressions (10.11)], $\gamma$ is the half angle of the die or the mandrel, and $\mu$ stands for the friction coefficient as in the stationary processes.

The physical interpretation of the characteristics is the same as in the case of flat rings. The characteristics of the first family represent the loading paths of material particles in the $r, b$-plane. The relevant values found along them describe the loading and deformation history of a particle of the material. The characteristics of the second family ($b = $ const) represent various stages of the advancing process, defined by the instantaneous value of the radius $b$ of the free edge. The magnitudes $h$, $\sigma_r$, and $\sigma_\theta$ obtained along these characteristics represent the stress and wall-thickness distribution along the radius for a fixed $b$.

Certain practical problems, as for instance the example of the fina stage of tube drawing considered in the next section, reduce to the solution of a characteristic boundary value problem. In numerous problems we have, however, the mixed boundary value problems. Two basic types of such problems are shown in Fig. 186. Along the characteristic $AB$ of the first family, representing in the $r, b$-plane the stress-free edge of the tube, the stress $\sigma_r$ must be zero, which for the case of expanding of the end of a tube (Fig. 186a) gives the boundary value $\omega = 5\pi/3$, and for sinking (Fig. 186b) the value $\omega = 2\pi/3$. These boundary values are obtained directly from (10.11) keeping in mind that $a_r = c(2\varepsilon_r + \varepsilon_\theta)$. Since the free edge is stressed by uniaxial tension or compression ($\sigma_j = \pm\sigma_{\text{pl}}$), the incompressibility condition requires that $2\varepsilon_r + \varepsilon_\theta = 0$. Thus $\omega$ is known along $AB$, and this allows to integrate the differential relation (10.17) along that line, because the function $\lambda(\omega)$ is now replaced by a constant value. This leads to the simple relation $h = h_0/\sqrt{b}$ for both values of $\omega$. This relation is identical with that for flat rings. Recall that to obtain it, we assumed without loss of generality that the initial radius of the tube is equal to unity; thus $r_0 = 1$. Moreover, we have $v = 1$ on the free edge, according to the assumption introduced at the beginning of the section. Thus all the relevant functions along the characteristic $AB$ are known. The non-characteristic vertical line $AC$ corresponds to the radius $r_0$, where the particles of the tube enter the deforming region. Thus along $AC$ the thickness $h$ is known, for it must be of the magnitude $h_0$ of the thickness of the undeformed tube. All these data constitute a mixed boundary value problem. Solving it, we can find all the required values in the triangle $ABC$. All numerical computations can be performed in the same way as in Sections 9.7 and 9.9.

The Cauchy boundary value problem will not be considered here, because although one can imagine forming processes such that the solution reduces to this problem, none of these are known to be of practical significance.

As a working example, let us consider the process of expanding the end of a tube. We assume isotropic hardening with the linear law, corresponding to the relation $k = k_0 + c\varepsilon_i$. As in previous examples, the numerical computations will be performed for $c = k_0$. Friction on the tube-mandrel interface is disregarded. Figure 187 shows the net of characteristics which only slightly differs from that obtained for a non-hardening material. The variation of the wall thickness is shown in Fig. 188. The influence of the strain-hardening is more visible than in the previous examples, particularly for large ratios of the diameter of the expanded

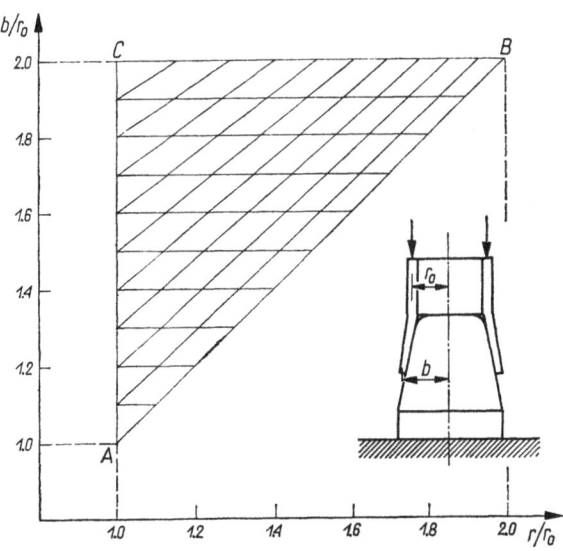

Fig. 187. The mesh of characteristics for the process of expanding the end of a tube

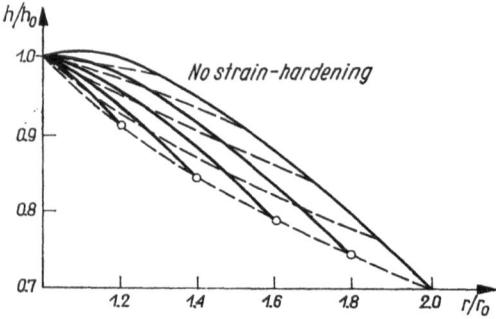

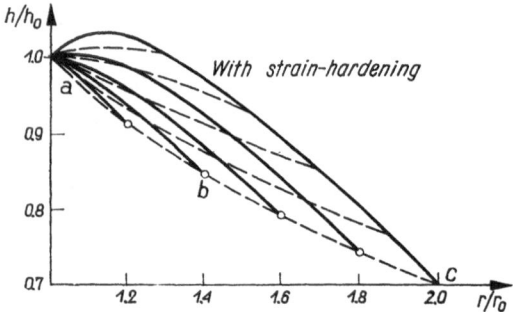

Fig. 188. Variation in wall thickness at different stages expanding the end of a tube

flange to the diameter of the tube. Full lines denote the variation of the thickness along the radius, for definite radii of the free edge *b*. Dashed lines show the history of variation of the thickness for various portions of the tube entering successively the deforming zone. In particular, the dashed line *abc* represents the history of variation of the wall thickness at the free edge. Figure 189 shows the value of the $\sigma_r$ stress component directed along the cone generatrix. The continuous lines also in this figure show the variation of stresses along the radius for fixed values of the radius *b*. The force *P* needed to expand the flange is shown in Fig. 190 in relation to its value at the beginning of the process. It is seen that the influence of strain-hardening is not very strong, particularly for relatively small extensions of the flange.

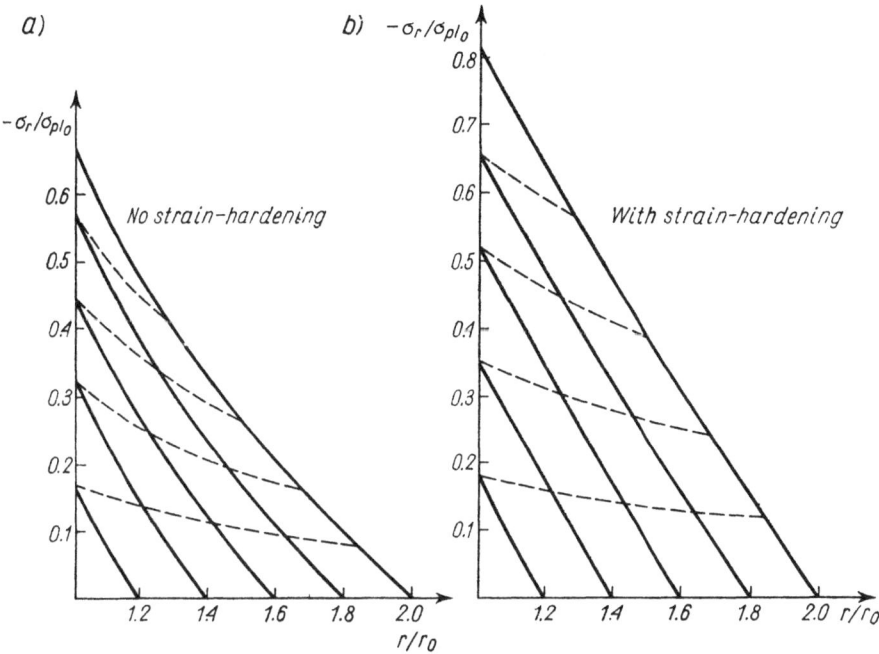

Fig. 189. Variation in the stress component directed along the cone generatrix

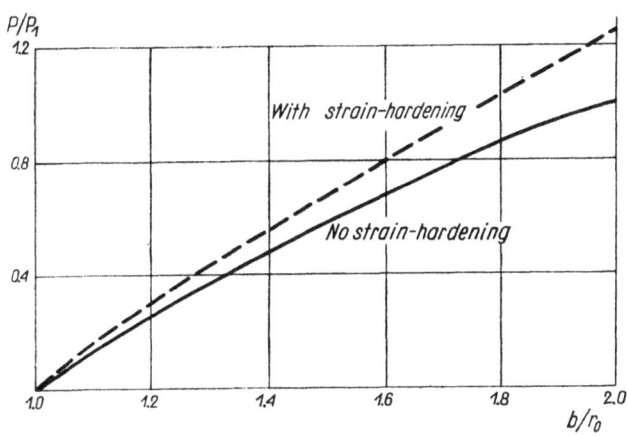

Fig. 190. The relationship between the total force and the flange extension

### 10.5   Non-stationary process as the final stage of
###            a stationary process

A non-stationary process may in numerous plastic forming operations constitute the final stage of a process of the steady-state run. The method of solution for this final stage is analogous to those described in foregoing examples. In specific problems the magnitudes obtained from the solution of the preceding stationary stage of the process constitute the initial values for the solution of the final stage.

Let us consider, as a typical example, the final stage of the tube drawing. It begins at the moment when the end section of the tube in Fig. 180 reaches the edge of the die. A more advanced stage of this non-stationary process is shown in Fig. 191. We are concerned now with a typical char-

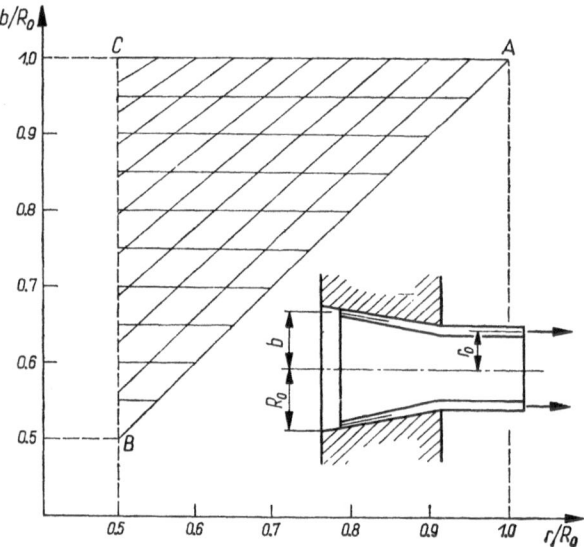

Fig. 191. The mesh of characteristics for the final stage of the tube drawing process

acteristic problem, for which the net of characteristics is schematically shown in Fig. 165b. The computations must be of course stopped in the present case at the vertical line drawn from the point $C$, because at the radius $r = r_0$ the material leaves the deforming region. Along the characteristic line $AB$, corresponding to the free edge of the tube, we have $\omega = 2\pi/3$, since at this edge $\sigma_r = 0$ and $\sigma_\theta = -\sigma_{p1}$. Moreover, on $AB$

the radial velocity component may always be assumed to be equal to unity. Hence we have there $v = 1$. The thickness on $AB$ is given by the relation $h = h_0/\sqrt{b}$, where $h_0$ is the wall thickness of the tube before deformation. We may assume without loss of generality that the initial radius of the stress free edge is of the magnitude $b_0 = R_0 = 1$. Along the characteristic line $AC$, representing the state along the generatrix at the initial moment of the non-stationary stage, the wall thickness $h$, the function $\omega$, and the radial velocity component $v$ must of course vary in the same manner as in the stationary stage preceding the final stage. These data suffice to solve the problem within the entire field of characteristics. Computations are performed by assuming the isotropic strain-hardening hypothesis and the same linear relation $k = k_0 + c\varepsilon_i$ with $c = k_0$, as in the previous examples. The friction forces on the tube-die interface are disregarded. As the starting point for the procedure of successive approximations, the solution for a non-hardening material was used. Figure 191 shows the mesh of characteristics for such a solution. For the case where the strain-hardening is accounted for, the mesh of characteristics showed but little change. The variation of the wall thickness along the radius for material with no strain-hardening is shown in Fig. 192a and for strain-hardening material in Fig. 192b. Continuous lines show the variation of the thickness with the radius for fixed values of the stress free edge radius $b$. The line $AB$ corresponds to $b/R_0 = 1$. Hence this line must be identical with the corresponding line obtained in the foregoing solution for the stationary process (see Fig. 182). Dashed lines represent trajectories of the material particles. In particular, the line $AC$ shows how the thickness of the free edge varies with decreasing radius of this edge. It is seen that the influence of strain-hardening is rather small, and for the kinematic strain-hardening rule it will be even less, similarly as in the case of the stationary stage of the process. However, a considerable difference is observed in stresses. Figure 193 shows the stresses $\sigma_r$ for the material with and without strain-hardening, respectively. Continuous lines represent also here the variation of stress with the radius for fixed values of $b$. Dashed lines correspond to the history of variation of the stress component $\sigma_r$ for a material particle. In Fig. 194 there is shown the wall-thickness distribution in the end-part of the drawn tube obtained from the solution with no strain-hardening. The assumption of strain-hardening gives slightly different distribution of the thickness.

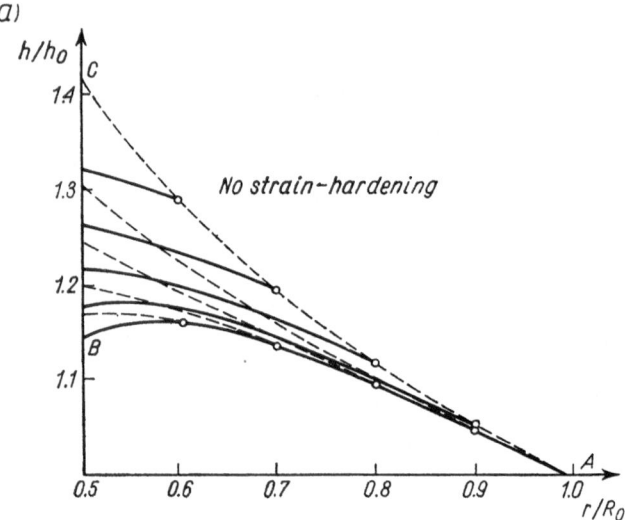

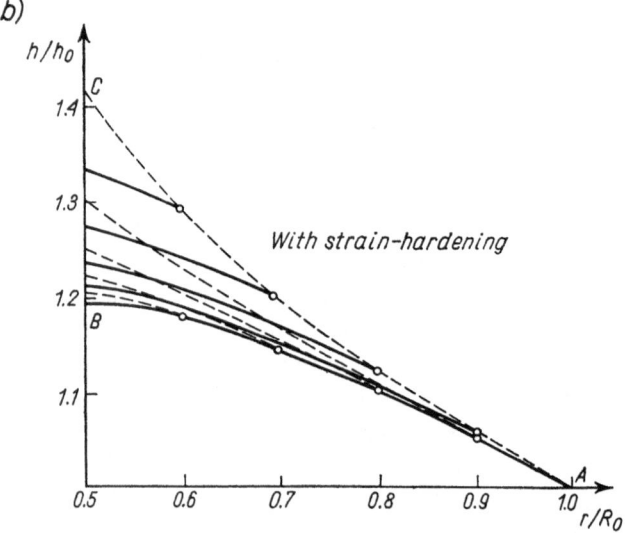

Fig. 192. Variation in wall thickness at different stages of the final stage of the tube drawing process

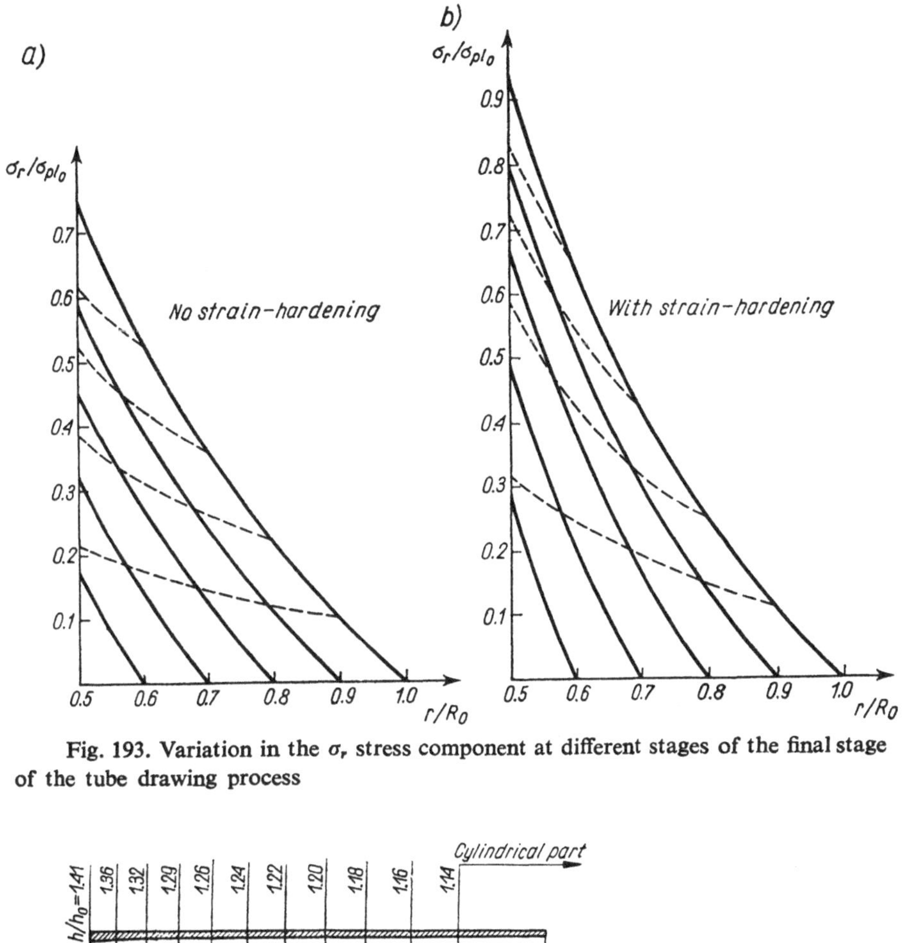

a)

b)

No strain-hardening

With strain-hardening

Fig. 193. Variation in the $\sigma_r$ stress component at different stages of the final stage of the tube drawing process

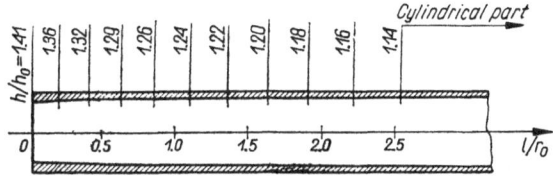

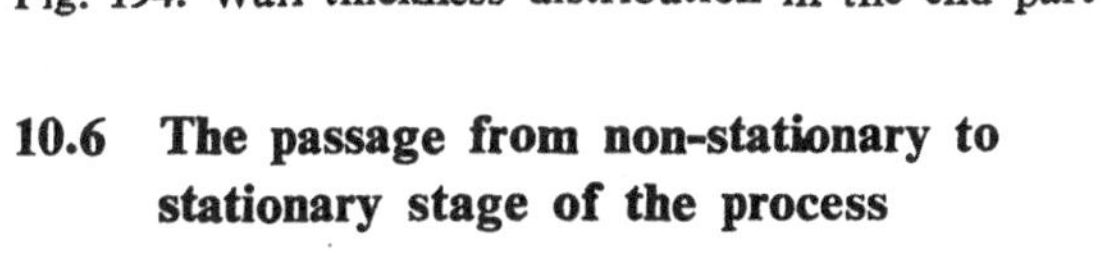

Fig. 194. Wall thickness distribution in the end part of the tube

## 10.6   The passage from non-stationary to stationary stage of the process

Each stationary process begins as a non-stationary one, strictly speaking, the steady state stage of any process is reached after an infinite time. In practice, however, as we shall demonstrate this below, real conditions do

301

not differ after a short time much from stationary ones. Solutions of problems of passage to the stationary stage and determination of the period of time after which this stage is reached, are of theoretical importance, and they are also not without practical significance. Unfortunately, in most cases such solutions are connected with mathematical difficulties. However, for the tube forming processes considered in this chapter such solutions, accounting also for the strain-hardening effect and friction on the tube-tool interface are available [126, 127].

As a typical example let us consider the stage following the non-stationary expansion of a tube (Fig. 187). This stage begins when the stress free edge reaches the cylindrical part of the mandrel (Fig. 195). As

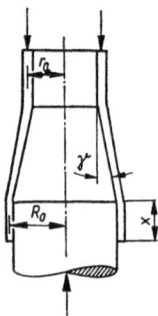

Fig. 195. The expansion of a tube diameter

a time scale we can now take the length $x$ of the expanded cylindrical part of the tube, assuming that the end rim of the tube moves with unit velocity $c = 1$. It is to be expected that as the length $x$ increases, the distribution of the wall thickness along the conical part of the mandrel will approach the distribution obtained for the stationary process. The equation of equilibrium (10.3), the yield condition (10.4), and the relation between the radial velocity component $v$ and stress components (10.15) do not change. Only in the incompressibility condition (10.16) we must substitute $x$ for $b$

$$\frac{1}{h}\left(\frac{\partial h}{\partial x}+v\frac{\partial h}{\partial r}\right)+\frac{\partial v}{\partial r}+\frac{v}{r}=0. \tag{10.19}$$

The differential equation of the first family of characteristics will now have the form $dr - v\,dx = 0$. The characteristics of the second family are determined by the equation $x = $ const. The differential equations which must be satisfied along characteristic lines are the same as in Section 10.4. Thus they have the form (10.17) and (10.18).

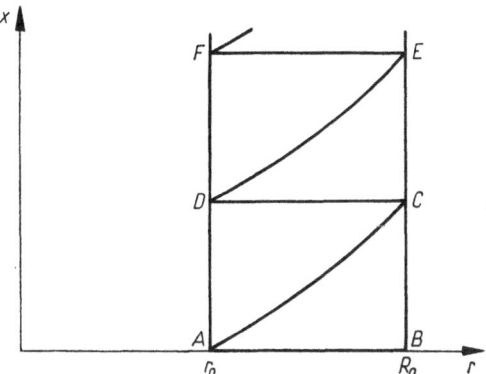

Fig. 196. Boundary and initial value problems for the analysis of transition from the non-stationary to the stationary stage of tube expanding

Let us consider the boundary and the initial value conditions in the $r, x$-plane (Fig. 196). The segment $AB$ of the characteristic line $x = 0$ belonging to the second family corresponds to the instant when the free edge reaches the cylindrical part of the mandrel. Thus we know from the solution of the previous stage (see Section 10.4, Figs 187, 188, 189) all the relevant functions along $AB$. Along the vertical line $r = R_0$, which represents the boundary between the expanded cylindrical part of the tube and its conical part; we have $\sigma_r = 0$, hence $\omega = 5\pi/3$, and the radial velocity component $v = \sin\gamma$, since we have assumed that the velocity of the end section of the tube is $c = 1$. These data constituting a mixed boundary value problem for the equations of characteristics, enable us to determine the solution in the entire region $ABC$ in the $r, x$-plane. Next, along the vertical line $r = r_0$ we have $h = h_0$, which together with the values of $h$ and $v$ already found along the characteristic $AC$ of the first family, constitutes a mixed boundary value problem in the curvilinear triangle $ACD$. In this way we can extend the solution to an arbitrarily large value of $x$. Figure 197 shows the variation of the wall thickness $h$ along the conical portion of the mandrel for various values of $x' = x\sin\gamma$. The solution for the magnitudes of the angle $\gamma$ used in practice, very quickly approaches the solution for the stationary process shown by the upper dashed line and obtained in the

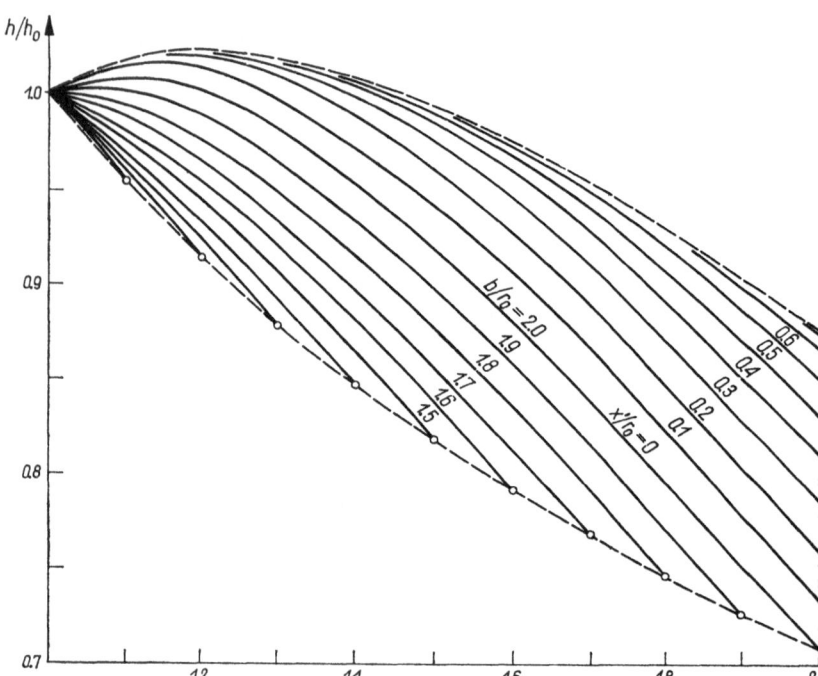

Fig. 197. Variation in wall thickness at different stages of the process of tube expanding

manner described in Section 10.2 For $x' = 0.8r_0$ the thickness distribution along the radius differs very little from that for the stationary process. Numerical computations have been done for a non-hardening material. Figure 197 incorporates the results already obtained for the preceding stage of the process in which the free edge of the tube is in contact with the conical part of the tool (compare Fig. 188a).

Other cases of passage from the non-stationary stage of the process to the stationary one may be solved in the same manner. For example, we may mention here the sinking of a tube, as shown in Fig. 186b.

# Drawing and stretchforming of thin-walled shells of arbitrary double curvature

## 11.1 Introduction

We shall consider in this chapter the theory of those processes of plastic forming of shells of arbitrary double curvature, in which one of the shell surfaces is in contact with the die. This theory is a generalization of the theory of plastic forming of axially symmetric shells discussed in the foregoing chapter. Examples of plastic forming of shells of arbitrary double curvature are shown in Fig. 198.

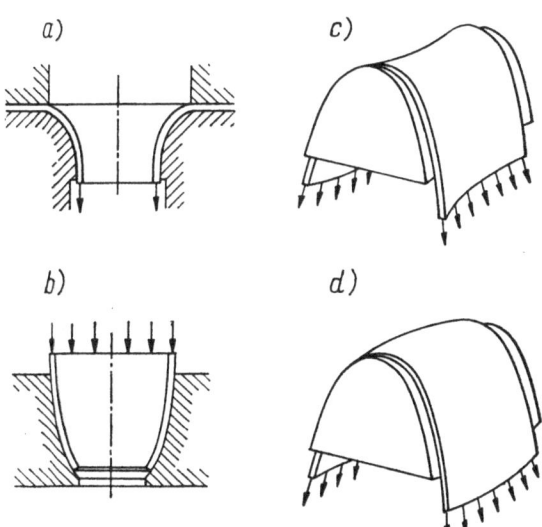

Fig. 198. Examples of plastic forming of shells of arbitrary double curvature

In order to make this chapter sufficiently self-contained the necessary rudiments of the theory of surfaces will be briefly presented below. Any surface can be described in a Cartesian coordinate system $x, y, z$ by an equation

$$F(x, y, z) = 0, \tag{11.1}$$

or in a parametric form, representing coordinates of the surface points as functions of two independent parameters $\alpha$ and $\beta$,

$$
\begin{aligned}
x &= \varphi(\alpha, \beta), \\
y &= \psi(\alpha, \beta), \\
z &= \omega(\alpha, \beta).
\end{aligned} \tag{11.2}
$$

The equation of a surface may also be represented in the vectorial form, where each point of the surface is defined by the radius-vector from a fixed pole $O$

$$\mathbf{r} = \mathbf{r}(\alpha, \beta),$$

or in the equivalent form

$$\mathbf{r} = x(\alpha, \beta)\mathbf{i} + y(\alpha, \beta)\mathbf{j} + z(\alpha, \beta)\mathbf{k}. \tag{11.3}$$

If the surface is defined by (11.2) or (11.3) and if one of the parameters is fixed, e.g. if $\beta = \beta_0$, the point $\mathbf{r}(\alpha, \beta)$ will move along a certain curve contained in the surface, while the other parameters will assume different values. Such a curve will be defined by the relation $\mathbf{r} = \mathbf{r}(\alpha, \beta_0)$. If now the parameter $\beta$ assumes different constant values $\beta_1, \beta_2, ..., \beta_n$, a family of curves lying on the surface is obtained. To each point of these curves corresponds a different value of the parameter $\alpha$. Thus each line of that family may be treated as the *coordinate line* $\alpha$ since along it $\beta = $ const. Similarly, if the parameter $\alpha$ assumes constant values $\alpha_1, \alpha_2, ..., \alpha_n$, we obtain a family of curves along which the $\beta$-parameter changes. These curves may be called the *lines of the coordinate* $\beta$. The two families constitute a curvilinear net of coordinates on the surface. The location of an arbitrary point $M$ on the surface can be defined by two coordinates $\alpha_M$ and $\beta_M$.

If through a given point $M$ of the surface a number of curves lying on the surface are plotted, then the tangents to these lines form a plane. Such a plane is defined as the tangent plane to the surface at $M$. In the tangent plane at $M$ two vectors $\mathbf{r}_1 = \partial \mathbf{r}/\partial \alpha$ and $\mathbf{r}_2 = \partial \mathbf{r}/\partial \beta$ can be plotted, both being tangent at this point to the coordinate lines $\alpha$ and $\beta$, respectively. If the scalar product of the vectors $\mathbf{r}_1$ and $\mathbf{r}_2$ is equal to zero ($\mathbf{r}_1 \cdot \mathbf{r}_2 = 0$), or in other words, if

$$\frac{\partial x}{\partial \alpha}\frac{\partial x}{\partial \beta} + \frac{\partial y}{\partial \alpha}\frac{\partial y}{\partial \beta} + \frac{\partial z}{\partial \alpha}\frac{\partial z}{\partial \beta} = 0, \tag{11.4}$$

the coordinate lines of the two families are orthogonal. In our considerations only orthogonal systems of coordinate lines on the surface will be assumed and therefore all the formulae derived further are valid for such systems only.

The distance $ds$ between two very close points on the surface results from the formula

$$ds^2 = dx^2 + dy^2 + dz^2, \tag{11.5}$$

where $dx, dy$ and $dz$ represent increments of the Cartesian coordinates between the two points. They can be expressed by means of the curvilinear coordinates $\alpha$ and $\beta$

$$dx = \frac{\partial x}{\partial \alpha}d\alpha + \frac{\partial x}{\partial \beta}d\beta,$$

$$dy = \frac{\partial y}{\partial \alpha}d\alpha + \frac{\partial y}{\partial \beta}d\beta,$$

$$dz = \frac{\partial z}{\partial \alpha}d\alpha + \frac{\partial z}{\partial \beta}d\beta.$$

Substituting these expressions in (11.5) and taking into account (11.4), we obtain

$$ds^2 = A^2 d\alpha^2 + B^2 d\beta^2, \tag{11.6}$$

where $A\,d\alpha = ds_\alpha$ and $B\,d\beta = ds_\beta$ represent the lengths of elementary arcs along the lines $\alpha$ and $\beta$, respectively. The equation (11.6) constitutes the so-called *first quadratic form* of the surface and represents one of the basic equations of the theory of surfaces. The coefficients $A$ and $B$, which are known as the *coefficients of the first quadratic form*, are expressed by the relations

$$A^2 = \left(\frac{\partial x}{\partial \alpha}\right)^2 + \left(\frac{\partial y}{\partial \alpha}\right)^2 + \left(\frac{\partial z}{\partial \alpha}\right)^2,$$

$$B^2 = \left(\frac{\partial x}{\partial \beta}\right)^2 + \left(\frac{\partial y}{\partial \beta}\right)^2 + \left(\frac{\partial z}{\partial \beta}\right)^2. \tag{11.7}$$

These formulae constitute the foundations of the so-called *internal geometry of surfaces*.

The radius of curvature of a surface at a specific point $M$ in an arbitrary direction is defined as the radius of curvature of the curve obtained by intersecting the surface by a normal plane containing that specific point and the given direction. The length of the radius $R$ is given by

$$\frac{1}{R} = \frac{L\,d\alpha^2 + 2M\,d\alpha\,d\beta + N\,d\beta^2}{A^2 d\alpha^2 + B^2 d\beta^2}. \tag{11.8}$$

The expression in the numerator on the right-hand side is called the *second quadratic form*. The coefficients $L$, $M$, and $N$ are known as the *coefficients of the second quadratic form*. They are defined by the relations

$$L = \frac{1}{AB} \begin{vmatrix} \dfrac{\partial^2 x}{\partial \alpha^2} & \dfrac{\partial^2 y}{\partial \alpha^2} & \dfrac{\partial^2 z}{\partial \alpha^2} \\[2mm] \dfrac{\partial x}{\partial \alpha} & \dfrac{\partial y}{\partial \alpha} & \dfrac{\partial z}{\partial \alpha} \\[2mm] \dfrac{\partial x}{\partial \beta} & \dfrac{\partial y}{\partial \beta} & \dfrac{\partial z}{\partial \beta} \end{vmatrix}, \qquad N = \frac{1}{AB} \begin{vmatrix} \dfrac{\partial^2 x}{\partial \beta^2} & \dfrac{\partial^2 y}{\partial \beta^2} & \dfrac{\partial^2 z}{\partial \beta^2} \\[2mm] \dfrac{\partial x}{\partial \alpha} & \dfrac{\partial y}{\partial \alpha} & \dfrac{\partial z}{\partial \alpha} \\[2mm] \dfrac{\partial x}{\partial \beta} & \dfrac{\partial y}{\partial \beta} & \dfrac{\partial z}{\partial \beta} \end{vmatrix},$$

$$M = \frac{1}{AB} \begin{vmatrix} \dfrac{\partial^2 x}{\partial \alpha \, \partial \beta} & \dfrac{\partial^2 y}{\partial \alpha \, \partial \beta} & \dfrac{\partial^2 z}{\partial \alpha \, \partial \beta} \\[2mm] \dfrac{\partial x}{\partial \alpha} & \dfrac{\partial y}{\partial \alpha} & \dfrac{\partial z}{\partial \alpha} \\[2mm] \dfrac{\partial x}{\partial \beta} & \dfrac{\partial y}{\partial \beta} & \dfrac{\partial z}{\partial \beta} \end{vmatrix}, \tag{11.9}$$

where $x, y, z$ are given by (11.2).

The magnitude of the radius $R$ of curvature, determined by (11.8), depends on the direction of the normal plane. Two of the radii of curvature at a point are of special significance, namely the largest and the smalest. Such radii represent the principal radii of curvature, and their directions— the principal directions of curvature. Curves on the surface, which at any point have the direction of a principal curvature, are called *lines of curvature*. In practical applications it is most convenient to choose the coordinate system on the surface coinciding with the lines of curvature. In such a case the coefficient $M$ of the second quadratic form is equal to zero,

$$M = 0, \tag{11.10}$$

and the principal radii of curvature can be obtained from the formulae

$$R_\alpha = \frac{A^2}{L} \text{ in } \alpha\text{-direction},$$

$$R_\beta = \frac{B^2}{L} \text{ in } \beta\text{-direction}.$$

(11.11)

Below we give some examples of parametric equations of certain surfaces.

1. Ellipsoid:

$$x = a\frac{\cos\beta}{\operatorname{ch}\alpha}, \ y = b\frac{\sin\beta}{\operatorname{ch}\alpha}, \ z = -c\operatorname{th}\alpha.$$

2. Hyperboloid of two sheets:

$$x = -a\frac{\cos\beta}{\operatorname{sh}\alpha}, \ y = -b\frac{\sin\beta}{\operatorname{sh}\alpha}, \ z = -c\operatorname{coth}\alpha.$$

3. Elliptic paraboloid:
$$x = ae^\alpha\cos\beta, \ y = be^\alpha\sin\beta, \ z = \tfrac{1}{2}e^{2\alpha}.$$

4. Hyperboloid of one sheet:

$$x = a\frac{\cos\beta}{\cos\alpha}, \ y = b\frac{\sin\beta}{\cos\alpha}, \ z = c\tan\alpha.$$

5. Hyperbolic paraboloid:
$$x = ae^\alpha\operatorname{ch}\beta, \ y = be^\alpha\operatorname{sh}\beta, \ z = \tfrac{1}{2}e^{2\alpha}.$$

In practical applications of greater significance are various ring-shaped surfaces with a non-circular line of central points of cross-sections (Fig. 199). Examples of parametric equations of such surfaces are given below.

6. Elliptic line of central points, cross-section elliptic:

$$x = \frac{a^2\cos\beta}{\sqrt{a^2\cos^2\beta+b^2\sin^2\beta}} - \frac{c^2\cos\alpha\cos\beta}{\sqrt{c^2\cos^2\alpha+d^2\sin^2\alpha}},$$

$$y = \frac{b^2\sin\beta}{\sqrt{a^2\cos^2\beta+b^2\sin^2\beta}} - \frac{c^2\cos\alpha\sin\beta}{\sqrt{c^2\cos^2\alpha+d^2\sin^2\alpha}},$$

$$z = \frac{d^2\sin\alpha}{\sqrt{c^2\cos^2\alpha+d^2\sin^2\alpha}}.$$

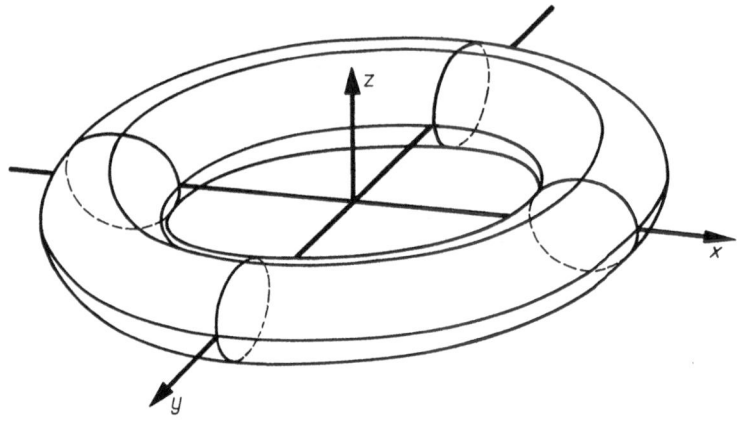

Fig. 199. A ring-shaped surface with non-circular line of central points of cross-sections

7. Oval line of central points, cross-section elliptic:

$$x = \frac{a^2\cos\beta}{\sqrt{a^2\cos^2\beta+b^2\sin^2\beta}} - \frac{c^2\cos\alpha\cos\beta}{\sqrt{c^2\cos^2\alpha+d^2\sin^2\alpha}} + c\cos\beta,$$

$$y = \frac{b^2\sin\beta}{\sqrt{a^2\cos^2\beta+b^2\sin^2\beta}} - \frac{c^2\cos\alpha\sin\beta}{\sqrt{c^2\cos^2\alpha+d^2\sin^2\alpha}} + c\sin\beta,$$

$$z = \frac{d^2\sin\alpha}{\sqrt{c^2\cos^2\alpha+d^2\sin^2\alpha}}.$$

Substituting in these equations $c = d$, one obtains equations for surfaces of circular cross-section.

### 11.2   Basic relations

The basic relations of the theory of plastic forming of thin shells of arbitrary double curvature have been given by A. A. Ilyushin [59]. He considered the general case where the material of the shell is compressed from both sides. Since for such a general case no methods of solution of even specific problems are as yet available, our considerations will be limited to the cases where only one of the surfaces of the shell is in contact with the die.

In order to obtain the solution for the stress and velocity fields in

310

the shell, one has to determine at each point the three stress components $\sigma_\alpha$, $\sigma_\beta$, $\tau$ (Fig. 200), the contact pressure $p$ between the deformed sheet and the die, the velocity components $v_\alpha$ and $v_\beta$ (Fig. 201), and the wall thickness $h$.

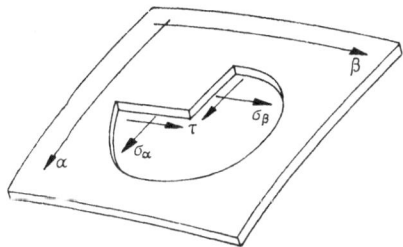

Fig. 200. The components of stress tensor

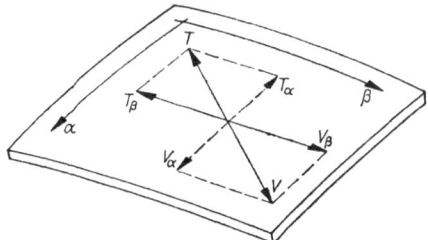

Fig. 201. The components of velocity vector $v_\alpha$, $v_\beta$ and the components of friction force $T_\alpha$, $T_\beta$

Consider now the system of governing equations. The friction forces between the die and the material sliding over its surface are accounted by the equation of equilibrium. Assume that the friction force $T$ per unit contact area is $T = \mu p$, where $\mu$ is the friction coefficient assumed to have a constant value over the entire contact area, and that at each point the direction of the friction force is opposite to the direction of the velocity vector. The vector $T$ can be resolved into two components $T_\alpha$ and $T_\beta$. From Fig. 201 one obtains

$$T_\alpha = \mu p \frac{v_\alpha}{v}, \quad T_\beta = \mu p \frac{v_\beta}{v}, \quad \text{where } v = \sqrt{v_\alpha^2 + v_\beta^2}.$$

Let us derive now the equations of equilibrium for thin-walled shells of arbitrary double curvature. Let *CDFE* (Fig. 202) represent an element

311

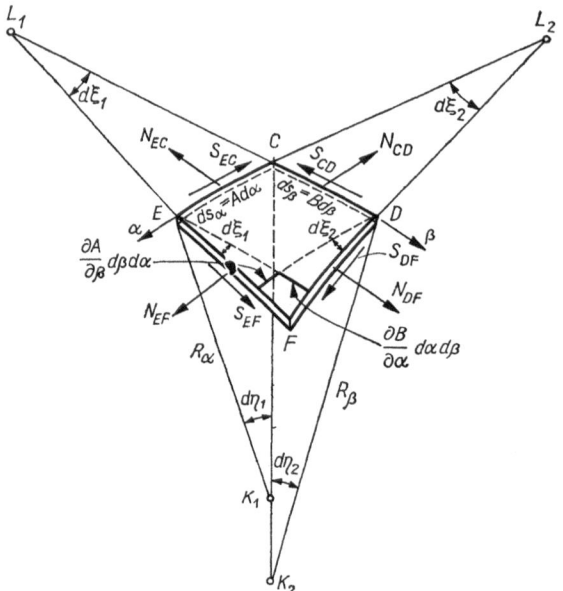

Fig. 202. Equilibrium of a small element of the shell

of the shell bounded by two pairs of coordinate lines $\alpha$, $\alpha + d\alpha$ and $\beta$, $\beta + d\beta$. The edges of the element are of the following lengths

$$\overline{CE} = A d\alpha, \quad \overline{DF} = \left(A + \frac{\partial A}{\partial \beta} d\beta\right) d\alpha,$$

$$\overline{CD} = B d\beta, \quad \overline{EF} = \left(B + \frac{\partial B}{\partial \alpha} d\alpha\right) d\beta. \tag{11.12}$$

Let $d\eta_1$ and $d\eta_2$ denote the angles contained between the normals to the shell at $E$ and $C$ and at $C$ and $D$, respectively. Similarly, let $d\xi_1$ and $d\xi_2$ denote the angles in the plane tangent to the shell surface, formed by the tangents to the lines of curvature at $C$, $E$, and $D$. The values of these angles can be obtained from Fig. 202.

$$d\eta_1 = \frac{A d\alpha}{R_\alpha}, \qquad d\eta_2 = \frac{B d\beta}{R_\beta}, \tag{11.13}$$

$$d\xi_1 = \frac{1}{B} \frac{\partial A}{\partial \beta} d\alpha, \quad d\xi_2 = \frac{1}{A} \frac{\partial B}{\partial \alpha} d\beta. \tag{11.14}$$

For the sake of brevity let us introduce the notation

$$N_\alpha = \sigma_\alpha h, \; N_\beta = \sigma_\beta h, \; S_\alpha = S_\beta = S = \tau h. \tag{11.15}$$

Along the edges of the element *CDFE* we have:

edge *CD*: $N_{CD} = N_\alpha$, $\quad\quad\quad\quad S_{CD} = S$,

edge *EF*: $N_{EF} = N_\alpha + \dfrac{\partial N_\alpha}{\partial \alpha}\, d\alpha$, $\quad S_{EF} = S + \dfrac{\partial S}{\partial \alpha}\, d\alpha$,

edge *EC*: $N_{EC} = N_\beta$, $\quad\quad\quad\quad S_{EC} = S$,

edge *DF*: $N_{DF} = N_\beta + \dfrac{\partial N_\beta}{\partial \beta}\, d\beta$, $\quad S_{DF} = S + \dfrac{\partial S}{\partial \beta}\, d\beta$.

The positive directions of these unit forces are shown in Fig. 202. We obtain any resultant force as the product of the unit force and the length of the respective edge

edge *CD*: normal force $N_\alpha B d\beta$,

   tangent force $SB d\beta$,

edge *EF*: normal force $\left[ N_\alpha B + \dfrac{\partial}{\partial \alpha}(N_\alpha B)\, d\alpha \right] d\beta$,

   tangent force $\left[ SB + \dfrac{\partial}{\partial \alpha}(SB)\, d\alpha \right] d\beta$,

edge *EC*: normal force $N_\beta A\, d\alpha$,

   tangent force $SA\, d\alpha$,

edge *DF*: normal force $\left[ N_\beta A + \dfrac{\partial}{\partial \beta}(N_\beta A)\, d\beta \right] d\alpha$,

   tangent force $\left[ SA + \dfrac{\partial}{\partial \beta}(SA)\, d\beta \right] d\alpha$.

Projecting these forces on $CL_1$ and $CL_2$ and on normal $CK_2$ and taking, moreover, into account external forces, i.e. the friction forces $T_\alpha$, $T_\beta$ and the contact pressure $p$, we obtain after certain simplifications

$$\frac{\partial}{\partial \alpha}(N_\alpha B)\, d\alpha\, d\beta + \frac{\partial}{\partial \beta}(SA)\, d\alpha\, d\beta -$$

$$- N_\beta A\, d\alpha\, d\xi_2 + SB\, d\beta\, d\xi_1 - T_\alpha AB\, d\alpha\, d\beta = 0,$$

$$\frac{\partial}{\partial \beta}(N_\beta A)\,d\alpha\,d\beta + \frac{\partial}{\partial \alpha}(SB)\,d\alpha\,d\beta -$$

$$- N_\alpha B d\beta\,d\xi_1 + SA\,d\alpha\,d\xi_2 - T_\beta AB\,d\alpha\,d\beta = 0,$$

$$N_\alpha B d\beta\,d\eta_1 + N_\beta A d\alpha\,d\eta_2 = pAB\,d\alpha\,d\beta.$$

Substituting (11.13) and (11.14) at place of $d\xi_1$, $d\xi_2$, $d\eta_1$ and $d\eta_2$ and replacing forces by the stress components obtained from (11.15), we obtain the final form of the differential equations of equilibrium

$$h(\sigma_\alpha - \sigma_\beta)\frac{\partial B}{\partial \alpha} + B\frac{\partial(h\sigma_\alpha)}{\partial \alpha} + 2h\tau\frac{\partial A}{\partial \beta} + A\frac{\partial(h\tau)}{\partial \beta} - AB\mu p\frac{v_\alpha}{v} = 0,$$

$$h(\sigma_\beta - \sigma_\alpha)\frac{\partial A}{\partial \beta} + A\frac{\partial(h\sigma_\beta)}{\partial \beta} + 2h\tau\frac{\partial B}{\partial \alpha} + B\frac{\partial(h\tau)}{\partial \alpha} - AB\mu p\frac{v_\beta}{v} = 0, \quad (11.16)$$

$$\frac{\sigma_\alpha}{R_\alpha} + \frac{\sigma_\beta}{R_\beta} = \frac{p}{h}.$$

Let us derive now the formulae for the strain rates $\epsilon_\alpha$ and $\epsilon_\beta$ along the coordinate lines $\alpha$ and $\beta$ and for the strain rate component $\epsilon_{\alpha\beta}$. Each of these strain rate components may be considered as composed of two parts. The first, connected with the variation of the flow velocity along the coordinate lines, can be derived in the same way as in the plane flow theory

$$\epsilon_\alpha' = \frac{\partial v_\alpha}{\partial s_\alpha} = \frac{1}{A}\frac{\partial v_\alpha}{\partial \alpha},$$

$$\epsilon_\beta' = \frac{\partial v_\beta}{\partial s_\beta} = \frac{1}{B}\frac{\partial v_\beta}{\partial \beta}, \quad (11.17)$$

$$\epsilon_{\alpha\beta}' = \frac{1}{2}\left(\frac{\partial v_\alpha}{\partial s_\beta} + \frac{\partial v_\beta}{\partial s_\alpha}\right) = \frac{1}{2}\left(\frac{1}{B}\frac{\partial v_\alpha}{\partial \beta} + \frac{1}{A}\frac{\partial v_\beta}{\partial \alpha}\right).$$

The second part results from the fact that in the curvilinear coordinate system the distance between the coordinate lines is different at each point of the mesh. An elementary segment $CE$ (Fig. 203) of the length $ds_\alpha = A\,d\alpha$ is shifted along the $\beta$-line with the velocity $v_\beta$. After passing the distance $ds_\beta$ its length increment is $(\partial A/\partial\beta)\,d\alpha\,d\beta$. Thus the strain increment is equal to $d\epsilon_\alpha'' = (1/A)\,(\partial A/\partial\beta)\,d\beta$. The increment $d\beta$ can be expressed by the time increment $dt$ by means of the relation $v_\beta dt = B d\beta$. Thus

$$\epsilon_\alpha'' = \frac{d\epsilon_\alpha''}{dt} = \frac{v_\beta}{AB}\frac{\partial A}{\partial \beta} \quad \text{and similarly} \quad \epsilon_\beta'' = \frac{d\epsilon_\beta''}{dt} = \frac{v_\alpha}{AB}\frac{\partial B}{\partial \alpha}. \quad (11.18a)$$

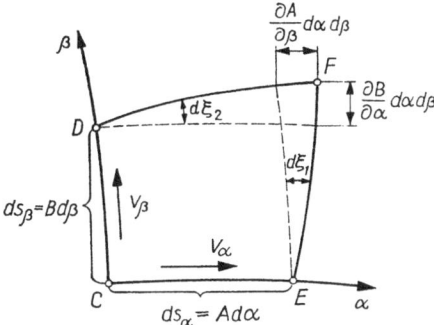

Fig. 203. A small element of the shell

When the sector $CE$ is shifted to a new position along a $\beta$-line, the right angle which it makes with the $\beta$-direction changes to $\frac{1}{2}\pi - d\xi_2$, where $d\xi_2$ is defined by (11.14). Differentiating the second of the relations (11.14) with respect to time and remembering that $d\beta/dt = v_\beta/B$, we obtain

$$\frac{d\xi_2}{dt} = \frac{v_\beta}{AB}\frac{\partial B}{\partial \alpha}, \text{ and similarly } \frac{d\xi_1}{dt} = \frac{v_\alpha}{AB}\frac{\partial A}{\partial \beta}.$$

The rate of the angle variation at $F$ is equal to the sum $-[d\xi_1/dt + d\xi_2/dt]$. Thus for the rate of distortion which is half this sum we obtain

$$\epsilon_{\alpha\beta}'' = -\frac{1}{2AB}\left(\frac{\partial B}{\partial \alpha}v_\beta + \frac{\partial A}{\partial \beta}v_\alpha\right). \tag{11.18b}$$

Thus finally the strain rates are expressed by the following formulae

$$\epsilon_\alpha = \frac{1}{A}\frac{\partial v_\alpha}{\partial \alpha} + \frac{v_\beta}{AB}\frac{\partial A}{\partial \beta},$$

$$\epsilon_\beta = \frac{1}{B}\frac{\partial v_\beta}{\partial \beta} + \frac{v_\alpha}{AB}\cdot\frac{\partial B}{\partial \alpha}, \tag{11.19}$$

$$\epsilon_{\alpha\beta} = \frac{A}{2B}\frac{\partial}{\partial \beta}\left(\frac{v_\alpha}{A}\right) + \frac{B}{2A}\frac{\partial}{\partial \alpha}\left(\frac{v_\beta}{B}\right),$$

obtained by adding (11.17) and (11.18).

Let us derive now the expression for the condition of incompressibility which has the general form

$$\epsilon_\alpha + \epsilon_\beta + \epsilon_h = 0.$$

315

The average strain rate across the thickness $\epsilon_h$ can be obtained from the expression for the thickness increment

$$\mathrm{d}h = \frac{\partial h}{\partial t}\,\mathrm{d}t + \frac{\partial h}{\partial s_\alpha}\,\mathrm{d}s_\alpha + \frac{\partial h}{\partial s_\beta}\,\mathrm{d}s_\beta$$

and, consequently,

$$\epsilon_h = \frac{1}{h}\frac{\mathrm{d}h}{\mathrm{d}t} = \frac{1}{h}\left(\frac{\partial h}{\partial t} + \frac{v_\alpha}{A}\frac{\partial h}{\partial \alpha} + \frac{v_\beta}{B}\frac{\partial h}{\partial \beta}\right).$$

Hence the incompressibility condition takes the form

$$\frac{\partial h}{\partial t} + \frac{1}{AB}\frac{\partial}{\partial \alpha}(hBv_\alpha) + \frac{1}{AB}\frac{\partial}{\partial \beta}(hAv_\beta) = 0. \tag{11.20}$$

The remaining relations depend on the choice of the basic yield criterion, i.e. whether the Huber–Mises criterion or the Tresca criterion is accepted.

The Huber–Mises yield criterion can now be written in the form [compare (9.3)]

$$\sigma_\alpha^2 - \sigma_\alpha\sigma_\beta + \sigma_\beta^2 + 3\tau^2 = 3k^2. \tag{11.21}$$

The flow rule associated with (11.21) is obtained from the general relations (3.16), bearing in mind that now $\sigma_m = \frac{1}{3}(\sigma_\alpha + \sigma_\beta)$

$$\frac{\epsilon_\alpha}{2\sigma_\alpha - \sigma_\beta} = \frac{\epsilon_\beta}{2\sigma_\beta - \sigma_\alpha} = \frac{\epsilon_{\alpha\beta}}{3\tau}.$$

Substituting here the relations (11.19), one obtains

$$\begin{aligned}
\frac{1}{2\sigma_\alpha - \sigma_\beta}&\left(\frac{1}{A}\frac{\partial v_\alpha}{\partial \alpha} + \frac{1}{AB}\frac{\partial A}{\partial \beta}v_\beta\right) \\
&= \frac{1}{2\sigma_\beta - \sigma_\alpha}\left(\frac{1}{B}\frac{\partial v_\beta}{\partial \beta} + \frac{1}{AB}\frac{\partial B}{\partial \alpha}v_\alpha\right) \\
&= \frac{1}{6\tau}\left[\frac{A}{B}\frac{\partial}{\partial \beta}\left(\frac{v_\alpha}{A}\right) + \frac{B}{A}\frac{\partial}{\partial \alpha}\left(\frac{v_\alpha}{B}\right)\right].
\end{aligned} \tag{11.22}$$

For the Tresca yield criterion in case where $\sigma_1\sigma_2 \leqslant 0$, or in other words, if $\sigma_\alpha\sigma_\beta \leqslant \tau^2$, we can write [compare (9.5)]

$$(\sigma_\alpha - \sigma_\beta)^2 + 4\tau^2 = 4k^2. \tag{11.23}$$

The Tresca yield condition is represented in the principal stress plane by a hexagon shown in Fig. 163. The edges $AB$ and $DE$ of the hexagon correspond to the condition (11.23). The strain rate "vector" is, according to the associated flow law, orthogonal to the yield surface. Thus the associated flow law written in terms of principal strain rates $\epsilon_1$ and $\epsilon_2$ has now the simple form

$$\epsilon_1 = -\epsilon_2. \tag{11.24}$$

Relation (11.24) implies drastic restrictions concerning the deformation mode. Writing the incompressibility condition in terms of principal strain rates

$$\epsilon_1 + \epsilon_2 + \epsilon_3 = 0,$$

where $\epsilon_3 = \epsilon_h$, we obtain from (11.24) that $\epsilon_h = 0$. This means that during the entire process of plastic deformation we shall have $h = $ const; in other words, the thickness of the deformed sheet will not change. The incompressibility condition in the general form (11.20) will be satisfied automatically.

The flow law associated with (11.23) can be expressed by the velocity components $v_\alpha$ and $v_\beta$ in the form

$$
\frac{1}{\sigma_\alpha - \sigma_\beta} \left( \frac{1}{A} \frac{\partial v_\alpha}{\partial \alpha} + \frac{1}{AB} \frac{\partial A}{\partial \beta} v_\beta \right)
$$
$$
= \frac{1}{\sigma_\beta - \sigma_\alpha} \left( \frac{1}{B} \frac{\partial v_\beta}{\partial \beta} + \frac{1}{AB} \frac{\partial B}{\partial \alpha} v_\alpha \right)
$$
$$
= \frac{1}{4\tau} \left[ \frac{A}{B} \frac{\partial}{\partial \beta} \left( \frac{v_\alpha}{A} \right) + \frac{B}{A} \frac{\partial}{\partial \alpha} \left( \frac{v_\beta}{B} \right) \right], \tag{11.25}
$$

equivalent to (11.24). Equations (11.25) have been obtained by analogy with (9.7).

If in the Tresca yield condition the principal stresses have the same signs ($\sigma_1 \sigma_2 \geqslant 0$), or, in other words, if $\sigma_\alpha \sigma_\beta \geqslant \tau^2$, then we have [see (9.8)]

$$(\sigma_\alpha - \sigma_\beta)^2 + 4\tau^2 = 4[\sigma_{p1} - \tfrac{1}{2}|\sigma_\alpha + \sigma_\beta|]^2. \tag{11.26}$$

The equations of equilibrium (11.16), the condition of incompressibility (11.20), the yield criterion, and the two equations resulting from the associated flow rule constitute a system of seven equations with seven sought for functions $\sigma_\alpha$, $\sigma_\beta$, $\tau$, $p$, $v_\alpha$, $v_\beta$, and $h$. The stress field is strongly coupled in this system with the velocity field since the wall thickness $h$, changing its value during the process of deformation, appears in the equations of equilibrium, and moreover, the flow velocities appear in the expression for friction forces on the contact surface. Thus the solution of this system is difficult, but in some cases it can be obtained explicitly. In other cases it is necessary to make certain simplifying assumptions. For the Huber–Mises yield condition we know a solution only under the assumption that the wall thickness is constant. However, this conclusion resulted directly from the flow law (11.24) associated with the Tresca yield criterion, which indicates that also for the Huber–Mises condition the assumption that the thickness $h$ does not change will not lead in most cases to considerable errors.

Similarly as in the theory of plane stress (Chapter 9), one can solve the problem under the assumption that wall thickness $h$ varies along the surface, thus $h = h(\alpha, \beta)$, although it is independent of time. Equations of characteristics for such a case have been derived in [123]. Their applicability is rather limited and therefore, following the author's works [121, 122] we shall consider here only the case $h = \text{const}$.

### 11.3   Characteristics of the stress field for the Huber–Mises yield criterion

Under assumption that the thickness $h$ is constant the problem is reduced to determination of three stress components $\sigma_\alpha$, $\sigma_\beta$ and $\tau$, the contact pressure $p$, and the two velocity components $v_\alpha$ and $v_\beta$. We have for this a set of six equations composed of the equations of equilibrium in which, at present, the friction forces will be neglected [compare (11.16)],

$$(\sigma_\alpha - \sigma_\beta)\frac{\partial B}{\partial \alpha} + B\frac{\partial \sigma_\alpha}{\partial \alpha} + 2\tau\frac{\partial A}{\partial \beta} + A\frac{\partial \tau}{\partial \beta} = 0,$$

$$(\sigma_\beta - \sigma_\alpha)\frac{\partial A}{\partial \beta} + A\frac{\partial \sigma_\beta}{\partial \beta} + 2\tau\frac{\partial B}{\partial \alpha} + B\frac{\partial \tau}{\partial \alpha} = 0, \qquad (11.27)$$

$$\frac{\sigma_\alpha}{R_\alpha} + \frac{\sigma_\beta}{R_\beta} = \frac{p}{h},$$

the Huber–Mises yield condition (11.21) and the relations (11.22). The incompressibility condition (11.20) cannot be satisfied if the assumption $h = \text{const}$ is introduced.

It should be emphasized here that if at the beginning of the deformation process the thickness is uniformly distributed, the present theory represents the exact approach to the analysis of the incipient plastic flow.

This system of governing equations displays the very important feature that it contains a subsystem composed of the first two equations of equilibrium (11.27) and the yield criterion (11.21) containing only three of the sought for unknowns: $\sigma_\alpha$, $\sigma_\beta$ and $\tau$. Thus the problem is in this sense statically determinate. This means that the stress field can be found independently and then checked whether it is compatible with all the kinematic conditions of the specific problem considered. The contact pressure $p$ can be obtained from the third equation of equilibrium (11.27).

Let us introduce, similarly as in the plane stress theory [compare (9.10)], an auxiliary function $\omega(\alpha, \beta)$ which is related to the principal stresses by

$$\begin{aligned}
\tfrac{1}{2}(\sigma_1 + \sigma_2) &= \sqrt{3}\,k\cos\omega, \\
\tfrac{1}{2}(\sigma_1 - \sigma_2) &= k\sin\omega.
\end{aligned} \tag{11.28}$$

Then the stress components can be expressed as follows by means of the function $\omega$ and the angle $\varphi$ between the direction of the larger principal stress $\sigma_1$ and the $\beta$-direction

$$\begin{aligned}
\sigma_\alpha &= k\left(\sqrt{3}\cos\omega - \sin\omega\cos 2\varphi\right), \\
\sigma_\beta &= k\left(\sqrt{3}\cos\omega + \sin\omega\cos 2\varphi\right),
\end{aligned} \qquad \tau = k\sin\omega\sin 2\varphi. \tag{11.29}$$

The equations (11.29) satisfy identically the yield condition (11.21) for any value of $\omega$ and $\varphi$.

Substituting (11.29) into the first two equations of equilibrium (11.27), we obtain the basic system of two quasi-linear partial differential equations with two sought for functions $\omega$ and $\varphi$

$$\begin{aligned}
-B\left(\sqrt{3}\sin\omega + \cos\omega\cos 2\varphi\right)\frac{\partial\omega}{\partial\alpha} &+ A\cos\omega\sin 2\varphi\frac{\partial\omega}{\partial\beta} \\
+ 2B\sin\omega\sin 2\varphi\frac{\partial\varphi}{\partial\alpha} &+ 2A\sin\omega\cos 2\varphi\frac{\partial\varphi}{\partial\beta} \\
= 2\sin\omega\cos 2\varphi\frac{\partial B}{\partial\alpha} &- 2\sin\omega\sin 2\varphi\frac{\partial A}{\partial\beta},
\end{aligned}$$

$$\begin{aligned}
B\cos\omega\sin 2\varphi\frac{\partial\omega}{\partial\alpha} &- A\left(\sqrt{3}\sin\omega - \cos\omega\cos 2\varphi\right)\frac{\partial\omega}{\partial\beta} \\
+ 2B\sin\omega\cos 2\varphi\frac{\partial\varphi}{\partial\alpha} &- 2A\sin\omega\sin 2\varphi\frac{\partial\varphi}{\partial\beta} \\
= -2\sin\omega\sin 2\varphi\frac{\partial B}{\partial\alpha} &- 2\sin\omega\cos 2\varphi\frac{\partial A}{\partial\beta}.
\end{aligned} \tag{11.30}$$

319

This system can be solved by means of the method of characteristics (see Appendix 2). Following the procedure described in the Appendix we arrive at the differential equations of characteristics

$$\frac{d\alpha}{d\beta} = \frac{B}{A} \frac{\sqrt{3}\sin\omega\sin 2\varphi \pm \sqrt{3-4\cos^2\omega}}{\sqrt{3}\sin\omega\cos 2\varphi - \cos\omega}. \qquad (11.31)$$

The upper sign (plus) in the numerator on the right-hand side refers to the first family of characteristics, while the lower sign (minus) corresponds to the second family.

Along the characteristic lines the following differential equations must be satisfied

$$d\varphi \mp \frac{\sqrt{3-4\cos^2\omega}}{2\sin\omega} d\omega = \frac{1}{A}\frac{\partial B}{\partial \alpha}d\beta - \frac{1}{B}\frac{\partial A}{\partial \beta}d\alpha. \qquad (11.32)$$

The upper sign (minus) and the lower sign (plus) refer to the first and to the second families of characteristics respectively.

Equations (11.31) and (11.32) indicate that the basic system (11.30) does not always have real characteristics. Hence it need not always be of the hyperbolic type. It is of the hyperbolic type only if the expression under the root $(3-4\cos^2\omega)$ is positive. In other words, only for $\pi/6 < \omega < 5\pi/6$ or $7\pi/6 < \omega < 11\pi/6$. For other values of $\omega$ the system (11.30) is of the elliptic type and the method of characteristics cannot be used. For the particular values $\omega = \pi/6$, $5\pi/6$, $7\pi/6$, $11\pi/6$ the system becomes of the parabolic type, having one family of characteristics. The range within which the basic system (11.30) is hyperbolic is shown by heavy lines in Fig. 160 where the ellipse represents the Huber–Mises yield condition in the principal stress plane. Our further considerations will be limited to the cases where the equations are of the hyperbolic type.

Let us recall the auxiliary notation (9.16) used by V. V. Sokolovsky in his study of the plane stress theory. Employing this, we represent the equations of characteristics in the compact form

$$\frac{d\alpha}{d\beta} = \frac{B}{A}\tan(\varphi + \psi),$$

$$(11.33a)$$

$$d\varphi + d\chi = \frac{1}{A}\frac{\partial B}{\partial \alpha}d\beta - \frac{1}{B}\frac{\partial A}{\partial \beta}d\alpha$$

320

for the first family of characteristics, and in the form

$$\frac{\mathrm{d}\alpha}{\mathrm{d}\beta} = \frac{B}{A}\tan(\varphi-\psi),$$

$$\mathrm{d}\varphi - \mathrm{d}\chi = \frac{1}{A}\frac{\partial B}{\partial \alpha}\mathrm{d}\beta - \frac{1}{B}\frac{\partial A}{\partial \beta}\mathrm{d}\alpha$$

(11.33b)

for the second family.

The partial derivatives of the coefficients $A$ and $B$, appearing in the equations of characteristics (11.33), are known functions of the coordinates $\alpha$ and $\beta$ if the shape of the shell is given. Thus the problem of integration of the basic system of equations (11.30) has been reduced to the integration of ordinary differential equations (11.33) along the characteristic lines. The integration procedure for particular cases consists in consecutively solving respective boundary value problems.

The characteristics form on the shell a non-orthogonal mesh of lines making the angles $2\psi$, in general different at each point. Note that the characteristic lines for the Huber–Mises yield condition cannot be identified with the slip-lines as is the case for the Tresca yield condition.

Equation (11.33) for the generalized plane stress have been derived in the author's work [122]. Putting in them $A = B = 1$, and replacing $\beta$ and $\alpha$ by the Cartesian coordinates $x$ and $y$, respectively, we arrive at the equations of characteristics (9.18) for the standard (true) plane stress which are due to V. V. Sokolovsky.

Consider now the basic boundary value problems for the equations (11.33).

### 1 Cauchy boundary value problem

Along a sector $AB$ of a regular curve (Fig. 204) lying on the surface of the shell the values of the two functions $\omega$ and $\varphi$, and consequently also the values of the auxiliary functions $\psi$ and $\chi$ are given. The functions $\omega$ and $\varphi$ are continuous along $AB$, their first derivatives existing everywhere. The required property of the sector $AB$ is that it intersects each of the characteristics at most once. These conditions are sufficient to find the solution in the entire curvilinear triangle $ABC$, formed on the surface of the shell by $AB$ and the segments of characteristics passing through the terminal points of the sector $AB$. The equations of characteristics can be integrated by means of the method of finite differences. For that purpose the dif-

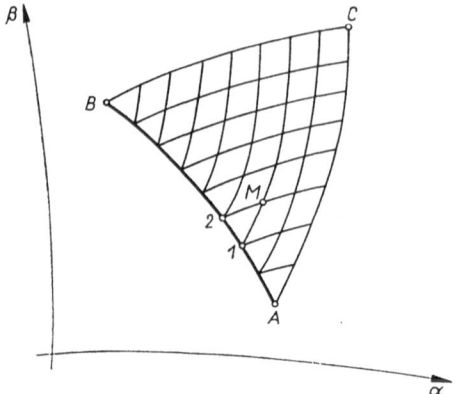

Fig. 204. The Cauchy boundary value problem

ferentials must be replaced in (11.33) by finite differences. The recurrence formulae for all sought for magnitudes at a nodal point $M$ of the mesh of characteristics will be derived below under the assumption that the characteristic passing through the points $1$ and $M$ belongs to the first family (11.33a) and the characteristic passing through $2$ and $M$ belongs to the second family (11.33b) and, moreover, that at the points $1$ and $2$ all required magnitudes are known. To start the numerical procedure an appropriate number of points on $AB$ should be chosen. Since all functions are assumed to be known along $AB$, all necessary values at these points can easily be found. The coordinates $\alpha_M$ and $\beta_M$ of the point $M$ can be obtained from the equations

$$\alpha_M - \alpha_1 = (\beta_M - \beta_1)\left[\frac{B}{A}\tan(\varphi+\psi)\right]_1,$$

$$\alpha_M - \alpha_2 = (\beta_M - \beta_2)\left[\frac{B}{A}\tan(\varphi-\psi)\right]_2,$$

(11.34)

where the suffix notations 1 and 2 indicate that a magnitude with the suffix has been taken at the point $1$ or $2$.

Then the values of $\varphi_M$ and $\chi_M$ can be found from the equations

$$\varphi_M + \chi_M = \varphi_1 + \chi_1 + \left(\frac{1}{A}\frac{\partial B}{\partial \alpha}\right)_1(\beta_M - \beta_1) - \left(\frac{1}{B}\frac{\partial A}{\partial \beta}\right)_1(\alpha_M - \alpha_1),$$

$$\varphi_M - \chi_M = \varphi_2 - \chi_2 + \left(\frac{1}{A}\frac{\partial B}{\partial \alpha}\right)_2(\beta_M - \beta_2) - \left(\frac{1}{B}\frac{\partial A}{\partial \beta}\right)_2(\alpha_M - \alpha_2).$$

(11.35)

Having found all the quantities at $M$, we may, in order to improve the accuracy, repeat the calculations introducing in (11.34) the mean values

$$\frac{1}{2}\left\{\left[\frac{B}{A}\tan(\varphi+\psi)\right]_1 + \left[\frac{B}{A}\tan(\varphi+\psi)\right]_M\right\} \text{ instead of } \left[\frac{B}{A}\tan(\varphi+\psi)\right]_1,$$

and

$$\frac{1}{2}\left\{\left[\frac{B}{A}\tan(\varphi-\psi)\right]_2 + \left[\frac{B}{A}\tan(\varphi-\psi)\right]_M\right\} \text{ instead of } \left[\frac{B}{A}\tan(\varphi-\psi)\right]_2.$$

If the so obtained second approximation of the coordinate values $\alpha_M$ and $\beta_M$ is sufficiently close to their first approximation, the computations can be stopped. If, however, the two approximations give remarkably different values of $\alpha_M$ and $\beta_M$, calculations should be repeated, until the two consecutive approximations become sufficiently close one to the other. In practice the second approximation is in most cases sufficiently accurate.

Also in equations (11.35), having previously found $\alpha_M$ and $\beta_M$ from (11.34), one can introduce at once the mean values, for example,

$$\frac{1}{2}\left[\left(\frac{1}{A}\frac{\partial B}{\partial\alpha}\right)_1 + \left(\frac{1}{A}\frac{\partial B}{\partial\alpha}\right)_M\right] \text{ instead of } \left(\frac{1}{A}\frac{\partial B}{\partial\alpha}\right)_1$$

and similarly with other expressions.

## 2 *Characteristic boundary value problem*

Along two given segments $AB$ and $AC$ of characteristics belonging to different families (comp. Fig. 53) all necessary quantities, i.e. $\omega$, $\varphi$, $\psi$, and $\chi$, are given. They are continuous along these lines and their first derivatives exist. In practice such a situation usually results from preceding calculations. These data are sufficient to find all the values inside the curvilinear quadrangle $ABDC$, bound by two pairs of characteristics $AB$, $CD$ and $AC$, $BD$. The numerical procedure is the same as in the case of the Cauchy problem. Equations (11.34) and (11.35) are valid also in the present case. An appropriate number of points should be chosen on both segments $AB$ and $AC$. The two points adjacent to the corner at $A$ form a situation identical with that shown in Fig. 204 for the three points *1*, *2*, and $M$. Thus the numerical computations have to be started from the points adjacent to the corner $A$.

## 3   *Mixed boundary value problems*

In practical applications there occurs most frequently the kind of mixed boundary problem, where along a segment $AB$ of a non-characteristic line $\alpha = \alpha(\beta)$ the value of the angle $\varphi$ is known, and moreover along a characteristic line $AC$ intersecting $AB$ at $A$, all necessary quantities are given (compare Fig. 55). These data are sufficient to solve the problem in the curvilinear triangle $ABC$ formed by segments of the two initial lines $AB$, $AC$ and the segment $BC$ of the characteristic of the other family. All numerical calculations can be done by means of the equations (11.34) and (11.35), except for the points lying on the non-characteristic line $AB$. Assume that the characteristic $AC$ belongs to the second family. Then the coordinates $\alpha_M$ and $\beta_M$ of the point $M$ at which the characteristic of the first family starting from the point $1$ intersects $AB$ can be calculated from the equation

$$\alpha_M - \alpha_1 = (\beta_M - \beta_1)\left[\frac{B}{A}\tan(\varphi+\psi)\right]_1,$$

and the equation $\alpha = \alpha(\beta)$ of the line $AB$.

Now the value $\chi_M$ can be calculated directly from the first equation (11.35) since $\varphi_M$ is given along $AB$. In this case the accuracy of computations may be improved by introducing in the second approximation the mean values of the appropriate quantities.

### 11.4   Characteristics of the velocity field associated with the Huber–Mises yield criterion

Having found the stress field, we can determine the velocity components $v_\alpha$ and $v_\beta$ from (11.22) since the stress components $\sigma_\alpha$, $\sigma_\beta$, and $\tau$ are already known and can be treated as variable coefficients appearing in the terms containing derivatives of velocities $v_\alpha$ and $v_\beta$ or the velocities themselves. Equations (11.22) can be written in the more suitable form

$$6\tau\frac{1}{B}\frac{\partial v_\beta}{\partial \beta} - (2\sigma_\beta - \sigma_\alpha)\frac{1}{B}\frac{\partial v_\alpha}{\partial \beta} - (2\sigma_\beta - \sigma_\alpha)\frac{1}{A}\frac{\partial v_\beta}{\partial \alpha}$$

$$= -6\tau\frac{1}{AB}\frac{\partial B}{\partial \alpha}v_\alpha - (2\sigma_\beta - \sigma_\alpha)\frac{1}{AB}\left(\frac{\partial B}{\partial \alpha}v_\beta + \frac{\partial A}{\partial \beta}v_\alpha\right), \qquad (11.36_1)$$

$$- (2\sigma_\alpha - \sigma_\beta) \frac{1}{B} \frac{\partial v_\beta}{\partial \beta} + (2\sigma_\beta - \sigma_\alpha) \frac{1}{A} \frac{\partial v_\alpha}{\partial \alpha}$$

$$= (2\sigma_\alpha - \sigma_\beta) \frac{1}{AB} \frac{\partial B}{\partial \alpha} v_\alpha - (2\sigma_\beta - \sigma_\alpha) \frac{1}{AB} \frac{\partial A}{\partial \beta} v_\beta. \tag{11.36$_2$}$$

The system (11.36) contains the two unknowns $v_\alpha$ and $v_\beta$. It can easily be shown that it is of the hyperbolic type in the same range of the auxiliary function $\omega$ as the analogous system (11.30) for stresses. The equations of characteristics of the system (11.36) in that range of $\omega$ have the form

$$\frac{d\alpha}{d\beta} = \frac{B}{A} \tan(\varphi \pm \psi),$$

where $\psi$ is the auxiliary function defined by (9.16). Thus the velocity characteristics coincide with the stress characteristics. This means that the velocity field may be found on the basis of the previously obtained mesh of characteristics for the stress field. Thus a solution of the velocity field is obtained by integrating along the characteristics the following differential equations

$$\cot(\varphi \pm \psi) dv_\beta + dv_\alpha = \frac{1}{A} \left( \frac{\partial B}{\partial \alpha} v_\beta + \frac{\partial A}{\partial \beta} v_\alpha \right) d\beta +$$

$$+ \frac{1}{B} \left[ \cot(\varphi + \psi) \cot(\varphi - \psi) \frac{\partial B}{\partial \alpha} v_\alpha - \frac{\partial A}{\partial \beta} v_\beta \right] d\alpha, \tag{11.37}$$

which have been obtained according to the standard procedure (see Appendix 2). The positive sign in the function $\cot(\varphi \pm \psi)$ on the left-hand side corresponds to the first family, while the negative sign refers to the second family of characteristics.

To obtain the velocity solution to a specific problem, one must successively solve a number of appropriate boundary value problems for the eqs. (11.37). Below are given the basic recurrence formulae for the Cauchy boundary value problem, i.e. for the problem when along a segment $AB$ of a non-characteristic line the two velocity components $v_\alpha$ and $v_\beta$ are known (compare Fig. 204). For the sake of brevity let us introduce the auxiliary notation

$$\Phi = = \frac{1}{A} \left( \frac{\partial B}{\partial \alpha} v_\beta + \frac{\partial A}{\partial \beta} v_\alpha \right),$$

$$\Psi = \frac{1}{B} \left[ \cot(\varphi + \psi) \cot(\varphi - \psi) \frac{\partial B}{\partial \alpha} v_\alpha - \frac{\partial A}{\partial \beta} v_\beta \right].$$

325

As in the case of the numerical procedure used for the computations of the stress field, finite differences will be introduced instead of differentials in the relations (11.37). If the velocity components $v_\alpha$ and $v_\beta$ are known at the two neighbouring points *1* and *2* of the mesh of characteristics, the velocity components $v_{\alpha M}$ and $v_{\beta M}$ at the third point $M$ may be calculated from the equations

$$\cot(\varphi+\psi)_{1M}v_{\beta M}+v_{\alpha M}$$
$$= v_{\beta 1}\cot(\varphi+\psi)_{1M}+v_{\alpha 1}+\Phi_{1M}(\beta_M-\beta_1)+\Psi_{1M}(\alpha_M-\alpha_1),$$
$$\cot(\varphi-\psi)_{2M}v_{\beta M}+v_{\alpha M} \tag{11.38}$$
$$= v_{\beta 2}\cot(\varphi-\psi)_{2M}+v_{\alpha 2}+\Phi_{2M}(\beta_M \mapsto \beta_2)+\Psi_{2M}(\alpha_M-\alpha_2),$$

where the suffixes $1M$ and $2M$ indicate that the quantity has been taken as the mean of its values at the points *1* and $M$ or *2* and $M$, respectively. For example, $\Phi_{2M} = \frac{1}{2}(\Phi_2+\Phi_M)$. Note that the values of $\varphi$ and $\psi$ are known at the nodal points of the mesh from the static solution. The relations (11.37) were derived in a previous work [122]. Putting $A = B = 1$ and replacing $\beta$ and $\alpha$ by the Cartesian coordinates $x$ and $y$, respectively, one obtains the relations (9.20) for the velocity field for the standard plane stress conditions.

### 11.5  Characteristics of the stress field for the Tresca yield criterion

Consider first the important case when the principal stresses have opposite signs. The Tresca yield criterion takes the form (11.23) and it constitutes with the two first equations of equilibrium (11.27) (we neglect for the present the friction forces on the die-plastic layer interface) a system of three equations with three sought for functions $\sigma_\alpha$, $\sigma_\beta$, $\tau$. The condition $h = $ const, so radically simplifying the solution, is now not an approximation, as it was in the case of the theory for the Huber–Mises yield condition (Section 11.3), but it results directly from the associated flow law (see Section 11.2).

To satisfy the yield condition (11.23) identically, let us express the stress components in terms of an auxiliary function $\chi$ and the angle $\varphi$ between the direction of the greater principal stress and the $\beta$-direction.

$$\sigma_\alpha = \sigma_0+2k\chi-k\cos 2\varphi,$$
$$\sigma_\beta = \sigma_0+2k\chi+k\cos 2\varphi, \qquad \tau = k\sin 2\varphi. \tag{11.39}$$

$\sigma_0$ represents an arbitrary constant, whose value should be chosen before the procedure of solving a specific problem is started. Usually one assumes $\sigma_0 = 0$, but in some cases it is more advantageous to assume a non-zero value for $\sigma_0$. The expressions (11.39) satisfy identically the yield criterion (11.23). Substituting (11.39) into the equilibrium equations (11.27), we obtain a set of two partial differential, quasi-linear equations in the two unknown functions $\chi$ and $\varphi$

$$B\frac{\partial \chi}{\partial \alpha} + B\sin 2\varphi \frac{\partial \varphi}{\partial \alpha} + A\cos 2\varphi \frac{\partial \varphi}{\partial \beta}$$
$$= \cos 2\varphi \frac{\partial B}{\partial \alpha} - \sin 2\varphi \frac{\partial A}{\partial \beta},$$
$$A\frac{\partial \chi}{\partial \beta} + B\cos 2\varphi \frac{\partial \varphi}{\partial \alpha} - A\sin 2\varphi \frac{\partial \varphi}{\partial \beta}$$
$$= -\sin 2\varphi \frac{\partial B}{\partial \alpha} - \cos 2\varphi \frac{\partial A}{\partial \beta}.$$

(11.40)

This system is always of the hyperbolic type. Hence it possesses two families of real characteristics. Proceeding in the standard way (see Appendix 2), we obtain the following differential equations

$$\frac{\mathrm{d}\alpha}{\mathrm{d}\beta} = \frac{B}{A}\tan\left(\varphi + \frac{\pi}{4}\right),$$
$$\mathrm{d}\varphi + \mathrm{d}\chi = \frac{1}{A}\frac{\partial B}{\partial \alpha}\mathrm{d}\beta - \frac{1}{B}\frac{\partial A}{\partial \beta}\mathrm{d}\alpha,$$

(11.41a)

of the characteristics of the first family, and

$$\frac{\mathrm{d}\alpha}{\mathrm{d}\beta} = \frac{B}{A}\tan\left(\varphi - \frac{\pi}{4}\right),$$
$$\mathrm{d}\varphi - \mathrm{d}\chi = \frac{1}{A}\frac{\partial B}{\partial \alpha}\mathrm{d}\beta - \frac{1}{B}\frac{\partial A}{\partial \beta}\mathrm{d}\alpha$$

(11.41b)

for the second family.

The partial derivatives of the coefficients $A$ and $B$ appearing in the equations of characteristics (11.41), are known functions of the coordinates $\alpha$ and $\beta$. Thus the problem of integration of the basic system of partial differential equations (11.40) has been reduced to the integration of the ordinary differential equations (11.41) along the characteristic lines. The characteristics of the two families form an orthogonal net on the surface of the shell, because at a generic point characteristics of the first family make an angle of $\varphi + \pi/4$, and characteristics of the second family—an angle of $\varphi - \pi/4$ with $\beta$-direction. Thus they intersect each other at the angle of $\pi/2$. Since the characteristics coincide with the lines of maximum shear stress, they may be regarded as the slip-lines.

The equations (11.41) have been derived in the author's previous work [122]. Putting in them $A = B = 1$, and replacing $\beta$ and $\alpha$ by the Cartesian coordinates $x$ and $y$, respectively, we obtain equations of characteristics (9.24) for the standard true plane stress.

One can formulate for equations (11.41) the same boundary value problems as for equations (11.33) obtained under the assumption of the Huber–Mises yield criterion. All recurrence formulae given in Section 11.3 hold valid in the present case provided the functions $\tan(\varphi + \psi)$ and $\tan(\varphi - \psi)$ are replaced by the functions $\tan(\varphi + \pi/4)$ and $\tan(\varphi - \pi/4)$ respectively. Now the function $\chi$ has evidently a different meaning.

The problem in the case where the principal stresses are of the same sign can be analysed in the same way as in Chapter 9 for the standard plane stress. The governing system of equations is of the parabolic type, having the one family of characteristics derived in [122]. These equations are not given here since they are little used in practice.

### 11.6   Characteristics of the velocity field associated with the Tresca yield condition

Our considerations will be limited here to problems in which the principal stresses have opposite signs, thus to the problems corresponding to the segments $AB$ and $DE$ of the Tresca hexagon in the principal stress plane (Fig. 163). Let us recall that for such states of stress the associated flow rule implies that the thickness of the plastically deformed sheet does not change during the deformation process. The incompressibility condition (11.20) is automatically satisfied. Thus we have a consistent theory not requiring any additional assumptions. When the Huber–Mises yield condition was accepted, the assumption that the thickness does not change, which had to be introduced, did not allow to accept the incompressibility condition. One must, however, remember that the Tresca yield condition, and certainly the flow law associated with it, only approximately describe the real behaviour of metals, which is much better described by the Huber–Mises theory.

Equations (11.25) may be presented in the more suitable form

$$(\sigma_\beta - \sigma_\alpha)\left(\frac{1}{B}\frac{\partial v_\alpha}{\partial \beta} + \frac{1}{A}\frac{\partial v_\beta}{\partial \alpha}\right) - 4\tau\frac{1}{B}\frac{\partial v_\beta}{\partial \beta}$$

$$= \frac{1}{AB}\left[4\tau\frac{\partial B}{\partial \alpha}v_\alpha + (\sigma_\beta - \sigma_\alpha)\left(\frac{\partial B}{\partial \alpha}v_\beta + \frac{\partial A}{\partial \beta}v_\alpha\right)\right], \qquad (11.42)$$

$$\frac{1}{A}\frac{\partial v_\alpha}{\partial \alpha} + \frac{1}{B}\frac{\partial v_\beta}{\partial \beta} = -\frac{1}{AB}\left(\frac{\partial B}{\partial \alpha}v_\alpha + \frac{\partial A}{\partial \beta}v_\beta\right).$$

They contain the two unknowns $v_\alpha$ and $v_\beta$, for the stresses should be found from the static solution before the kinematic solution is started. It can easily be shown that the system (11.42) is always of the hyperbolic type and hence it has two families of characteristics. Proceeding in the standard manner, we find the differential equations of characteristics

$$\frac{d\alpha}{d\beta} = \frac{B}{A}\tan\left(\varphi \pm \frac{\pi}{4}\right).$$

Comparing them with (11.41), we conclude that also in the present case the velocity characteristics coincide with the stress characteristics. In any specific problem the velocity field may be found by numerical integration of the differential relations

$$dv_\alpha - dv_\beta \tan(\varphi \pm \tfrac{1}{4}\pi)$$

$$= \frac{1}{A}\left[2\tan 2\varphi\frac{\partial B}{\partial \alpha}v_\alpha + \left(\frac{\partial B}{\partial \alpha}v_\beta + \frac{\partial A}{\partial \beta}v_\alpha\right)\right]d\beta -$$

$$-\frac{1}{B}\left[\frac{\partial B}{\partial \alpha}v_\alpha + \frac{\partial A}{\partial \beta}v_\beta\right]d\alpha, \qquad (11.43)$$

which must be satisfied along the characteristic lines. Relations (11.43) have been derived according to the standard procedure (see Appendix 2). The upper negative sign in the expression $\tan(\varphi \mp \tfrac{1}{4}\pi)$ corresponds to the first, and the lower positive sign, to the second family of characteristics.

## 11.7  Stretchforming of thin-walled shells

Consider an example of stretchforming of a shallow shell. A sheet of the plastic metal is stretched over a rigid die whose surface constitutes a portion of an ellipsoid of revolution with the main axis of symmetry chosen

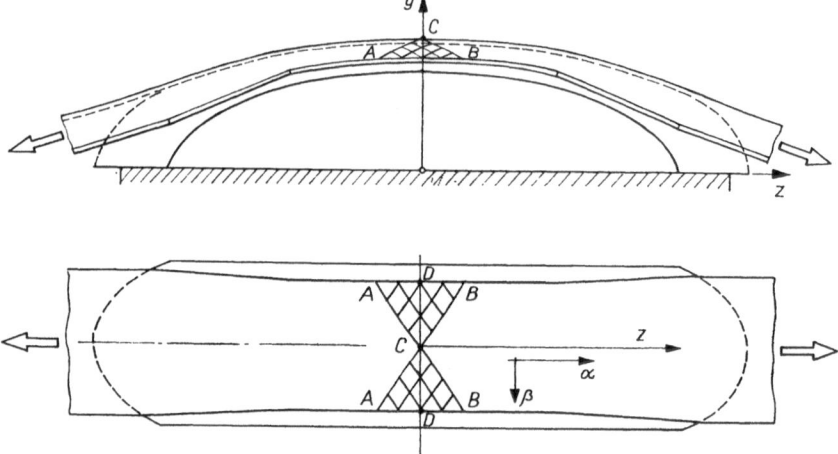

Fig. 205. Stretchforming of a shallow shell

as the $z$-axis of the coordinate system (Fig. 205). The main axis of the ellipsoid has the length $b$, while the other is of the length $a$. Assume an orthogonal system of coordinates $\alpha$, $\beta$ with the lines $\alpha = $ const coinciding with the parallels, and the $\beta = $ const lines, with the meridians of the ellipsoid.

The parametric equations of the shell surface have the form

$$x = r(\alpha)\cos\beta,$$
$$y = r(\alpha)\sin\beta,$$
$$z = \frac{b^2\cos\alpha}{\sqrt{a^2\sin^2\alpha+b^2\cos^2\alpha}},$$

where $r(\alpha)$ stands for the radius of the parallel $\alpha = $ const and is defined by the formula

$$r(\alpha) = \frac{a^2\sin\alpha}{\sqrt{a^2\sin^2\alpha+b^2\cos^2\alpha}}.$$

The coefficients of the first quadratic form can be obtained from the general relations (11.7)

$$A = \frac{ds_\alpha}{d\alpha} = \varrho_\alpha = \frac{r'(\alpha)}{\cos\alpha},$$

$$B = \frac{ds_\beta}{d\beta} = r.$$

Assume that the stress free edges of the stretched sheet are formed by intersection of the ellipsoid with the two planes

$$x = \pm\tfrac{1}{2}a.$$

In spite of the axial symmetry of the die, the problem is not axially symmetric and has to be solved according to the general theory. Assuming the Huber–Mises yield condition (11.21), one can show that the zone of incipient plastic flow will be localized in the region of the smallest curvature. In the present example the plastic zone has the form of two curvilinear triangles $ABC$. Boundaries of the two triangles were obtained numerically by solving the Cauchy boundary value problem, starting from the data along the stress free edge $AB$ of the sheet. Along $AB$ one of the principal stresses is zero, whereas the second equals to the yield point in simple tension and its direction is tangent to $AB$. Hence from (9.15) we obtain $\omega = \pi/3$ along $AB$. The angle $\varphi$ between the direction of the larger principal stress and the $\beta$-direction, is along $AB$ obviously equal to the angle at which the edge of the sheet intersects the parallel circles. This angle, determined from geometrical considerations, may be expressed by the formula

$$\varphi = \cos^{-1} \frac{2a\cos\alpha\sin\beta}{\sqrt{3a^2 - (b^2 - a^2)\cos^2\alpha}},$$

where for $\alpha$ and $\beta$ we have to write the coordinates of a specific point of the edge. As we have shown in the general theory, these data suffice to obtain the solution in the entire region $ABC$.

Numerical computations have been performed by means of the recurrence formulae (11.34) and (11.35) for the particular case $a = 0.4b$. The maximum tensile stress at the central point $C$ is $\sigma_\alpha^{(c)} = 1.015\sigma_{\text{pl}}$. Thus stresses are almost uniformly distributed along $D$–$D$. Such a result indicates that in analysing the stretchforming process of shallow shells of positive Gaussian curvature one can, with sufficient degree of accuracy, assume homogeneous stress distribution.

Note that this conclusion results directly if the Tresca yield criterion is assumed. Since the two principal stresses have evidently positive signs, the problem is of the parabolic type, and thus the deforming region will be limited to a single characteristic through the points $D$. The normal velocity suffers a jump across that characteristic, and the incipient plastic flow consists in local necking of the sheet along $D$–$D$.

Local necking may occur also in solutions obtained for the Huber–Mises yield criterion if the shell is not as shallow as that shown in Fig. 205.

331

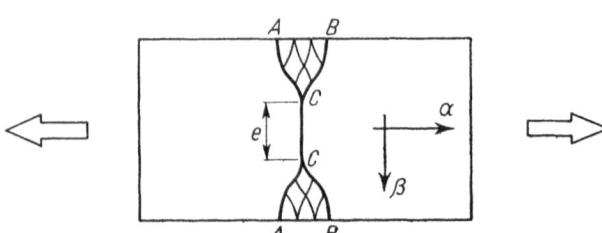

Fig. 206. Local necking in a stretched shell

At a certain distance from the edge the characteristics of the two families can merge as shown in Fig. 206. The state of stress along the segment $CC$ corresponds to the point $A$ of the Huber–Mises ellipse (Fig. 160). Along $CC$ the stresses are $\sigma_\alpha = 2k$ and $\sigma_\beta = k$. The auxiliary function $\omega$ equals $\pi/6$. For that value of $\omega$ the basic system of equations (11.30) becomes parabolic and it has one family of characteristics. Thus $CC$ represents a stress characteristic. Besides plastic yielding in the two zones $ABC$, deformation can occur due to necking of the sheet along $CC$. Such a mode of necking represents a particular case of the necking discussed in Chapter 9.

It is interesting to note that all features of the stress and velocity fields in stretchforming of shells of positive Gaussian curvature are identical with those for the flat notched strips loaded by a tensile force (see [51]).

For shells with negative Gaussian curvature (see Fig. 198c) the solving procedure remains unchanged. Note, however, that for these shells the principal stresses are everywhere of opposite signs. Thus the problem is always hyperbolic for both yield conditions. The mesh of characteristics will be of the same type as that shown in Fig. 205.

### 11.8 Drawing through a die, Huber–Mises yield condition

Assuming the Huber–Mises yield condition let us analyse the stress and velocity field in the neighbourhood of an oval hole in a sheet undergoing deformation by drawing through a die (Fig. 207).

Let the uniformly distributed drawing stress at the exit be $\sigma_0 = \sigma_{pl} = \sqrt{3}k$. The uniformly distributed drawing velocity is directed vertically. This formulation of the problem may be regarded as an approximate approach to the complex problem of drawing shells from a sheet material.

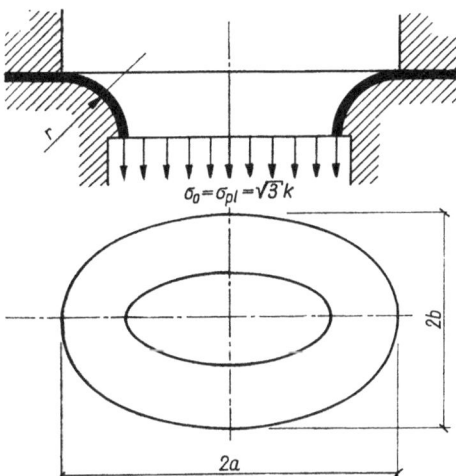

Fig. 207. Drawing through an oval die

An orthogonal coordinate system, coinciding with the lines of principal curvatures will be adopted. The parametric equations of the die surface are

$$x = \frac{a^2\cos\beta}{\sqrt{a^2\cos^2\beta+b^2\sin^2\beta}} - r\cos\alpha\cos\beta,$$

$$y = \frac{b^2\sin\beta}{\sqrt{a^2\cos^2\beta+b^2\sin^2\beta}} - r\cos\alpha\sin\beta,$$

$$z = r\sin\alpha.$$

The coefficients of the first quadratic form can be found from the relations (11.7)

$$A = r,$$

$$B = \frac{a^2b^2}{\sqrt{(a^2\cos^2\beta+b^2\sin^2\beta)^3}} - r\cos\alpha.$$

Owing to symmetry, it is sufficient to find the solution in the quadrant of the shell where the coordinates $x, y, z$ are positive (Fig. 208). Relations (11.29) indicate that on the edge, $AB$, which is the line $\alpha = 0$, we have $\varphi = \pi/2$ and $\omega = \pi/3$. Thus the knowledge of the boundary values along $AB$ is sufficient to solve the Cauchy problem for stresses in the curvilinear quadrangle $ABCD$, following the procedure defined by Fig. 204, and the recurrence equations (11.34) and (11.35). Now, after having found the

333

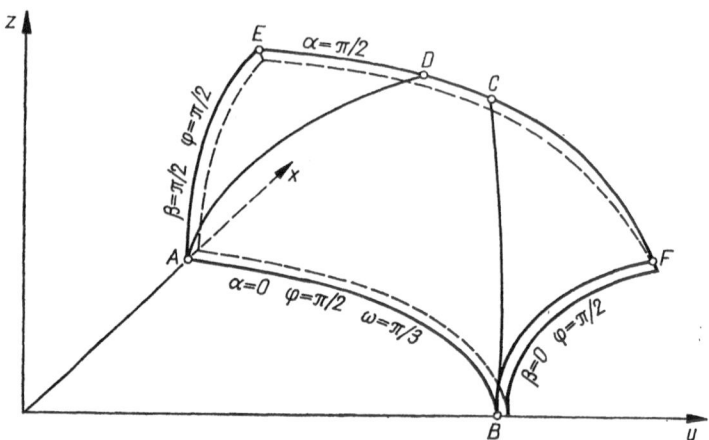

Fig. 208. Boundary value problems for computations of the stress field

values of functions $\varphi$ and $\omega$ along the characteristics $BC$ and $AD$, we can find the stresses in the curvilinear triangles $BCF$ and $ADE$. On account of symmetry, we have $\varphi = \pi/2$ along the lines $AE$ and $BF$. Since $\varphi$ and $\omega$ are known along the characteristics $AD$ and $BC$, and $\varphi$ is known also along the non-characteristic line $AE$ or $BF$, a typical mixed boundary value problem occurs in the two triangles. Solving numerically these mixed problems, we obtain the values of $\varphi$ and $\omega$ in the entire region $ABFE$. The values of stress components can be found by means of the expressions (11.29). Numerical results of computations are given in [122] and in the Polish edition of the present book. Figure 209 shows the mesh of characteristics. Computations have been performed for the particular case where $a = 1.0$, $b = 0.6$, and $r = 0.2$. The nodal points are labelled by two numbers, the first of which stands for the number of the characteristic of the first family, and the second represents the number of the characteristic of the second family intersecting each other at this node.

The magnitude of the pressure exerted by the sheet against the die is determined from the third equation of equilibrium (11.27) on substituting for the radii of curvature the expressions

$$R_\alpha = r \quad \text{and} \quad R_\beta = \frac{a^2 b^2}{\cos \alpha (a^2 \cos^2 \beta + b^2 \sin^2 \beta)^{3/2}} - r,$$

obtained from relations (11.11). The values of the contact pressure $p$ at the nodal points of the mesh of characteristics are given in Table 2 (p. 336).

334

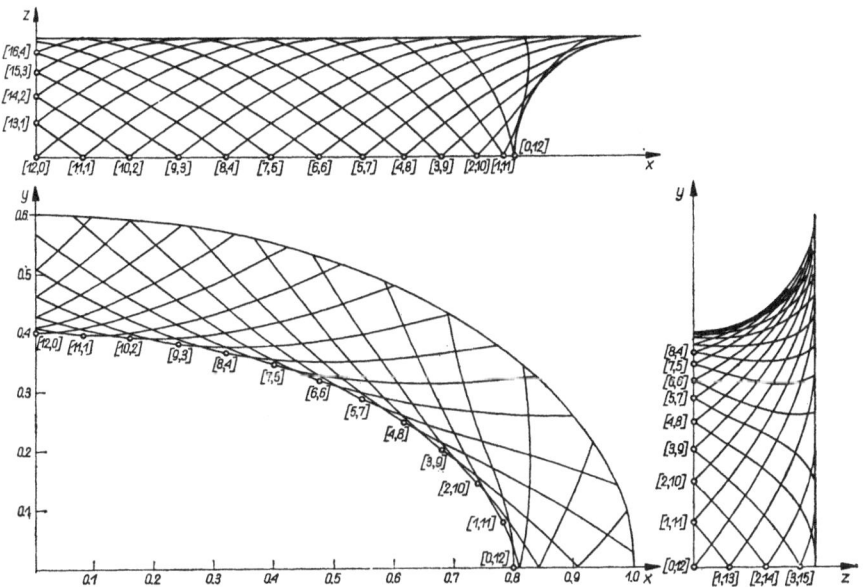

Fig. 209. The mesh of characteristics for the drawing problem shown in Fig. 207

The velocity field has been found under the assumption that along the edge $AB$, $v_\alpha = v_0 = \text{const}$ and $v_\beta = 0$. These data define uniquely the Cauchy boundary value problem for the differential equations (11.37) for the velocity field. Numerical computations have been carried out by means of the recurrence formulae (11.38). The values of the velocity components at the nodal points are collated in Table 3 (pp. 338–339).

## 11.9    Drawing through a die, Tresca yield condition

To give an example of an application of the theory for the Tresca yield condition, let us find the stress distribution in the vicinity of the exit section in the case of drawing of a shell through a hyperboloidal die. As in the foregoing example, the uniformly distributed tensile stress $\sigma_0$ is assumed to represent external loading in the exit section. In the orthogonal coordinate

335

**Table 2.** The pressure $\dfrac{p}{kh}$ for the example of drawing (Figs. 207 and 209)

Characteristics of the first family

| Characteristics of the second family | 0 | 1 | 2 | 3 | 4 | 5 | 6 | 7 | 8 | 9 | 10 | 11 | 12 | 13 | 14 | 15 |
|---|---|---|---|---|---|---|---|---|---|---|---|---|---|---|---|---|
| 0 | | | | | | | | | | | | | | | | |
| 1 | | | | | | | | | | | | | | | | |
| 2 | | | | | | | | | | | | | | | | |
| 3 | | | | 8.66 | | | | | | | | | 8.66 | 8.61 | 8.51 | 8.31 |
| 4 | | | | 8.66 | | | | | | | 8.66 | 8.66 | 8.65 | 8.55 | 8.40 | 8.15 |
| 5 | | | | 8.66 | | | | | 8.66 | 8.66 | 8.65 | 8.65 | 8.61 | 8.51 | 8.31 | 7.98 |
| 6 | | | | 8.66 | | | | 8.66 | 8.65 | 8.65 | 8.61 | 8.61 | 8.55 | 8.40 | 8.15 | |
| 7 | | | | 8.66 | | | 8.66 | 8.65 | 8.61 | 8.61 | 8.55 | 8.55 | 8.51 | 8.28 | 7.93 | |
| 8 | | | | 8.66 | 8.66 | 8.66 | 8.65 | 8.61 | 8.55 | 8.55 | 8.48 | 8.51 | 8.40 | 8.10 | | |
| 9 | | | 8.66 | 8.66 | 8.65 | 8.65 | 8.61 | 8.55 | 8.47 | 8.48 | 8.36 | 8.39 | 8.26 | 7.85 | | |
| 10 | | | 8.66 | 8.66 | 8.61 | 8.61 | 8.55 | 8.47 | 8.31 | 8.33 | 8.19 | 8.24 | 8.06 | | | |
| 11 | 8.66 | 8.66 | 8.68 | 8.66 | 8.55 | 8.55 | 8.41 | 8.26 | 8.06 | 8.15 | 7.89 | 7.99 | 7.80 | | | |
| 12 | | 8.77 | 8.81 | 8.70 | 8.37 | 8.38 | 8.13 | 7.84 | 7.51 | 7.79 | 7.48 | 7.70 | | | | |
| 13 | | 9.15 | 9.16 | 8.62 | 7.79 | 8.01 | 7.51 | 6.99 | 6.50 | 7.12 | | | | | | |
| 14 | | | 9.12 | 8.15 | 6.47 | 6.81 | 5.52 | 4.95 | | | | | | | | |
| 15 | | | | 6.63 | 4.41 | 4.76 | | | | | | | | | | |

system $\alpha, \beta$ coinciding with the lines of principal carvature, the parametric equations of the shell have the form (Fig. 210)

$$x = a\frac{\cos\beta}{\cos\alpha},$$

$$y = b\frac{\sin\beta}{\cos\alpha},$$

$$z = c\cot\alpha.$$

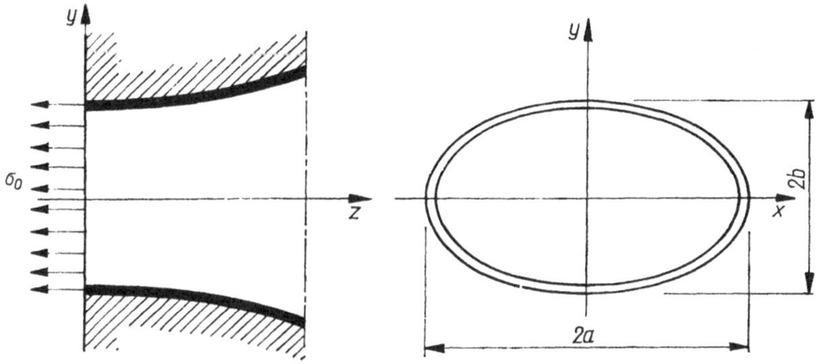

Fig. 210. Drawing through a hyperboloidal die

Expressions for the coefficients of the first quadratic form are obtained from the general formulae (11.7)

$$A = \frac{\sqrt{\sin^2\alpha(a^2\cos^2\beta + b^2\sin^2\beta) + c^2}}{\cos^2\alpha},$$

$$B = \frac{\sqrt{a^2\sin^2\beta + b^2\cos^2\beta}}{\cos\alpha}.$$

Owing to symmetry, it is sufficient to solve the problem for the quadrant of the shell which is bounded by the positive parts of the coordinate axes $x, y, z$ (Fig. 211). Along the edge $AB$, which coincides with the line $\alpha = 0$, we have $\sigma_\alpha = \sigma_0$, $\sigma_\beta = \sigma_0 - 2k$ and $\tau = 0$. Thus from (11.39) we find the values $\chi = -0.5$ and $\varphi = \pi/2$ along $AB$, assuming that the constant $\sigma_0$ appearing in (11.39) has the value of the tensile stress $\sigma_0$ acting on $AB$. These data define the Cauchy boundary value problem in the curvilinear triangle $ABC$ on the shell surface. In the triangle $BCD$ we have a mixed boundary value problem, for the $BD$ line represents the line

**Table 3.** Velocities for the example

| | | | | Characteristics | | | |
|---|---|---|---|---|---|---|---|
| | | 0 | 1 | 2 | 3 | 4 | 5 |
| 0 | $-v_\alpha/v_0$<br>$-v_\beta/v_0$ | | | | | | |
| 1 | $-v_\alpha/v_0$<br>$-v_\beta/v_0$ | | | | | | |
| 2 | $-v_\alpha/v_0$<br>$-v_\beta/v_0$ | | | | | | |
| 3 | $-v_\alpha/v_0$<br>$-v_\beta/v_0$ | | | | | | |
| 4 | $-v_\alpha/v_0$<br>$-v_\beta/v_0$ | | | | | | |
| 5 | $-v_\alpha/v_0$<br>$-v_\beta/v_0$ | | | | | | |
| 6 | $-v_\alpha/v_0$<br>$-v_\beta/v_0$ | | | | | | |
| 7 | $-v_\alpha/v_0$<br>$-v_\beta/v_0$ | | | | | | 1.00<br>0.00 |
| 8 | $-v_\alpha/v_0$<br>$-v_\beta/v_0$ | | | | | 1.00<br>0.00 | 1.00<br>0.00 |
| 9 | $-v_\alpha/v_0$<br>$-v_\beta/v_0$ | | | | 1.00<br>0.00 | 1.00<br>0.00 | 0.99<br>0.00 |
| 10 | $-v_\alpha/v_0$<br>$-v_\beta/v_0$ | | | 1.00<br>0.00 | 1.00<br>0.00 | 0.99<br>0.00 | 0.97<br>0.01 |
| 11 | $-v_\alpha/v_0$<br>$-v_\beta/v_0$ | | 1.00<br>0.00 | 1.00<br>0.00 | 0.98<br>0.00 | 0.95<br>0.01 | 0.92<br>0.02 |
| 12 | $-v_\alpha/v_0$<br>$-v_\beta/v_0$ | 1.00<br>0.00 | 1.00<br>0.00 | 0.97<br>0.00 | 0.93<br>0.02 | 0.88<br>0.02 | 0.83<br>0.04 |
| 13 | $-v_\alpha/v_0$<br>$-v_\beta/v_0$ | | 0.95<br>0.00 | 0.88<br>0.01 | 0.82<br>0.03 | 0.76<br>0.05 | 0.70<br>0.05 |
| 14 | $-v_\alpha/v_0$<br>$-v_\beta/v_0$ | | | 0.79<br>0.00 | 0.70<br>0.03 | 0.63<br>0.04 | 0.57<br>0.04 |
| 15 | $-v_\alpha/v_0$<br>$-v_\beta/v_0$ | | | | 0.59<br>0.00 | 0.52<br>0.02 | |

Characteristics of the second family

of drawing (Figs. 207 and 209)

of the first family

| 6 | 7 | 8 | 9 | 10 | 11 | 12 | 13 | 14 | 15 |
|---|---|---|---|---|---|---|---|---|---|
| | | | | | | 1.00 0.00 | | | |
| | | | | | 1.00 0.00 | 1.00 0.00 | 0.99 0.00 | | |
| | | | | 1.00 0.00 | 1.00 0.00 | 0.99 0.00 | 0.99 0.00 | 0.97 0.00 | |
| | | | 1.00 0.00 | 1.00 0.00 | 0.99 0.00 | 0.98 0.00 | 0.97 0.00 | 0.95 0.00 | 0.94 0.00 |
| | | 1.00 0.00 | 1.00 0.00 | 0.99 0.00 | 0.98 0.00 | 0.97 0.00 | 0.95 0.00 | 0.93 0.00 | 0.90 0.00 |
| | 1.00 0.00 | 1.00 0.00 | 0.99 0.00 | 0.98 0.00 | 0.97 0.00 | 0.95 0.00 | 0.92 0.01 | 0.90 0.01 | 0.87 0.01 |
| 1.00 0.00 | 1.00 0.00 | 0.99 0.00 | 0.98 0.00 | 0.96 0.00 | 0.95 0.01 | 0.92 0.01 | 0.89 0.01 | 0.87 0.01 | |
| 1.00 0.00 | 0.99 0.00 | 0.98 0.00 | 0.96 0.01 | 0.94 0.01 | 0.91 0.01 | 0.88 0.01 | 0.86 0.01 | | |
| 0.99 0.00 | 0.98 0.00 | 0.95 0.01 | 0.93 0.01 | 0.91 0.01 | 0.88 0.02 | 0.85 0.02 | | | |
| 0.97 0.00 | 0.95 0.01 | 0.92 0.01 | 0.90 0.01 | 0.87 0.02 | 0.84 0.02 | | | | |
| 0.93 0.01 | 0.91 0.01 | 0.87 0.02 | 0.85 0.02 | 0.82 0.03 | | | | | |
| 0.88 0.02 | 0.85 0.03 | 0.81 0.03 | 0.78 0.03 | | | | | | |
| 0.79 0.04 | 0.76 0.05 | 0.73 0.05 | | | | | | | |
| 0.66 0.05 | 0.64 0.06 | | | | | | | | |

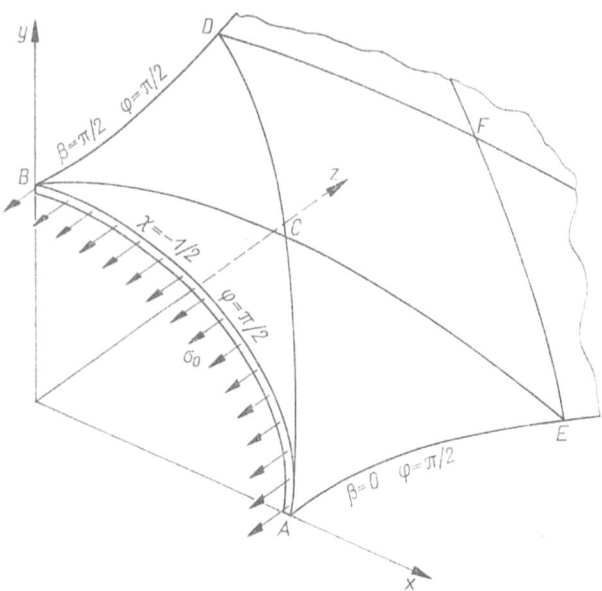

Fig. 211. Boundary value problems for the example shown in Fig. 210

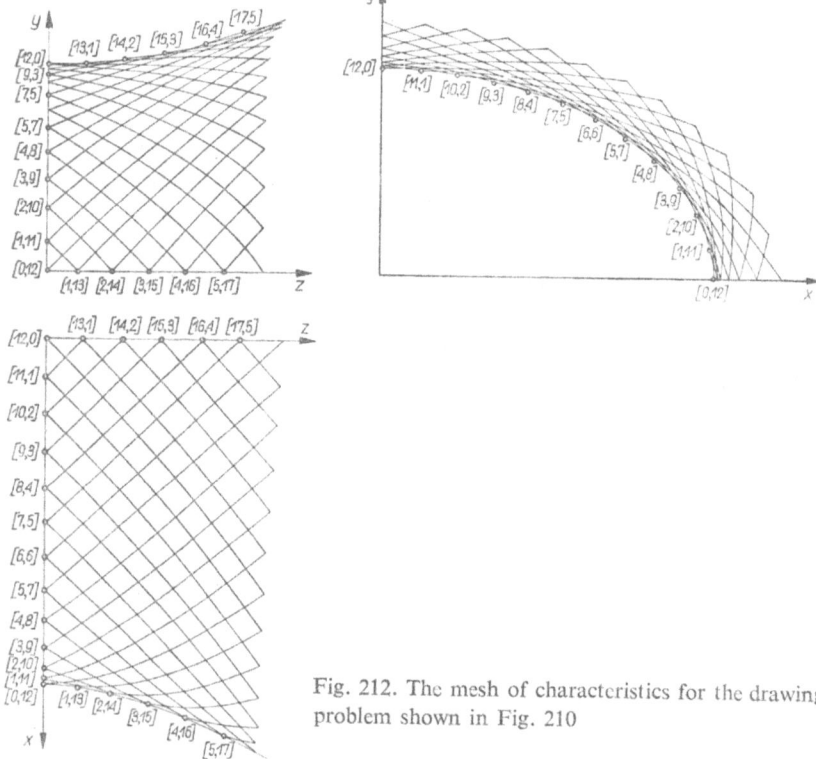

Fig. 212. The mesh of characteristics for the drawing problem shown in Fig. 210

$\beta = \pi/2$ and along it, owing to symmetry, we have $\varphi = \pi/2$. Similarly in the curvilinear triangle $ACE$ the data define the mixed problem since along $AE$ the values $\beta = 0$ and $\varphi = \pi/2$ are given. In the curvilinear quadrangle $DCEF$ we have the characteristic problem. Repeating this procedure one can extend the solution further and further, finding the values of $\chi$ and $\varphi$ for the entire quadrant of the shell. Figure 212 shows the mesh of characteristics, for the particular case $a = 1.0$, $b = 0.6$ and $c = 1.0$.

In the present case the characteristics coincide with the slip-lines. Numerical results are given in [121] and in the Polish edition of this book.

## 11.10   Drawing through a die, influence of the friction on the die-sheet interface

Retaining the previous assumption that the wall thickness does not change and taking into account the friction forces between the die and the sheet, we may write the first two equations of equilibrium (11.16) as follows:

$$(\sigma_\alpha - \sigma_\beta)\frac{\partial B}{\partial \alpha} + B\frac{\partial \sigma_\alpha}{\partial \alpha} + 2\tau\frac{\partial A}{\partial \beta} + A\frac{\partial \tau}{\partial \beta} - \frac{1}{h}AB\mu p\frac{v_\alpha}{v} = 0,$$

$$(\sigma_\beta - \sigma_\alpha)\frac{\partial A}{\partial \beta} + A\frac{\partial \sigma_\beta}{\partial \beta} + 2\tau\frac{\partial B}{\partial \alpha} + B\frac{\partial \tau}{\partial \alpha} - \frac{1}{h}AB\mu p\frac{v_\beta}{v} = 0.$$

$$(11.44)$$

Since there is evident coupling of the stress and velocity fields, caused by the last terms in the equations of equilibrium, the standard procedure used in the foregoing sections cannot be applied. However, the problem can be solved by means of a successive approximations procedure [122]. In processes such as drawing, the velocity field can depend only slightly on the friction conditions, because in a specified period of time a strictly defined volume of the material must pass through the die. Hence for the first approximation we may accept the solution in which friction forces are neglected. The velocity field resulting from such a solution enables us to start the procedure of successive approximations. Thus in the second approximation we assume that the velocity field is known and coinciding with that obtained in the first approximation. The problem will be analysed here for the Huber–Mises yield condition and the associated flow rule. It can be solved in an analogous way also for the Tresca yield condition.

The Huber–Mises yield condition (11.21) is identically satisfied when the stress components are expressed by means of the relations (11.29).

Substituting (11.29) into the equations of equilibrium (11.44), we obtain a system of quasi-linear partial differential equations with two sought for functions $\omega$ and $\varphi$. This system differs from the previous system (11.30) by the additional term $(AB/kh)\mu p v_\alpha/v$, which appears on the right-hand side of the first equation, and the term $(AB/kh)\mu p v_\beta/v$ in the second equation. The left-hand sides of the two systems are the same. Thus the differential equations of the characteristics

$$\frac{d\alpha}{d\beta} = \frac{B}{A}\tan(\varphi\pm\psi),$$

which one obtains by equating to zero the characteristic determinant of the system are the same in both cases. But the differential relations which must be satisfied along these characteristics take now another form,

$$d\varphi\pm d\chi = \left[\frac{1}{A}\frac{\partial B}{\partial\alpha} + \frac{\mu p}{kh}\left(\frac{v_\alpha}{v}\cos 2\varphi - \frac{v_\beta}{v}\sin 2\varphi\right)\frac{B}{2\sin\omega}\right]d\beta -$$
$$-\left[\frac{1}{B}\frac{\partial A}{\partial\beta} - \frac{\mu p}{kh}\left(\frac{v_\alpha}{v}\sin 2\varphi + \frac{v_\beta}{v}\cos 2\varphi\right)\frac{A}{2\sin\omega}\right]d\alpha, \qquad (11.45)$$

where the positive sign on the left-hand side refers to the first, and the negative sign, to the second family of characteristics.

The solution of particular problems consists in successive numerical computations of the boundary value problems. Having found the stresses in the second approximation, we can obtain the corrected velocity field. The equations of characteristics for velocities (11.37) remain unaltered. Then the third approximation can be computed, and so on. In practice the second approximation gives sufficiently accurate results.

For example, we solved the previous problem from Fig. 207. The friction coefficient $\mu = 0.2$ was deliberately taken evidently too large, in order to emphasize the influence of friction on the stress and velocity distributions. In Fig. 213 the dashed lines show the mesh of characteristics obtained according to the present procedure, accounting for the friction forces. The mesh of characteristics obtained previously for the perfectly smooth contact (see Fig. 209) is shown in Fig. 213 by continuous lines. Generally, the pattern of the mesh remains unaltered, but the nodal points have been slightly shifted. This fact allows us to compare directly without interpolation the values of the contact pressure $p$, which for the present case have been collated in Table 4 (p. 343). Velocities are given in the Table 5 (p. 344–345). It is readily seen that the values of the contact pressure

**Table 4.** The values of the contact pressure $p/kh$ for the example shown in Fig. 213

|  | Characteristics of the first family | | | | | | | | | | | | | | |
|---|---|---|---|---|---|---|---|---|---|---|---|---|---|---|---|
| | 0 | 1 | 2 | 3 | 4 | 5 | 6 | 7 | 8 | 9 | 10 | 11 | 12 | 13 | 14 |
| 0 | | | | | | | | | | | | 8.66 | 8.66 | | |
| 1 | | | | | | | | | | | 8.66 | 8.41 | 8.41 | 8.26 | |
| 2 | | | | | | | | | | 8.66 | 8.41 | 8.27 | 8.26 | 7.97 | 7.64 |
| 3 | | | | | | | | | 8.66 | 8.42 | 8.27 | 7.98 | 7.98 | 7.64 | 7.21 |
| 4 | | | | | | | | 8.66 | 8.42 | 8.28 | 7.99 | 7.67 | 7.67 | 7.22 | |
| 5 | | | | | | | 8.66 | 8.43 | 8.29 | 8.01 | 7.68 | 7.26 | 7.22 | | |
| 6 | | | | | | 8.66 | 8.45 | 8.31 | 8.04 | 7.70 | 7.28 | | | | |
| 7 | | | | | 8.66 | 8.47 | 8.34 | 8.08 | 7.74 | 7.32 | | | | | |
| 8 | | | | 8.66 | 8.50 | 8.38 | 8.12 | 7.79 | 7.32 | | | | | | |
| 9 | | | 8.66 | 8.56 | 8.44 | 8.17 | 7.82 | 7.34 | | | | | | | |
| 10 | | 8.66 | 8.67 | 8.59 | 8.28 | 7.83 | 7.23 | | | | | | | | |
| 11 | 8.66 | 8.88 | 8.69 | 8.50 | 7.85 | 7.00 | | | | | | | | | |
| 12 | | 9.37 | 9.20 | 8.02 | 6.29 | | | | | | | | | | |
| 13 | | | 8.50 | 5.85 | | | | | | | | | | | |
| 14 | | | | | | | | | | | | | | | |

Characteristics of the second family

**Table 5.** The values of velocities

|  |  |  | Characteristics | | | | | |
|---|---|---|---|---|---|---|---|---|
|  |  |  | 0 | 1 | 2 | 3 | 4 | 5 |
| Characteristics of the second family | 0 | $-v_\alpha/v_0$ $-v_\beta/v_0$ |  |  |  |  |  |  |
|  | 1 | $-v_\alpha/v_0$ $-v_\beta/v_0$ |  |  |  |  |  |  |
|  | 2 | $-v_\alpha/v_0$ $-v_\beta/v_0$ |  |  |  |  |  |  |
|  | 3 | $-v_\alpha/v_0$ $-v_\beta/v_0$ |  |  |  |  |  |  |
|  | 4 | $-v_\alpha/v_0$ $-v_\beta/v_0$ |  |  |  |  |  |  |
|  | 5 | $-v_\alpha/v_0$ $-v_\beta/v_0$ |  |  |  |  |  |  |
|  | 6 | $-v_\alpha/v_0$ $-v_\beta/v_0$ |  |  |  |  |  |  |
|  | 7 | $-v_\alpha/v_0$ $-v_\beta/v_0$ |  |  |  |  |  | 1.00 0.00 |
|  | 8 | $-v_\alpha/v_0$ $-v_\beta/v_0$ |  |  |  |  | 1.00 0.00 | 1.00 0.00 |
|  | 9 | $-v_\alpha/v_0$ $-v_\beta/v_0$ |  |  |  | 1.00 0.00 | 1.00 0.00 | 0.99 0.00 |
|  | 10 | $-v_\alpha/v_0$ $-v_\beta/v_0$ |  |  | 1.00 0.00 | 1.00 0.00 | 0.98 0.00 | 0.96 0.01 |
|  | 11 | $-v_\alpha/v_0$ $-v_\beta/v_0$ |  | 1.00 0.00 | 1.00 0.00 | 0.98 0.00 | 0.95 0.01 | 0.92 0.02 |
|  | 12 | $-v_\alpha/v_0$ $-v_\beta/v_0$ | 1.00 0.00 | 1.00 0.00 | 0.97 0.01 | 0.94 0.01 | 0.90 0.02 | 0.86 0.03 |
|  | 13 | $-v_\alpha/v_0$ $-v_\beta/v_0$ |  | 0.96 0.00 | 0.91 0.01 | 0.83 0.04 | 0.78 0.08 |  |
|  | 14 | $-v_\alpha/v_0$ $-v_\beta/v_0$ |  |  | 0.77 0.00 | 0.65 0.06 |  |  |

## 11.10 Drawing through a die, influence of the friction

for the example shown in Fig. 213

of the first family

| 6 | 7 | 8 | 9 | 10 | 11 | 12 | 13 | 14 |
|---|---|---|---|---|---|---|---|---|
|  |  |  |  |  |  | 1.00 0.00 |  |  |
|  |  |  |  |  | 1.00 0.00 | 1.00 0.00 | 0.99 0.00 |  |
|  |  |  |  | 1.00 0.00 | 1.00 0.00 | 0.99 0.00 | 0.99 0.00 | 0.97 0.00 |
|  |  |  | 1.00 0.00 | 1.00 0.00 | 0.99 0.00 | 0.99 0.00 | 0.97 0.00 | 0.95 0.00 |
|  |  | 1.00 0.00 | 1.00 0.00 | 0.99 0.00 | 0.99 0.00 | 0.97 0.00 | 0.95 0.00 |  |
|  | 1.00 0.00 | 1.00 0.00 | 0.99 0.00 | 0.99 0.00 | 0.96 0.00 | 0.95 0.00 |  |  |
| 1.00 0.00 | 1.00 0.00 | 0.99 0.00 | 0.98 0.00 | 0.96 0.00 | 0.94 0.00 |  |  |  |
| 1.00 0.00 | 0.99 0.00 | 0.98 0.00 | 0.96 0.00 | 0.94 0.00 |  |  |  |  |
| 0.99 0.00 | 0.98 0.00 | 0.96 0.00 | 0.94 0.00 |  |  |  |  |  |
| 0.97 0.00 | 0.95 0.00 | 0.93 0.01 |  |  |  |  |  |  |
| 0.94 0.01 | 0.92 0.01 |  |  |  |  |  |  |  |
| 0.88 0.02 |  |  |  |  |  |  |  |  |

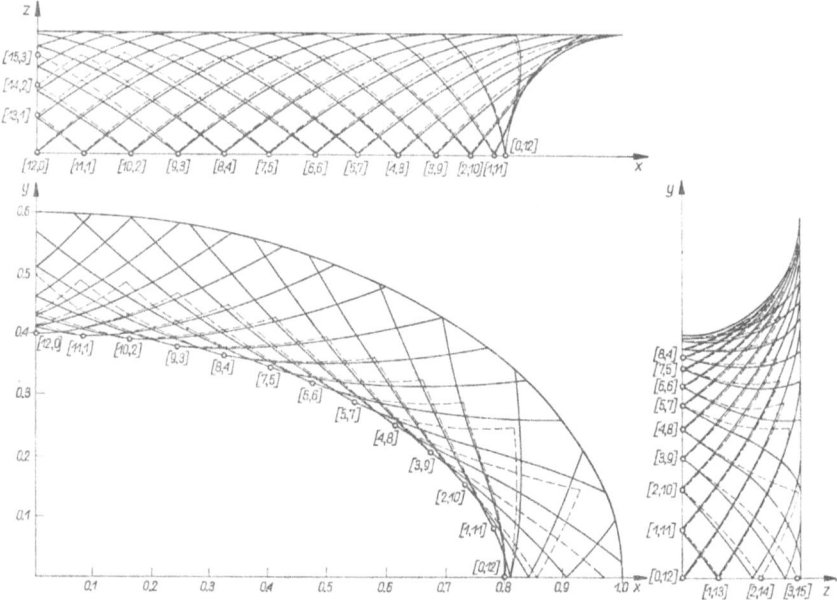

Fig. 213. The mesh of characteristics for the drawing problem shown in Fig. 207. Dashed lines correspond to the solution with friction. Continuous lines represent solution without friction

are remarkably different in the two solutions (compare the Tables 2 and 4). Depending on the lengths of the radii of curvature at a generic point, the contact pressure may be in the present solution greater or smaller than that in the previous solution where we assumed perfectly smooth contact. Comparison of Tables 3 and 5 shows that the velocity field is only slightly influenced by friction forces, which reassures our initial assumption enabling us to apply the procedure of successive approximations.

# Appendix 1

## 1 Suffix notation

To solve specific practical problems we usually use the Cartesian coordinate system $x, y, z$. The stress components are denoted as shown in Fig. 13. The normal stresses are denoted by one suffix, while the tangential components by two suffixes. Similarly one denotes the components of other tensors, for instance of the strain increment tensor or the strain rate tensor. Such a notation of tensor components is not convenient in a general discussion of equations of motion, in relations between stress and strain rate components, in expressions for the rate of energy dissipation, etc. These relations can be written in a compact form if the general suffix notation is introduced, with one suffix for vectors and two suffixes for tensors.

In suffix notation the Cartesian coordinates are denoted as follows: $x = x_1$, $y = x_2$, $z = x_3$. Now the components of the stress and strain rate tensors may be written in the following form

$$\begin{bmatrix} \sigma_{11} & \sigma_{12} & \sigma_{13} \\ \sigma_{21} & \sigma_{22} & \sigma_{23} \\ \sigma_{31} & \sigma_{32} & \sigma_{33} \end{bmatrix} \quad \text{and} \quad \begin{bmatrix} \epsilon_{11} & \epsilon_{12} & \epsilon_{13} \\ \epsilon_{21} & \epsilon_{22} & \epsilon_{23} \\ \epsilon_{31} & \epsilon_{32} & \epsilon_{33} \end{bmatrix},$$

or briefly $\sigma_{ij}$ and $\epsilon_{ij}$, where $i$ and $j$ are used as suffixes assuming the appropriate values $1, 2$, and $3$. The condition of symmetry of tensors, which in technical texts take the form $\tau_{xy} = \tau_{yx}$, $\tau_{yz} = \tau_{zy}$, $\tau_{zx} = \tau_{xz}$ may be now written in the form of one equality $\sigma_{ij} = \sigma_{ji}$, $\epsilon_{ij} = \epsilon_{ji}$, and similarly for other tensors.

The vector defined by three components $v_x, v_y, v_z$ in suffix notation is denoted as $v_i$, where $i$ takes the values $1, 2$, and $3$.

## 2  *Summation convention*

An expression of the type

$$\sigma_{11}+\sigma_{22}+\sigma_{33} = \sum_{i=1}^{3} \sigma_{ii} \tag{a}$$

shows the abbreviating role of the $\sum$-notation. According to the summation convention the summation sign $\sum$ may be dropped and (a) becomes

$$\sigma_{11}+\sigma_{22}+\sigma_{33} = \sigma_{ii}.$$

Thus the summation convention implies summation with respect to the repeated letter suffix. As an example of application of this convention in more complex cases, one can take the expression for the rate of work of external forces $X_i$ acting on the surface of a deforming body

$$X_i v_i = X_1 v_1 + X_2 v_2 + X_3 v_3,$$

where the vector $v_i$ represents the velocity of displacement of the point at which the force $X_i$ is applied.

To give another example, let us write the expression for the rate of internal energy dissipation

$$\sigma_{ij}\epsilon_{ij} = \sigma_{11}\epsilon_{11}+\sigma_{12}\epsilon_{12}+\sigma_{13}\epsilon_{13}+ $$
$$+\sigma_{21}\epsilon_{21}+\sigma_{22}\epsilon_{22}+\sigma_{23}\epsilon_{23}+ $$
$$+\sigma_{31}\epsilon_{31}+\sigma_{32}\epsilon_{32}+\sigma_{33}\epsilon_{33}.$$

The convention does not apply to non-repeated suffixes, which are called free suffixes. For example, consider the expression

$$\sigma_{ij}\sigma_{jk} = \sigma_{i1}\sigma_{1k}+\sigma_{i2}\sigma_{2k}+\sigma_{i3}\sigma_{3k},$$

in which $i$ and $k$ are free suffixes. Assigning to them the values $1$, $2$, and $3$ in all possible combinations we represent in the compact form $\sigma_{ij}\sigma_{jk}$, nine different sums.

The use of brackets is explained by the following examples

$$(\sigma_{ii})^2 = (\sigma_{11}+\sigma_{22}+\sigma_{33})^2,$$
$$\sigma_{ii}^2 = \sigma_{11}^2+\sigma_{22}^2+\sigma_{33}^2.$$

The summation convention is particularly useful for a short representation of complex systems of equations. For example, the equation

$$p_{ij}q_{jk} = r_{ik}$$

stands for nine equations. The suffixes $i$ and $k$ are free. The free suffixes must be the same on both sides. Each equation of the set corresponds to one of the nine possible combinations of the values of free suffixes. Thus we have

$$p_{11}q_{11}+p_{12}q_{21}+p_{13}q_{31} = r_{11} \quad \text{for} \quad i = 1, \, k = 1,$$
$$p_{21}q_{13}+p_{22}q_{23}+p_{23}q_{33} = r_{23} \quad \text{for} \quad i = 2, \, k = 3.$$

The summation convention permits to write shortly also expressions involving derivatives. For example,

$$\mathrm{d}u = \frac{\partial u}{\partial x_1}\mathrm{d}x_1 + \frac{\partial u}{\partial x_2}\mathrm{d}x_2 + \frac{\partial u}{\partial x_3}\mathrm{d}x_3 = \frac{\partial u}{\partial x_i}\mathrm{d}x_i.$$

Here we have summation over the repeated suffix $i$. Similarly,

$$\frac{\partial v_i}{\partial x_i} = \frac{\partial v_1}{\partial x_1} + \frac{\partial v_2}{\partial x_2} + \frac{\partial v_3}{\partial x_3}.$$

Further examples where summation convention is applied, can be found in equations (2.1), (2.18b), and (3.32). Note, however, that in expressions such as [comp. (2.14)]

$$\epsilon_{ij} = \frac{1}{2}\left(\frac{\partial v_i}{\partial x_j}+\frac{\partial v_j}{\partial x_i}\right),$$

there is no summation, because in no term are the suffixes repeated. Both suffixes $i$ and $j$ are free, and therefore the expression represents simply nine equations corresponding to the various combinations of the values of $i$ and $j$.

# Appendix 2

*Characteristics*

We begin with the equation in one unknown function $u(x, y)$ and two independent variables $x$ and $y$

$$P(u, x, y)\frac{\partial u}{\partial x} + Q(u, x, y)\frac{\partial u}{\partial y} = R(u, x, y). \tag{1}$$

A number of practical problems of plastic flow reduce to the solution of equations of this type. In practical cases usually the value of the function $u$ is given along some line $C$, described by an equation $y = y(x)$. The solution consists in determining the function $u$ in the region adjacent to $C$. This kind of problem is known as the Cauchy boundary value problem. If we consider the problem in the three-dimensional space $u, x, y$, (Fig. A-1) the boundary conditions on $C$ can be represented by a line $U$, whose $u$-coordinates are the boundary values of $u(x, y)$. The integral surface $u(x, y)$ satisfying (1) must pass through $U$. This condition must be complemented by the knowledge of the slope of the integral surface along $U$, because only in such a case is the surface $u(x, y)$ uniquely defined. The slope is determined by the derivatives $\partial u/\partial x$ and $\partial u/\partial y$. In order to find these derivatives, let us complement (1) by the expression

$$du = \frac{\partial u}{\partial x}dx + \frac{\partial u}{\partial y}dy. \tag{2}$$

for the variation of $u$ along $C$.

Solving now equations (1) and (2) with respect to the derivatives $\partial u/\partial x$ and $\partial u/\partial y$, we obtain

$$\frac{\partial u}{\partial x} = \frac{\begin{vmatrix} R & Q \\ du & dy \end{vmatrix}}{\begin{vmatrix} P & Q \\ dx & dy \end{vmatrix}}, \qquad \frac{\partial u}{\partial y} = \frac{\begin{vmatrix} P & R \\ dx & du \end{vmatrix}}{\begin{vmatrix} P & Q \\ dx & dy \end{vmatrix}}.$$

It is seen that if the characteristic determinant in the denominator is zero, the derivatives $\partial u/\partial x$ and $\partial u/\partial y$ are not defined. This means that the line $C$, along which the boundary conditions determining the Cauchy problem

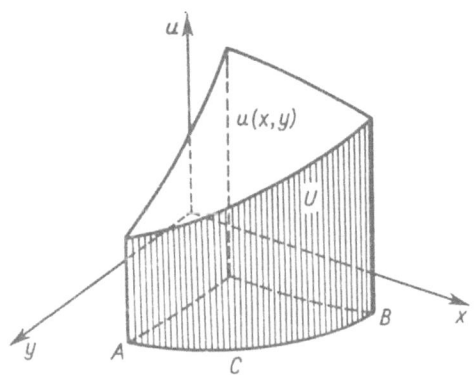

*a)*

Fig. A-1. The Cauchy boundary problem for equation (1)

are given, cannot have at any point the direction determined by the differential equation

$$\begin{vmatrix} P & Q \\ dx & dy \end{vmatrix} = 0 \text{ or in different way } \frac{dx}{P} = \frac{dy}{Q}. \tag{3}$$

Such a direction is referred to as a characteristic direction, and the lines of that direction, along which the additional condition

$$\begin{vmatrix} R & Q \\ du & dy \end{vmatrix} = 0, \text{ or equivalently } \begin{vmatrix} P & R \\ dx & du \end{vmatrix} = 0$$

is satisfied are called the *characteristics of* the equation (1). Thus the differential equations of characteristics are

$$\frac{dx}{P} = \frac{dy}{Q} = \frac{du}{R}. \tag{4}$$

Characteristics lie on the integral surface $u = u(x, y)$. Thus equations (4) are completely equivalent to the original differential equation (1). This property is of great practical significance since the integration of a partial differential equation reduces to the integration of ordinary differential equations (4) along some lines. In the general case integration is conducted numerically. In all equations differentials are replaced by finite differences and consecutive points of each characteristic are subsequently calculated step by step, starting from point on the initial line *C*. As we already know, this line cannot coincide with any of the characteristics. For each consecutive point *M* of a characteristic we find the coordinates $x_M, y_M$ and the value $u_M$ of the function *u*. If, however, the coefficients *P* and *Q* in (1) are functions of independent variables *x* and *y* only, thus if

$$P(x,y)\frac{\partial u}{\partial x}+Q(x,y)\frac{\partial u}{\partial y} = R(u,x,y), \qquad (5)$$

then the differential equation of characteristics has the form

$$\frac{dy}{dx} = \frac{Q(x,y)}{P(x,y)} = \Phi(x,y).$$

This equation can be integrated independently of the function *u*. In other words, we can find a characteristic in advance, and then determine the function *u* along it.

Consider now a system of two partial differential equations in two unknown functions *u, v* and two independent variables *x, y*

$$A_1(u,v,x,y)\frac{\partial u}{\partial x}+B_1(u,v,x,y)\frac{\partial v}{\partial x}+C_1(u,v,x,y)\frac{\partial u}{\partial y}+$$

$$+ D_1(u,v,x,y)\frac{\partial v}{\partial y} = F_1(u,v,x,y),$$

$$A_2(u,v,x,y)\frac{\partial u}{\partial x}+B_2(u,v,x,y)\frac{\partial v}{\partial x}+C_2(u,v,x,y)\frac{\partial u}{\partial y}+ \qquad (6)$$

$$+ D_2(u,v,x,y)\frac{\partial v}{\partial y} = F_2(u,v,x,y).$$

The Cauchy boundary value problem is defined when along some line *C*, whose equation is $y = y(x)$, the values of both functions *u* and *v* are given. These functions must be continuous along *C*. The integral surface in the four-dimensional space will be uniquely defined, if the derivatives $\partial u/\partial x, \ldots, \partial v/\partial y$ are determined on *C*. We can find these derivatives by complementing the set of equations (6) by expressions for the differentials of *u* and *v*

$$du = \frac{\partial u}{\partial x}dx+\frac{\partial u}{\partial y}dy,$$

$$dv = \frac{\partial v}{\partial x}dx+\frac{\partial v}{\partial y}dy. \qquad (7)$$

Solving now equations (6) and (7) with respect to the derivatives, we arrive at the conclusion that this cannot be done only when the characteristic determinant is zero,

$$\begin{vmatrix} A_1 & B_1 & C_1 & D_1 \\ A_2 & B_2 & C_2 & D_2 \\ dx & 0 & dy & 0 \\ 0 & dx & 0 & dy \end{vmatrix} = 0. \tag{8}$$

This equation can be transformed into a more convenient form

$$\begin{vmatrix} C_1 - A_1 \dfrac{dy}{dx} & D_1 - B_1 \dfrac{dy}{dx} \\ C_2 - A_2 \dfrac{dy}{dx} & D_2 - B_2 \dfrac{dy}{dx} \end{vmatrix} = 0. \tag{9}$$

This is a quadratic equation in $dy/dx$. Depending on the sign of the discriminant of this equation the system (6) belongs to different types. If the discriminant is negative, the derivatives $\partial u/\partial x, \ldots, \partial v/\partial y$ are defined along any line $C$. We say in such a case that the system (6) is *elliptic* and has *no real characteristics*. Such a system cannot be solved by means of the method of characteristics. If, however, the discriminant of the quadratic equation (9) is positive, we obtain two real roots for the derivative $dy/dx$

$$dy/dx = m_1 \quad \text{and} \quad dy/dx = m_2. \tag{10}$$

Returning to the Cauchy problem, we see that the line $C$, along which the boundary conditions are given, cannot have at any point the direction determined by equations (10).

The lines satisfying the differential equations (10) and the additional differential relations obtained from the equation

$$\begin{vmatrix} F_1 & B_1 & C_1 & D_1 \\ F_2 & B_2 & C_2 & D_2 \\ du & 0 & dy & 0 \\ dv & dx & 0 & dy \end{vmatrix} = 0 \tag{11}$$

are referred to as the characteristics of equations (6). The determinant in (11) is formed by replacing an arbitrary column in the characteristic determinant (8) by the column of free terms.

In practice it is more advantageous to determine these differential equations from their general form

$$\left( H + K \frac{dy}{dx} \right) du + L\, dv + M\, dx + N\, dy = 0, \tag{12}$$

where the coefficients $H, K, L, M, N$ are defined by the determinants

$$H = \begin{vmatrix} D_1 & A_1 \\ D_2 & A_2 \end{vmatrix}, \quad K = \begin{vmatrix} A_1 & B_1 \\ A_2 & B_2 \end{vmatrix}, \quad L = \begin{vmatrix} D_1 & B_1 \\ D_2 & B_2 \end{vmatrix},$$

$$M = \begin{vmatrix} F_1 & D_1 \\ F_2 & D_2 \end{vmatrix}, \quad N = \begin{vmatrix} B_1 & F_1 \\ B_2 & F_2 \end{vmatrix}.$$

By substituting both values of the derivative $\mathrm{d}y/\mathrm{d}x$ [see (10)] into (12) we obtain finally the equations of the two families of characteristics

$$\frac{\mathrm{d}y}{\mathrm{d}x} = m_1, \quad (H + Km_1)\mathrm{d}u + L\,\mathrm{d}v + M\,\mathrm{d}x + N\,\mathrm{d}y = 0,$$

$$\frac{\mathrm{d}y}{\mathrm{d}x} = m_2, \quad (H + Km_2)\mathrm{d}u + L\,\mathrm{d}v + M\,\mathrm{d}x + N\,\mathrm{d}y = 0.$$

$$(13)$$

# References

[1] J. M. ALEXANDER: On Complete Solutions for Frictionless Extrusion in Plane Strain, *Q. Appl. Math. 19* (1961) pp. 31–37.

[2] A. G. ATKINS and D. TABOR: Plastic Indentation in Metals with Cones, *J. Mech. Phys. Solids 13* (1965) pp. 149–164.

[3] A. BALTOV and A. SAWCZUK: A Rule of Anisotropic Hardening, *Acta Mechanica 1* (1965) pp. 81–92.

[4] Ch. S. BARRET: *Structure of Metals*, McGraw-Hill Book Comp., New York/London 1943.

[5] J. F. BELL: *The Experimental Foundations of Solid Mechanics*, Encyclopedia of Physics, Ed. S. Flügge, Vol. VI a/1, Festkörpermechanik I, Springer-Verlag. Berlin/Heidelberg/New York 1973.

[6] P. K. BERTSCH and W. N. FINDLEY: An Experimental Study of Subsequent Yield Surfaces; Corners, Normality, Bauschinger and Allied Effects, *J. Appl. Mech. 25* (1958) pp. 201–209.

[7] J. BIAŁKIEWICZ: Theoretical Analysis of a Tube Extrusion Process, (in Polish), *Mech. Teor. Stos. 13* (1975) pp. 251–259.

[8] J. BIAŁKIEWICZ and W. SZCZEPIŃSKI: On the Mechanics of the Forging Process in Dies, (in Polish), *Mech. Teor. Stos. 9* (1971) pp. 321–328.

[9] J. F. W. BISHOP: Calculations on Sheet-Drawing under Back Tension Through a Rough Wedge-Shaped Die, *J. Mech. Phys. Solids 2* (1953) pp. 39–42.

[10] J. F. W. BISHOP: On the Complete Solution to Problems of Deformation of Plastic-Rigid Material, *J. Mech. Phys. Solids 2* (1953) pp. 43–53.

[11] J. F. BRATT and O. KANAN: Determination of the Yield Condition in the Third Quadrant of the Stress Plane, *J. Appl. Mech. 33* (1966) p. 228.

[12] P. W. BRIDGMAN: The Stress Distribution at the Neck of a Tension Specimen, *Proc. ASM 32* (1944) pp. 553–574.

[13] P. W. BRIDGMAN: *Studies in Large Plastic Flow and Fracture*, Harvard University Press, Cambridge, Mass. 1964.

[14] G. I. BYKOVTSEV: On the Velocity Field in the Problem of Indentation of a Flat Punch into a Plastic Half-Space (in Russian), *Prikl. Mat. Mekh. 25* (1961) pp. 552–553.

[15] J. D. CAMPBELL and J. DUBY: The Yield Behaviour of Mild Steel in Dynamic Compression, *Proc. Roy. Soc. London, ser. A, 1204, 236* (1956) pp. 24–40.

355

## References

[16] J. D. CAMPBELL and C. J. MAIDEN: The Effect of Impact Loading on the Static Yield Strength of a Medium Carbon Steel, *J. Mech. Phys. Solids 6* (1957) pp. 53–62.

[17] D. S. CLARK, P. E. DUWEZ: The Influence of Strain Rate on Some Tensile Properties of Steel, *ASTM 50* (1950) pp. 560–575.

[18] I. F. COLLINS: Geometric Properties of Some Slip-Line Fields for Compression and Extrusion, *J. Mech. Phys. Solids 16* (1968) pp. 137–152.

[19] A. D. COX, G. EASON, and H. G. HOPKINS: Axially Symmetric Plastic Deformations in Solids, *Phil. Trans. Roy. Soc. of London 254, 1036* (1961) pp. 1–45.

[20] DAO-DUY-TIEN: Rockwell Hardness Test as a Problem in Plasticity, (in Polish), *Rozpr. Inżyn. 21* (1973) pp. 709–721.

[21] N. N. DAVIDENKOV and N. I. SPIRIDONOVA: Analysis of the State of Stress in the Neck of a Tension Test Specimen, *Proc. ASTM 46* (1946) p. 1147–1158.

[22] P. DEWHURST: Plane-Strain Indentation on a Smooth Foundation: A Range of Solutions for Rigid-Perfectly Plastic Strip, *Int. J. Mech. Sci. 16* (1974) pp. 923–930.

[23] L. DIETRICH: Contact Stresses in the Compression of a Truncated Plastic Cone, (in Polish), *Rozpr. Inżyn. 18* (1970) pp. 353–370.

[24] L. DIETRICH and W. SZCZEPIŃSKI: Plastic Yielding of Axially-Symmetric Bars with Non-Symmetric V-Notch, *Acta Mechanica 4* (1967) pp. 230–240.

[25] L. DIETRICH, K. TURSKI: Load-Carrying Capacity of Tensile Axially-Symmetric Bars Weakened by Series of V-Notches, (in Polish), *Mech. Teoret. Stos. 6* (1968) pp. 437–448.

[26] D. C. DRUCKER: A More Fundamental Approach to Plastic Stress-Strain Relations, *Proc. 1st U. S. Nat. Congr. Appl. Mech.* (1952), pp. 487–491.

[27] D. C. DRUCKER, H. J. GREENBERG and W. PRAGER: The Safety Factor of an Elastic-Plastic Body in Plane Strain, *J. Appl. Mech. 73* (1951) pp. 371–378.

[28] D. C. DRUCKER, W. PRAGER and H. J. GREENBERG: Extended Limit Design Theorems for Continuous Media, *Quart. J. Appl. Math. 9* (1952) pp. 381–389.

[29] B. A. DRUJANOV: On the Deformation of a Strip by Drawing Through a Curvilinear Die (in Russian), *Mekh. Tvierd. Tiela 1* (1966) pp. 125–131.

[30] B. A. DRUJANOV: *On the Rolling of a Strip under Condition of Maximum Shear*, (in Russian), in *Plastic Flow of Metals*, Nauka Publ., Moscow 1968, pp. 102–113.

[31] B. A. DRUJANOV: On Application of Rigid-Plastic Analysis to Certain Cold-Working Operations, *Mekh. Tvierd. Tiela 3* (1971) pp. 179–183.

[32] D. S. DUGDALE: Wedge Indentation Experiments with Cold-Worked Metals, *J. Mech. Phys. Solids 1* (1953) pp. 14–26.

[33] D. S. DUGDALE: Cone Indentation Experiments, *J. Mech. Phys. Solids 2* (1954) pp. 265–277.

[34] G. EASON and R. T. SHIELD: The Plastic Indentation of a Semi-Infinite Solid by a Perfectly Rough Circular Punch, *Zeitschr. Ang. Mat. Phys. 11* (1960) pp. 33–43.

[35] S. ERBEL: *Investigations of Properties of Strongly Deformed Metals*, (in Polish) Thesis, Technical University, Warsaw 1965.

[36] T. C. FIRBANK and P. R. LANCASTER: Plane-Strain Drawing Between Dies with a Circular Profile and Zero Exit Angle, *Int. Journ. Mech. Sci. 6* (1964) pp. 415–420.

[37] H. FORD: *Advanced Mechanics of Materials*, Longmans, London 1963.

356

# References

[38] H. Geiringer: Beitrag zum volständigen ebenen Plastizitätsproblem, *Proc. 3rd Intern. Congr. Appl. Mech.*, Stockholm 1930.

[39] A. P. Green: A Theoretical Investigation of the Compression of a Ductile Material Between Smooth Flat Dies, *Phil. Mag. 42* (1951) pp. 900–918.

[40] A. P. Green: The Plastic Yielding of Notched Bars due to Bending, *Quart. J. Appl. Math. 6* (1953) pp. 223–239.

[41] A. P. Green: On the Use of Hodographs in Problems of Plane Plastic Strain, *J. Mech. Phys. Solids 2* (1954) pp. 78–80.

[42] W. A. Green: Extrusion Through Smooth Square Dies of Medium Reduction, *J. Mech. Phys. Solids 10* (1962) pp. 225–233.

[43] O. D. Grigoriev: On the Condition of the Positive Rate of Energy Dissipation in Plane Plastic Flow Problems (in Russian), *Zhurn. Prikl. Mekh. Tekhn. Phys. 1* (1962) p. 164.

[44] J. Grundzweig, J. M. Longman and N. J. Petch: Calculations and Measurements on Wedge-Indentation, *J. Mech. Phys. Solids 2* (1954) pp. 81–86.

[45] A. Haar, Th. Kármán: Zur Theorie der Spannungszustände in plastischen und sandartigen Medien, *Nachr. Ges. Wiss. Göttingen Math.-Phys. Klasse* (1909) pp. 204–218.

[46] M. Hawrysz: *Plastic Flow Processes with Geometric Similarity*, Thesis, Wroclaw Technical University, Wrocław 1976.

[47] R. Hill: A Theoretical Analysis of the Stresses and Strains in Extrusion and Piercing, *J. Iron and Steel Inst. 158* (1948) pp. 177.

[48] R. Hill: Some Special Problems of Indentation and Compression in Plasticity, *Proc. 7th Intern. Congr. Appl. Mech.*, London 1948.

[49] R. Hill: The Plastic Yielding of Notched Bars under Tension, *Q. J. Mech. Appl. Math. 2* (1949) p. 40.

[50] R. Hill: *The Mathematical Theory of Plasticity*, Clarendon Press, Oxford 1950.

[51] R. Hill: On Discontinuous Plastic States, with Special Reference to Localized Necking in Thin Sheets, *J. Mech. Phys. Solids 1* (1952) pp. 19–30.

[52] R. Hill: On the Mechanics of Cuttings Metal Strips with Knife-Edged Tools, *J. Mech. Phys. Solids 1* (1953) pp. 265–270.

[53] R. Hill, E. H. Lee and S. J. Tupper: The Theory of Wedge Indentation of Ductile Materials, *Proc. Roy. Soc. A, 188* (1947) pp. 273.

[54] R. Hill, E. H. Lee and S. J. Tupper: A Method of Numerical Analysis of Plastic Flow in Plane Strain and its Application to the Compression of a Ductile Material between Rough Plates, *J. Appl. Mech. 18* (1951) pp. 46–52.

[55] R. Hill and S. J. Tupper: A New Theory of Plastic Deformation in Wire-Drawing, *J. Iron and Steel Inst. 159* (1948) p. 353.

[56] M. T. Huber: Specific Work of Deformation as a Measure of the Level of Straining, (in Polish), *Czasopismo Techniczne*, Lemberg 1904.

[57] Ju. I. Jagn and O. A. Shishmariev: Some Results of Elastic Limit Investigation of Plastically Tensioned Nickel Specimens, (in Russian), *Doklady Ak. Nauk USSR 119* (1958) pp. 46–48.

[58] A. A. Ilyushin: *Plasticity*, (in Russian), Moscow 1949.

357

# References

[59] A. A. ILYUSHIN: On Problems of the Theory of Flow of a Plastic Medium on Surfaces, (in Russian), *Prikl. Math. Mekh. 18* (1954) pp. 265–288.

[60] A. I. ISHLINSKY: General Theory of Plasticity with Linear Hardening, (in Russian), *Ukr. Math. Journ. 3* (1954) pp. 314–324.

[61] A. I. ISHLINSKY: Axially Symmetric Problem of Plasticity and the Brinell Hardness Test, (in Russian), *Prikl. Math. Mekh. 8* (1944) p. 201.

[62] H. J. IVEY: Plastic Stress-Strain Relations and Yield Surfaces for Aluminium Alloys, *J. Mech. Engig. Sci. 3* (1961) pp. 15–31.

[63] W. JOHNSON: Extrusion Through Square Dies of Large Reductions, *J. Mech. Phys. Solids 4* (1956) pp. 191–198.

[64] W. JOHNSON: Over Estimates of Load for Some Two-Dimensional Forging Operations, *Proc. 3rd U. S. Nat. Congr. Appl. Mech.* 1958, pp. 571–579.

[65] W. JOHNSON, G. L. BARAYA and R. A. C. SLATER: On Heat Lines or Lines of Thermal Discontinuity, *Intern. Journ. Mech. Sci. 6* (1964) pp. 409–414.

[66] W. JOHNSON and H. KUDO: *The Mechanics of Metal Extrusion*, Manchester University Press, Manchester 1962.

[67] W. JOHNSON and H. KUDO: *Plane-Strain Deep Indentation*, Proc. 5th Intern. Machine Tool Design and Research Conference, University of Birmingham, Pergamon Press, 1964, pp. 441–447.

[68] L. M. KACHANOV: *Foundations of the Theory of Plasticity*, North-Holland, Amsterdam/London 1971.

[69] J. I. KADASHEVITSH, V. V. NOVOZHILOV: Theory of Plasticity Accounting for the Residual Stresses, (in Russian) *Prikl. Math. Mekh. 22* (1958) pp. 78–89.

[70] S. KOBAYASHI and E. G. THOMSEN: Approximate Solutions to a Problem of Press Forging, *Trans. ASME, Series B. J. Eng. Ind., 81* (1959) pp. 217–227.

[71] W. KOITER: *General Theorems for Elastic-Plastic Solids*, in: Sneddon and Hill (Eds.), *Progress in Solid Mechanics*, vol. 1, North-Holland 1960.

[72] L. KRONSJÖ and P. B. MELLOR: Plane-Strain Extrusion Through Concave Dies, *Int. J. Mech. Sci. 8* (1966) pp. 515–524.

[73] K. KWASZCZYŃSKA, Z. MRÓZ: A Theoretical Analysis of Plastic Compression of Short Circular Cylinders, *Arch. Mech. Stos. 19* (1967) pp. 787–797.

[74] T. LEHMAN: Ein neuer Ansatz für plastische Formänderungen mit Kaltverfestigung, *ZAMM 38* (1958), pp. 295–297.

[75] E. LEVIN: Indentation Pressure of a Smooth Circular Punch, *Quart. Appl. Math. 13* (1955) pp. 133–137.

[76] M. LÉVY: Extrait du mémoire sur les équations générales des mouvements intérieurs des corps solides ductile au delà des limites où l'élasticité pourrait les ramener à leur premier état, *J. Mech. Purres Appl. 16* (1871) pp. 369–372.

[77] U. S. LINDHOLM: Some Experiments with the Split Hopkinson Pressure Bar, *J. Mech. Phys. Solids 12* (1964) pp. 317–335.

[78] H. LIPPMAN: Principal Line Theory of Axially-Symmetric Plastic Deformation, *J. Mech. Phys. Solids 10* (1962) pp. 111–122.

[79] F. J. LOCKETT: Indentation of a Rigid Plastic Material by a Conical Indenter' *J. Mech. Phys. Solids 11* (1963) pp. 345–355.

# References

[80] W. LODE: Versuche über den Einfluss der mittleren Hauptspanung auf das Fliessen der Metalle Eisen, Kupfer und Nickel, *Zeitschr. für Physik 36* (1926).

[81] A. E. H. LOVE: *A Treatise on the Mathematical Theory of Elasticity*, 4th edition, Dover, New York 1944.

[82] W. M. MAIR and H. PUGH: Effect of Prestrain on Yield Surfaces in Copper, *J. Mech. Engng. Sci. 6* (1964) pp. 150–163.

[83] Z. MARCINIAK: Influence of the Sign Change of the Load on the Strain Hardening Curve of a Copper Test Piece Subject to Torsion, *Arch. Mech. Stos. 6* (1961) pp. 743–752.

[84] Z. MARCINIAK: Analysis of the Process of Forming Axially Symmetrical Draw-pieces with a Hole at the Bottom, *Arch. Mech. Stos. 15* (1963) pp. 821–832.

[85] R. MARJANOVIĆ and W. SZCZEPIŃSKI: Yield Surfaces of the M-63 Brass Prestrained by Cyclic Biaxial Loading, *Arch. Mech. Stos. 26* (1974) pp. 311–320.

[86] K. J. MARSH and J. D. CAMPBELL: The Effect of Strain Rate on the Post-Yield Flow of Mild Steel, *J. Mech. Physics Solids 11* (1963) pp. 49–63.

[87] J. MIASTKOWSKI: On the Influence of the Loading History on the Yield Surface, (in Polish), *Mech. Teor. Stos. 4* (1966) pp. 5–16.

[88] J. MIASTKOWSKI· Analysis of the Memory Effect of Plastically Prestrained Material, *Arch. Mech. Stos. 20* (1968) pp. 261–276.

[89] J. MIASTKOWSKI, W. SZCZEPIŃSKI: An Experimental Study of Yield Surfaces of Prestrained Brass, *Int. J. Solids and Structures 1* (1965) pp. 189–194.

[90] R. VON MISES: Mechanik der Festen Körper in plastisch deformablem Zustand, *Göttingen Nachrichten*, 1913.

[91] Z. MRÓZ: Graphical Solution of Axially Symmetric Problems of Plastic Flow, *J. Appl. Math. Phys. ZAMP 18* (1967) pp. 219–236.

[92] A. NADAI: *Theory of Flow and Fracture of Solids*, McGraw-Hill, New York 1950.

[93] P. M. NAGHDI, F. ESSENBURG and W. KOFF: An Experimental Study of Initial and Subsequent Yield Surfaces in Plasticity, *J. Appl. Mech. 25* (1958) pp. 201–209.

[94] J. NAJAR: Plane Polar-Like Rapid Flow Problems for Perfectly Plastic Materials, *Journal de Mécanique 7* (1968) pp. 249–279.

[95] J. NAJAR: Simple Waves in Plane Flow of a Perfectly Plastic Material, *Arch. Mech. Stos. 20* (1968) pp. 445–459.

[96] W. PRAGER: *Probleme der Plastizitätstheorie*, Birkhauser Verlag, Basel und Stuttgart 1955.

[97] W. PRAGER, P. G. HODGE: *Theory of Perfectly Plastic Solids*, J. Wiley Inc., New York, Chapman & Hall, London 1951, p. 172.

[98] L. PRANDTL: Über die Härte plastischer Körper, *Götinger Nachrichten*, mat.-phys., 1920, p. 74.

[99] L. PRANDTL: Anwendungsbeispiele zu einem Henckyschen Satz über das plastische Gleichgewicht, *Zeitschr. Angew. Math. Mech. 3* (1923) pp. 401–406.

[100] N. W. PURCHASE and S. J. TUPPER: Experiments with a Laboratory Extrusion Apparatus under Conditions of Plane Strain, *J. Mech. Phys. Solids 1* (1953) pp. 277–283.

# References

[101] J. RYCHLEWSKI: On the Correctness of the Solutions of Perfectly Plastic Flow, *Bull. Acad. Polon. Sci., Série Sci. Techn., 11* (1963) pp. 225–232.

[102] J. SALENÇON: Théorie des charges limites: sur la solution du problème du poinçonnement d'une plaque par un poinçon rectangulaire rigid, *C. R. Acad. Sci. Paris 264, Série A* (1967) pp. 613–616.

[103] J. SALENÇON: Théorie des charges limites: poinçonnement d'une plaque par deux poinçon symmétrique en déformation plane, cas ou l'epaisseur de la plaque est inférieure à largeur des poinçons, *C. R. Acad. Sci. Paris 265, Série A|* (1967) pp. 869–872.

[104] A. A. SAPRIKIN, L. A. SHOFMAN: The Flow of the Metal by Forming of Elements with Varying Cross-Section, (in Russian), *Kuzn. Shtampovotchnoe Proizvodstvo 4* (1965) pp. 1–5.

[105] R. T. SHIELD: On the Plastic Flow of Metals under Conditions of Axial Symmetry, *Proc. Roy. Soc. 233 A*, No. *1193* (1955) pp. 267–287.

[106] R. T. SHIELD and H. ZIEGLER: On Prager's Hardening Rule, *ZAMP IXa* (1958) pp. 260–275.

[107] L. A. SHOFMAN: *Applications of Rigid-Plastic Model of the Material to the Calculations of Distortion and Resistance of Plastically Deformed Bodies*, (in Russian), Chapter VII, in the book: *Foundations of the Theory of Plastic Forming of Metals*, Mashgiz, Moscow 1959.

[108] M. SINGH: A Linearized Theory of Tube Drawing, *ZAMP 15* (1964) pp. 1–12.

[109] V. V. SOKOLOVSKY: Equations of Plastic Equilibrium in the Plane State of Stress, (in Russian), *Prikl. Math. Mekh. 13* (1949).

[110] V. V. SOKOLOVSKY: *Theory of Plasticity*, (in Russian), Izdat. Vysshaya Shkola, Moscow 1969.

[111] V. V. SOKOLOVSKY: Some Remarks on the Plane Problem of Plasticity, (in Russian), *Prikl. Math. Mekh. 18* (1954) pp. 762–763.

[112] V. V. SOKOLOVSKY: Drawing of a Thin-Walled Tube Through a Conical Die, (in Russian), *Prikl. Math. Mekh. 24* (1960) pp. 959–961.

[113] V. V. SOKOLOVSKY: Expansion of a Circular Hole in a Rigid-Plastic Plate, (in Russian) *Prikl. Math. Mekh. 25* (1961) pp. 548–552.

[114] V. V. SOKOLOVSKY: Some Remarks on Linearisation of the Equations of Plasticity, *Prikl.* (in Russian) *Math. Mekh. 25* (1961) pp. 931–932.

[115] V. V. SOKOLOVSKY: Stress and Velocity Fields for the Drawing of a Plastic Strip, (in Russian), *Inzhn. Zhurnal 2* (1962) pp. 298–304.

[116] V. V. SOKOLOVSKY: Complete Plane Problems of Plastic Flow, *J. Mech. Phys. Solids 10* (1962) pp. 353–364.

[117] V. V. SOKOLOVSKY: Stress and Velocity Fields for Indentation of a Plastic Body by a Punch, *Inzhn. Zhurnal 3* (1963) pp. 160–164.

[118] A. J. M. SPENCER: The Approximate Solution of Certain Problems of Axially Symmetric Plastic Flow, *J. Mech. Phys. Solids 12* (1964) pp. 231–243.

[119] N. P. SUH, R. S. LEE, C. R. ROGERS: The Yielding of Truncated Solid Cones under Quasi-Static and Dynamic Loading, *J. Mech. Phys. Solids 16* (1968) pp. 357–372.

[120] H. W. Swift: Stresses and Strains in Tube Drawing, *Phil. Mag. 40* (1949) pp. 883–902.

[121] W. Szczepiński: Stress Field Equations for Problems of Drawing and Stretch Forming of Thin-Walled Shells of Double Curvature, (in Polish), *Archiwum Budowy Maszyn 6* (1959) pp. 325–343.

[122] W. Szczepiński: The Equations of Stress and Velocity During the Drawing and Stretch Forming Process of Thin Shells with Double Curvature, *Arch. Mech. Stos. 12* (1960) pp. 565–581.

[123] W. Szczepiński: Steady-State Plastic Flow Processes with Strain-Hardening Experimentally Determined, *Arch. Mech. Stos. 13* (1961) pp. 377–388.

[124] W. Szczepiński: The Method of Successive Approximations of Some Strain-Hardening Solutions, Proc. *4th U. S. Nat. Congr. Appl. Mech., Berkeley 1962*, pp. 1131–1135.

[125] W. Szczepiński: On the Effect of Plastic Deformation on Yield Condition, *Arch. Mech. Stos. 15* (1963) pp. 275–296.

[126] W. Szczepiński: Axially Symmetric Plane Stress Problem of a Plastic Strain-Hardening Body, *Arch. Mech. Stos. 15* (1963) pp. 611–633.

[127] W. Szczepiński: Some Solutions of Plastic Forming, Accounting for Kinematic, Isotropic and Mixed Strain-Hardening, *Proc. IASS Symposium on Non-Classical Shell Problems, North-Holland Publ. Comp., Amsterdam, PWN—Polish Scientific Publishers, Warszawa 1964*, pp. 867–876.

[128] W. Szczepiński: Indentation of a Plastic Block by two Opposite Narrow Punches, *Bull. Acad. Polon. Sci., Serie Sci. Techn. 14* (1966) pp. 671–676.

[129] W. Szczepiński: Experimental Analysis of Some Non-Steady State Processes of Plastic Flow, (in Polish), *Mech. Teor. Stos. 5* (1967) pp. 309–323.

[130] W. Szczepiński, L. Dietrich, E. Drescher and J. Miastkowski: Plastic Flow of Axially-Symmetric Notched Bars Pulled in Tension, *Int. J. Solids Structures 2* (1966) pp. 543–554.

[131] W. Szczepiński, J. Miastkowski: An Experimental Study of the Effect of the Prestraining History on the Yield Surfaces of an Aluminium Alloy, *J. Mech. Phys. Solids 16* (1968) pp. 153–162.

[132] G. I. Taylor and H. Quinney: The Plastic Distortion of Metals, *Phil. Trans. Roy. Soc. A, 230* (1931) pp. 323–362.

[133] L. D. Tomlenov: *Mechanics of Plastic Forming of Metals* (in Russian), Mashgiz, Moscow 1963.

[134] E. G. Thomsen: A New Method for the Construction of Hencky-Prandtl Nets, *J. Appl. Mech. 24* (1957) pp. 81–84.

[135] E. G. Thomsen, C. T. Yang, S. Kobayashi: *Mechanics of Plastic Deformation in Metal Processing*, McMillan Co., New York 1965.

[136] K. Turski: Effect of Plastic Deformation on Subsequent Behaviour of Metal in a Different Loading Program, *Bull. Acad. Polonais Sci. 19* (1971) pp. 89–96.

[137] E. P. Unksov: *An Engineering Theory of Plasticity*, Butterworths, London 1961.

[138] V. A. Zhalin, D. D. Ivlev, V. C. Mishtshenko: On Indentation of Ring-Shaped Punch in Semi-Infinite Plastic Body, (in Russian), *Zhurn. Prikl. Mat. Tekhn. Phys.* (1961) pp. 153–154.

# Supplementary references

[1] W. A. BACKOFEN: *Deforming Processing*, Addison-Wesley, 1972.

[2] P. BAQUE, E. FELDER, J. HYAFIL, Y. DESCATHA: *Mise en forme des metaux: Calculs par la plasticité*, Dunod, Paris 1973.

[3] A. GELEJI: *Bildsame Formung der Metalle in Rechnung und Versuch*, Academie-Verlag, Berlin 1961.

[4] S. I. GUBKIN: *Plastic Deformation of Metals* (in Russian), Metallurgizdat, Moscow 1961.

[5] O. HOFFMAN and G. SACHS: *Introduction to the Theory of Plasticity for Engineers*, McGraw-Hill, New York 1953.

[6] W. JOHNSON and H. KUDO: *The Mechanics of Metal Extrusion*, Manchester University Press, 1962.

[7] W. JOHNSON and P. B. MELLOR, *Engineering Plasticity*, Van Nostrand Reinhold, London 1973.

[8] W. JOHNSON, R. SOWERBY and I. B. HADDOW: *Plane Strain Slip Line Fields: Theory and Bibliography*, E. Arnold Ltd., London 1970.

[9] G. W. ROWE: *Theory of Metal Processes*, E. Arnold Ltd., London 1965.

[10] L. A. SHOFMAN: *The Fundamentals of Die Forging and Press Forging Design* (in Russian), Mashgiz, Moscow 1961.

[11] E. G. THOMSEN, C. T. YANG, S. KOBAYASHI: *Mechanics of Plastic Deformation in Metal Processing*, McMillan Co., New York 1965.

[12] E. P. UNKSOV: Engineering Theory of Plasticity, Butterworths, London 1962.

# Author index

363

# Subject index

## Subject index